S*pringer* **M***onographs in* **M***athematics*

Jay Jorgenson
Serge Lang

The Heat Kernel and Theta Inversion on $SL_2(C)$

Springer

Jay Jorgenson
Department of Mathematics
City College of New York
New York, NY 10031
USA
jjorgenson@mindspring.com

Serge Lang
(*deceased*)

ISBN 978-1-4419-2282-3 e-ISBN 978-0-387-38032-2
DOI 10.1007/978-0-387-38032-2

Mathematics Subject Classification (2000): 11Fxx, 32Wxx

Softcover reprint of the hardcover 1st edition 2008

Printed on acid-free paper

9 8 7 6 5 4 3 2 1

springer.com

Preface

The draft of the present book was first completed in August 2005, one month prior to the passing of Serge Lang. As such, this book is the last completed in its entirety by Lang.

Beginning the early 1990's, Lang became fascinated with the prospect of using heat kernels and heat kernel analysis in analytic number theory. Specifically, we developed a program of study where one would define Selberg-type zeta functions associated to finite volume quotients of symmetric spaces, and we speculated that each such zeta function would admit a functional equation where lower level Selberg-type zeta functions would appear. In the case of the Riemann surface associated to $\mathrm{PSL}_2(\mathbf{Z})$, the Selberg zeta function has a functional equation which involves the Riemann zeta function. Lang and I began the work necessary to carry out our proposed analysis for quotients of $\mathrm{SL}_n(\mathbf{C})$ by $\mathrm{SL}_n(\mathbf{Z}[\mathbf{i}])$. As with other mathematical works undertaken by Lang, he wanted to develop the foundations himself. The present book is the result of establishing, as only Lang could, the case $n = 2$.

During the reviewing process, I have refrained from making any changes in order to preserve Lang's style of exposition.

New York
May 2008

Jay Jorgenson

Contents

Part V The Eisenstein–Cuspidal Affair

Introduction

In this book, we are concerned with the group $SL_2(\mathbf{C})$ for several reasons that reinforce each other.

- Like the book $SL_2(\mathbf{R})$, but more significantly (as will be apparent from what follows), it provides an introduction to the general theory of semisimple or reductive groups G, with symmetric space G/K (K maximal compact).
- For some fundamental questions, it is easier than $SL_2(\mathbf{R})$, because the complexification has the effect of "splitting" certain objects that are not "split" on $SL_2(\mathbf{R})$, for instance the spherical kernel, the Harish-Chandra c-function, and the heat kernel.
- Furthermore, the real case can have a treatment depending on the complex case; cf. additional comments on this at the end of the introduction to Chapter 19.
- SL_2 may be viewed as part of a general "ladder" of groups (the SL_n-ladder and other ladders), actually the bottom geometric step of these ladders, causing the center of interest to shift from a given individual step in the ladder to the way the infinitely many steps are related to each other. Thus SL_2 may be viewed as the first step of an inductive structure. We treat it in a way that prepares for the higher levels, involving semisimple or reductive Lie groups. The above considerations are leading toward a reorganization of the way the theory of these groups is developed.
- $SL_2(\mathbf{R})$ and $SL_2(\mathbf{C})$ (mod their unitary subgroups) are natural models for the universal covering space of Riemann surfaces and 3-manifolds with constant negative curvature. Low-dimensional topologists or geometers are especially interested in them from the topological–differential-geometric point of view. How the invariants stemming from this direction are related to the invariants coming from the spherical inversion is a question deserving attention for its own sake.

Spherical Inversion

For all these reasons, we find it appropriate to start with a self-contained treatment of spherical inversion on $\mathrm{SL}_2(\mathbf{C})$. However, both the notation and the logical structure of statements and proofs are expected to go over to the higher-dimensional case. Our treatment occurs as Chapter 1, and culminates in Chapter 2 with the heat kernel. We follow Gangolli's approach [Gan 68], with the modification that comes from using a new space of test functions, the Gaussians, which are dense in anything one wants (cf. [JoL 03]), and set the stage for our computation of Gangolli's explicit formula for the heat Gaussian in the complex case. The real case will follow using the Flensted-Jensen method [Fle 78] or a variation [JoLS 03].

Gaussians actually serve (at least) three purposes:

- One of them is to serve as test functions in a new general development of the theory of semisimple or reductive Lie groups, of which Chapter 1 is an example.
- Another is to provide the basis for explicit formulas in this theory, a matter that will be discussed in greater detail below.
- The third is to lead immediately into the heat kernel, which is Gaussian. A treatment of the higher-dimensional case following the pattern of this book is now in order, to accompany [JoL 02].

Fourier and Eigenfunction Expansions

We are then ready to go on with various eigenfunction expansions on the space $\Gamma\backslash G/K$, where K is the unitary subgroup, and $\Gamma = \mathrm{SL}_2(\mathbf{Z}[\mathbf{i}])$ plays a role analogous to $\mathrm{SL}_2(\mathbf{Z})$ in the theory of the upper half-plane. The main expansion involves eigenfunctions of Casimir (Laplacian) and heat convolution, both in a discrete L^2 setting, and a continuous setting analogous to that of Fourier inversion on $\mathbf{R}$, but more complicated on $\Gamma\backslash G$. The required eigenfunctions are called Eisenstein series. The analysis of eigenfunctions involves genuine Fourier expansions with respect to the subgroup U of unipotent elements, and the discrete subsubgroup $\Gamma_U = \Gamma \cap U$. The expansion on $\Gamma\backslash G$ originated in Roelcke [Roe 66], with Selberg as precursor in the context of Riemann surfaces for the Selberg zeta function [Sel 56], pp. 74–81. Going from $\Gamma_U\backslash U$ to $\Gamma\backslash G$ can be viewed as an inductive step, the higher level reflecting complications partly due to the noncompactness of $\Gamma\backslash G$, and partly due to the more complicated group-theoretic structure. The expansions are usually called spectral expansions, because starting with Roelcke, they are carried out in the context of the spectral theorem of functional analysis. We have completely avoided this context. For further comments, historical and otherwise, see the introductions of Chapters 10 and 11, and Section 10.1.

Gaussians and the Trace Formula

Both for the expansions and for the subsequent operations involving integration over $\Gamma\backslash G$ and a suitable integral transform yielding zeta objects (see below), we need explicit formulas for all the expressions involved. Integrating the Fourier–Eisenstein eigenfunction expansion over $\Gamma\backslash G$ leads to the trace formula. Selberg first worked with cocompact Γ, especially compact Riemann surfaces [Sel 56], although he does mention the Eisenstein series relevance in the noncocompact case. See also Gangolli [Gan 77]. Jim Arthur carried out a general theory in the noncompact case [Art 74], [Art 78], [Art 89]. See also Gangolli–Warner [GaW 80], and Elstrodt–Grunewald–Mennicke [EGM 98], especially Chapter 6, Theorem 5.1. The theory of the trace formula, especially in the noncocompact case, involves general formulas for test functions. We need the formulas (trace and eigenfunctions) with very explicit functions, at least including the heat kernel, and so notably Gaussians to include the heat Gaussian, which allow for explicit evaluations of all terms. This is central for some specific application (theta inversion formula and beyond with the Gauss transform).

The corresponding theory in the real case can most likely be obtained by reduction to the complex case via an integral transform (Flensted–Jensen method [FlJ 86] and [JoLS 03]). But independently of the real case, the complex case deserves attention for itself, in a complex ladder; cf. the comments below.

The heat Gaussian is the most important normalization of a Gaussian, by the Dirac property and the heat equation. These two conditions determine the heat kernel uniquely, and in particular determine uniquely the floating constants that occur in the context of test functions. Our goal in this book is to obtain an explicit formula analogous to a classical theta inversion formula. We do this by integrating the explicit eigenfunction expansion of the heat kernel to get an explicit version of the Selberg trace formula for this kernel. In the case of cocompact Γ, such a formula was obtained by Gangolli [Gan 68], starting with the heat kernel and its Γ-periodization. Our way of getting the formula starts as in Gangolli, but then differs in several respects both from Gangolli (who does not have to deal with the cuspidal phenomena because of cocompactness), and from the way others treat such dealings in connection with the eigenfunction expansion.

The General Path to Theta Inversion

Our path is as follows:

(1) Start with the heat kernel on G/K.
(2) Periodize it with respect to Γ.
(3) Expand it in an eigenfunction decomposition (series and integral).
(4) Regularize the terms in the expansion, and integrate its restriction to the diagonal over $\Gamma\backslash G$.

The eigenfunction expansion has both a discrete part in L^2 and a continuous part represented by an integral, corresponding to the noncompactness; think of the Fourier transform. The eigenfunctions of the continuous part are Eisenstein series. The corresponding integral may be called the Eisenstein integral. The above-mentioned regularization involves two steps:

(4a) Separate the Γ-periodization into two terms: the cuspidal term and the noncuspidal term.

(4b) Give an asymptotic expression for the cuspidal term and the Eisenstein integral in terms of a coordinate $Y \to \infty$ on $\Gamma\backslash G/K$, showing that the divergent part for the cuspidal term and the Eisenstein integral are exactly the same. These can then be cancelled, and the $\Gamma\backslash G$-integral of the remaining terms can be carried out explicitly. We call this the *Eisenstein–cuspidal affair*.

This yields the *theta inversion relation* (cf. Section 14.7)

$$\frac{e^{-2t}}{t^{1/2}}\sum a_j e^{-b_j/8t} + c_{\mathrm{cus}}\frac{e^{-2t}}{t^{3/2}} + \int_0^\infty F_{\mathrm{cus}}(r)\,\mathbf{f}_t(r)\,dr$$

$$= \sum_{n=1}^{\infty} c_n e^{-\lambda_n t} + c_0 + \int_{-\infty}^{\infty} F(r)\,e^{-\lambda}(r)\,t_{dr}.$$

As the reader can see, the formula involves theta series and two integrals that are continuous analogues of theta series and might be called theta integrals. The $-\lambda_n$ are eigenvalues of Casimir (Laplacian) on an orthonormal basis of a certain subspace of $L^2(\Gamma\backslash G)$, and $-\lambda(r) = -2(1+r^2)$ is the eigenvalue on the Eisenstein series E_s with s on the axis, $s = 1+\mathbf{i}r$. The functions F and F_{cus} will be determined explicitly.

If one carries out the above procedure on the circle $\mathbf{R}/2\pi\mathbf{Z}$, viewed as a quotient of $\mathbf{R}$ by the discrete group of translations $2\pi\mathbf{Z}$ (cf. [Lan 99]), then one obtains the classical theta inversion relation (Poisson summation formula)

$$\frac{1}{(4\pi t)^{1/2}}\sum_{n\in\mathbf{Z}} e^{-(2\pi n)^2/4t} = \frac{1}{2\pi}\sum_{n\in\mathbf{Z}} e^{-n^2}t.$$

No integral occurs because the circle is compact, and the integral reflects a noncompactness structure, as when one takes the Fourier transform for Fourier inversion on $\mathbf{R}$ itself. In the terminology of the general Lie context, we would say that the integral reflects a cuspidal structure.

For the conjectural eigenfunction expansion in higher dimension, in terms of the heat kernel, see [JoL 02], Section 4.6, formula **EFEX 1**, and the comments in Section 4.7.

The above four steps covered in this book then lead into a fifth step:

(5) Take an appropriate integral transform of the theta relation to get a zeta object (zeta function of some sort).

The development of this fifth step will follow what we did in [JoL 94] and [JoL 96]. We comment on this next at greater length.

Zetas

The theta inversion formula is indeed not the end. It is the beginning of the theory of zeta functions that can be developed from such a formula. There are immediately two possible integral transforms that can be applied to the formula.

(a) Riemann took the *Mellin transform* in the most classical case of $\zeta_{\mathbf{Q}}$, which amounts to following the above path on the circle viewed as a quotient of **R** by the discrete group of translations $2\pi\mathbf{Z}$. Taking the Mellin transform in the geometric context leads to what is called the spectral zeta function, and various analogues.

(b) We apply what we call the *Gauss transform* rather than the Mellin transform, as in [JoL 94] (reproduced in [LanJo 01], vol. V). The Gauss transform is a Laplace transform with a change of variables, defined on a function θ by the expression

$$\operatorname{Gauss}(\theta)(s) = 2s\int_0^\infty e^{-s^2t}\theta(t)\,dt.$$

We call this the **additive zeta function** determined by the theta.

Actually, since the function $\theta(t)$ has a singularity at 0, it is necessary to regularize this transform as in [JoL 94]. If one carries out the Gauss transform procedure in the case of a compact Riemann surface $\Gamma\backslash G/K$, with $G = \mathrm{SL}_2(\mathbf{R})$, one obtains the logarithmic derivative of the Selberg zeta function according to a theorem of McKean [McK 72]. If one carries out the above procedure over the reals and the usual $2\pi\mathbf{Z}$ periodization, using the Mellin transform in the fifth step, one obtains the usual Riemann zeta function. In both cases, there is also a fudge term. Using the Gauss transform instead of the Mellin transform gives the logarithmic derivative of the sine function, up to further fudge terms. Note that the Gauss transform of terms other than the main noncuspidal periodization of the heat kernel yields fudge terms in the functional equation.

It is relevant to note here a comment of Iwaniec [Iwa 95]. He writes down the functional equation of the Selberg zeta function in multiplicative form (10.40)

$$Z(s) = \Psi(s)Z(1-s),$$

and then says on pp. 169–170:

> If you wish, the Selberg zeta-function satisfies an analogue of the Riemann hypothesis. However, the analogy with the Riemann zeta-function is superficial. First of all, it fails badly when it comes to development into Dirichlet's series. Furthermore, the functional equation (10.40) resists any decent interpretation as a kind of Poisson's summation principle.

Readers can evaluate for themselves whether "the analogy with the Riemann zeta function is superficial." Whether decent or indecent, our procedure shows the way to get systematically a construction of zeta functions in the most general case of semisimple or reductive groups.

Note that there have been papers giving analogues of the Selberg zeta function via the trace formula in some cases, such as Gangolli [Gan 77] and Gangolli–Warner [GaW 80], especially the latter in the noncocompact case. Gangolli–Warner handle some problems arising from the trace formula by using what is called the Maass–Selberg relations and the truncation method. We avoid these by using the noncuspidal trace separately first to get the theta series proper; and second by using the Eisenstein–cuspidal affair to take care of other terms. Furthermore, Gangolli–Warner apply the trace formula to a test function out of which the Selberg zeta function arises, but this leads into other convergence problems. We apply the Gauss transform as the fifth step to an existing theta inversion relation. These are three of the main differences between our procedure and a previous development of Selberg-like functions.

For further comments, see the end of Section 11.1.

Ladders

Spaces of type G/K give rise to what we call ladders, as illustrated by the following example, the $\{\mathrm{SL}_n\}$ ladder. Let $G_n = \mathrm{SL}_n(\mathbf{C})$ (say), and $K_n =$ unitary subgroup. Each G_n/K_n is naturally embedded in G_{n+1}/K_{n+1} (totally geodesic). Thus we have a sequence of geometric objects that can be displayed vertically as a ladder:

$$\begin{gathered} G_{n+1}/K_{n+1} \\ G_n/K_n \\ G_{n-1}/K_{n-1} \end{gathered}$$

On the other hand, we have the ladder of associated additive zeta functions $L_n(s)$ according to the above-mentioned procedure. (The L notation suggests logarithmic derivative as well as classical L-functions.) The fudge terms (written additively in our setup so we don't say fudge factors) in the additive functional equation of our additive zeta function will conjecturally come mainly from the zeta functions of lower level in the ladder. Other fudge factors will include higher-dimensional versions of gamma functions. Thus we have a zeta ladder in parallel to the ladder of spaces. The spaces G_n/K_n are not compact, but can be compactified by the spaces G_m/K_m with $m < n$, especially arising from parabolic subgroups. Thus the occurrence of functions $L_m(s)$ as fudge terms for L_n reflects the geometric construction of compactification.

Connection with Analytic Number Theory

Furthermore, a new connection arises between classical number-theoretic objects and objects coming from geometry and analysis, because classical Dedekind zeta functions will occur as fudge factors of geometric zetas. For instance, the logarithmic derivative $\zeta'_{\mathbf{Q}(\mathrm{i})}/\zeta_{\mathbf{Q}(\mathrm{i})}$ comes in as a fudge term via the continuous part of the theta inversion. Thus classical Dedekind zeta functions will occur as fudge factors of geometric zetas. For example, the above term occurs as a fudge term for all levels above 2, i.e., for all $\mathrm{SL}_n(\mathbf{C})$. In particular, zeros or poles at a given lower level belonging to the fudge factors are the main zeros or poles of such lower levels. The classical terminology about "trivial zeros" is thus inappropriate, because the zeros of a Riemann–Dedekind zeta become zeros for higher-level zetas. More appropriately, they might be called **fudge zeros** for higher levels.

Furthermore, the theory in the complex case is not just a poor relative of the theory in the real case, as it has often been conceived. After all, $\zeta_{\mathbf{Q}}$ divides $\zeta_{\mathbf{Q}(\mathrm{i})}$, showing that the real case of the Riemann hypothesis follows from the complex case. This property could be viewed on a paar with the Flensted–Jensen "reduction to the complex case." The complex case of the geometric theory (say for semisimple or reductive Lie groups) deserves being carried out systematically for its own sake from its very beginnings, including the higher ground of the explicit inversion formula and its connection with classical analytic number theory via the ladder structure (among other things). This doesn't mean that the real case shouldn't also be carried out, but it should not obscure the clearer formalism of Gaussians before one superimposes the integral over the normals.

Still another type of connection comes up as in Chapter 13, in determining an asymptotic expansion for the integral of the cuspidal term of the heat kernel Γ-periodization, taken over the truncated fundamental domain. We do just what is necessary for present purposes, but a whole new area extending items of analytic number theory (what is called classically approximate formulas) to thetas coming from a geometric context, involving special functions, is arising. One may call this the **ladderization of approximate formulas**, or **ladderization of asymptotic expansions**, or **ladderization of the Hardy–Littlewood theorem** for $\zeta_{\mathbf{Q}}$ [HaL 21]. In other words, the possibility of extending the Hardy–Littlewood-type expansion to more general series, including generalized zeta and Eisenstein series, is opening up.

Connections with Geometry

Ladders should occur with certain types of spaces arising from geometry, in various ways.

(a) First, moduli spaces in algebraic geometry have a tendency to be of type $\Gamma\backslash G/K$ or to be naturally embeddable in some $\Gamma\backslash G/K$ (G reductive or semisimple, and K the unitary subgroup). For instance, one can already see the Siegel modular ladder corresponding to the groups Sp_{2g} (associated with abelian varieties of dimension g); the ladder of moduli spaces for K3-surfaces corresponding to the group $\mathrm{SO}_0\,(2, 19)$, and Calabi–Yau manifolds with their more complicated moduli structure, which may involve (b) below; the moduli ladder of forms of higher degree as in a paper of Jordan [Jor 1880]; etc. The moduli spaces of curves of genus greater than or equal to 2 are embeddable in the moduli spaces of their Jacobians because of Torelli's theorem (two curves are isomorphic if and only if their Jacobians are isomorphic). Similarly, moduli spaces of Calabi–Yau manifolds are embeddable in $\Gamma\backslash G/K$'s.

Thus analytic properties of Mellin(θ) or of Gauss(θ) associated with such spaces (will) get related to the algebraic geometry and differential geometry of such spaces in ways that will both complement those in the past and lead to new ways in the future. Being on a G/K with $G = \mathrm{SL}_2(\mathbf{C})$ eliminates some more complicated phenomena having to do with higher-dimensional geometric structures. In any case, the geometric ladder and the ladder of zeta functions reflect each other, thereby interlocking the theory of spaces coming from algebraic and differential geometry, with analysis and a framework whose origins to a large extent stem from analytic number theory. On the other hand, for some purposes, and in any case as a necessary preliminary for everything else, the purely analytic aspects have to be systematically available.

(b) There are other manifestations of ladderlike stratifications. A theorem of Griffiths [Gri 71] states that given a projective variety V over $\mathbf{C}$, there is a Zariski open subset that is a quotient of a bounded domain (of holomorphy), and this bounded domain is $\mathbf{C}^\infty$ isomorphic to a cell (Euclidean space). We propose to go further, namely, the Zariski open subset can be chosen so that its universal covering space is real-analytically a G/K with G semisimple or reductive, so the Zariski open subset is a $\Gamma\backslash G/K$ with discrete Γ. Thus any projective variety could be stratified by a $\Gamma\backslash G/K$. This gives rise to the possibility of considering the classification of varieties or manifolds via stratification structures by such $\Gamma\backslash G/K$. In particular, to what extent is there a minimal Zariski-closed subset of "natural" varieties such as moduli spaces (curves, Calabi–Yau, etc.) that one has to delete to get the complement expressible as a semisimple or reductive $\Gamma\backslash G/K$?

Topologists have concentrated on the classification problem via connected sums, but we find indications that the stratification structure and its connection with eigenexpansion analysis deserves greater attention. Thurston's conjecture for 3-manifolds fits into this ladder scheme.

Having a stratification as suggested above, one may then define an associated zeta function following the five steps listed previously, and relate the analytic properties of these zeta functions with the algebraic-differential geometry of the variety.

(c) The trace formula and the spectral zeta function in connection with index theorems have some history dating back to the 1970s and 1980s. We mention only a few papers: Atiyah–Bott–Patodi [AtBP 73], Atiyah–Donnelly–Singer [AtDS 83], Barbasch–Moscovici [BaM 83], Müller [Mul 83], [Mul 84], [Mul 87]. For a more complete bibliography, cf. Müller's Springer Lecture Notes [Mul 87]. These papers are partly directed toward index theorems and the connection with number-theoretic invariants, as in the proof of a conjecture of Hirzebruch in [AtDS 83] and [Mul 87]. In retrospect, we would interpret the Atiyah–Donnelly–Singer paper as working on two steps of a ladder, with the compactification of one space by another, and the index theorem being applied on this compact manifold. A reconsideration of the above-mentioned papers in light of the present perspective is now in order.

Readers can compare Müller's formula for the heat kernel [Mul 84], Theorem 4.8, and [Mul 87], (9.5), obtained in the context of functional analysis, with our explicit formula. Working as we do in the complex case, where it is possible to use an explicit Gaussian representation for the heat kernel, and using the Gauss transform rather than the Mellin transform, puts a very different slant on the whole subject, and allows us to go in a very different direction, starting with the explicit theta inversion relation and its Gauss transform. The present book simply provides the very first step in the simplest ladder we could think of, in a manner that is adapted to its extension to the other steps of the ladder, as well as other ladders.

Towers of Ladders

The structure goes still further. To concentrate on certain aspects of analysis in the simplest case of *G/K*, we already picked the number field **Q**(**i**) instead of **Q** itself, so we used **Z**[**i**] instead of **Z**. However, one may consider an arbitrary number field, and the Hilbert–Asai symmetric space associated with it [Asa 70], [Jol 99]. Thus we may go up a tower of number fields $\{F_m\}$ (finite extensions), and then ladders over these $\{G_{m,n}/K_{m,n}\}$, giving rise to a tower of ladders, so a quarter lattice combining even more extensively geometric structures with classical number-theoretic ones.

Going up a tower ipso facto introduces questions of number theory. Already for quadratic fields, the Eisenstein series give rise to a whole direction as in [EGM 85], [EGM 87], [EGM 98], with the theory of special values. However, the existence of a ladder over a fixed base such as Spec (**Z**) or Spec (**Z**[**i**]) has its own number-theoretic relevance, because the Riemann–Dedekind zeta should turn out to be a common factor of the zetas in all steps of the ladder, using Eisenstein series twisted by the heat kernel as in [Jol 02]. The case of SL_2 $(n = 2)$ is too small to have room for such a twist, visible only for $n \geqq 3$, but that the Riemann zeta appears in the functional equation of the Eisenstein series for $n = 2$ is classical. Our treatment prepares the ground for a continuation of [Jol 02], in what we view as an open-ended development.

Part I
Gaussians, Spherical Inversion, and the Heat Kernel

Chapter 1
Spherical Inversion on $SL_2(\mathbf{C})$

We are concerned with the group $G = SL_2(\mathbf{C})$. Like other groups that generalize it, the basic structure is the Iwasawa decomposition $G = UAK$, where:

U is the subgroup of upper triangular unipotent matrices;
A is the subgroup of diagonal matrices with positive diagonal components;
K is the unitary subgroup.

After giving the proof for this decomposition, we discuss characters on A, and then tabulate systematically integral formulas related to Haar measure on G, and various other decompositions (e.g., polar) that require the computation of a Jacobian in each case. In the general case of semisimple Lie groups, these are mostly due to Harish-Chandra [Har 58].

The Iwasawa decomposition leads into the Harish transform and Mellin transform, whose composite is the spherical transform, of Harish-Chandra, relating analytic (Fourier-type) expansions on G/K to those on A, which provides a Euclidean context.

We are especially interested in the heat Gaussian on G/K, corresponding to a normalized Gaussian on A. This heat Gaussian controls the general expansion of functions in various spaces. We choose the space of all Gaussians as a natural space of test functions for which we can prove the inversion formulas explicitly and very simply, at the level of elementary calculus. These test functions immediately provide the appropriate background for the heat kernel, which is characterized among them by simple conditions.

The construction of the heat kernel on general symmetric spaces G/K stems from Gangolli's fundamental paper [Gan 68]. He showed how the heat Gaussian on a general G/K is obtained as the inverse spherical transform of a Gaussian on $\check{\mathfrak{a}}$. We carry this out directly on the complex group $SL_2(\mathbf{C})$, when no great machinery is needed, and the formula for the heat Gaussian "splits."

Representation theory per se forms a motivating force for a whole establishment. We are motivated differently, namely by the development of zeta functions from theta inversion relations, including regularized products, regularized series, and explicit formulas. It has been realized in different contexts (Selberg trace formula

[Sel 56], Gangolli's construction of the heat kernel on general G/K's with cocompact discrete Γ) that theta inversion formulas can be viewed as part of a much larger context, stemming from the theory of semisimple Lie groups, symmetric spaces, and the heat kernel. The spherical inversion then provides an essential background for this direction. The essential feature here is the explicit form of the various formulas, notably our formulas for the orbital integral in Theorems 1.5.1 and 1.6.1, which are the backbone not just of spherical inversion theory, but of the explicit trace formula. The first explicit connection of this chapter with the trace formula will be found at the end of Chapter 5.

In turn, the explicit version of the trace formula yields theta inversion formulas, whence zeta functions via Mellin and especially Gauss transforms as in [JoL 94], when applied to the heat kernel. Although this chapter corresponds to the chapter in $SL_2(\mathbf{R})$ [Lan 75/85], but on $SL_2(\mathbf{C})$, our goal is actually not so much to do the spherical transform for its own sake as to apply it to the heat kernel, which was not mentioned in [Lan 75/85]. One may summarize an essential relationship by saying that *the t in the Poisson inversion formula is the same t as in the heat kernel (or heat Gaussian).*

We note that the literature on SL_2 ($\mathbf{C}$ or $\mathbf{R}$) uses mostly (if not entirely) what is called the Selberg transform. We find it much more valuable to plug in right away with the transforms used by Harish-Chandra. We find the latter clearer than the "Selberg transform" on SL_2, and they have the advantage of preparing the ground for inductive procedures to higher dimensions or rank of more general groups (semisimple, reductive).

Gelfand–Naimark first treated the representation theory in the complex case of the classical groups [GeN 50], and Harish-Chandra completed this for all complex groups, and then all real groups [Har 54], [Har 58], by means of the Harish-Chandra series, taking his motivation from linear differential equations. This method was followed in the standard references [Hel 84], [GaV 88], and also in [JoL 01] for SL_n. However, in Chapter XII of [JoL 01], for spherical inversion we suggested the possibility of an entirely different approach to the general case, having its origins in the Flensted–Jensen method of reduction to the complex case [FlJ 78], [FlJ 86]. This program is in the process of being carried out, using the normal transform and its relation to spherical inversion and the heat Gaussian, as described in the paper with A. Sinton [JLS 03]. In particular, the plan is to do the theory for $SL_n(\mathbf{C})$ first by the direct, simple, and explicit method as in the present chapter; then apply the normal transform to go from $SL_n(\mathbf{C})/K_n(\mathbf{C})$ to an arbitrary symmetric space G_1/K_1, which can be totally geodesically embedded in the symmetric space of $SL_n(\mathbf{C})$. This includes the real semisimple symmetric spaces embedded in their complexification as a special case. In any case, [JoLS 03] shows that $SL_n(\mathbf{C})$ is not only a particular (even significant) example for the theory of semisimple or reductive groups. It is a dominant object in this theory, controlling the others. Actually, the family $\{SL_n\}$ is not the only such family, for instance $\{Sp_n\}$ and $\{SO(p,q)\}$ also can play such a universal role, as discussed in [JoLS 03].

1.1 The Iwasawa Decomposition, Polar Decomposition, and Characters

The groups G, U, A, K are as mentioned above. Thus U is the group of 2×2 matrices of the form

$$u(x) = \begin{pmatrix} 1 & x \\ 0 & 1 \end{pmatrix} \quad \text{with } x \in \mathbf{C}.$$

The elements of A are of the form

$$a = \begin{pmatrix} a_1 & 0 \\ 0 & a_2 \end{pmatrix} \quad \text{with } a_1, a_2 > 0 \text{ and } a_2 = a_1^{-1}.$$

We also define the diagonal group $\mathbf{D}$ consisting of all elements

$$z = \begin{pmatrix} z_1 & 0 \\ 0 & z_1^{-1} \end{pmatrix} \quad \text{with } z_1 \neq 0 \text{ in } \mathbf{C}.$$

The elements k of K are the matrices satisfying (besides determinant 1)

$$k^{-1} = k^*, \text{ where } M^* = {}^t\bar{M} \text{ for a matrix } M.$$

Theorem 1.1.1. Iwasawa decomposition. *Every element $g \in G$ has a unique expression as a product*

$$g = uak \text{ with } u \in U,\ a \in A,\ k \in K.$$

Furthermore, let $\mathrm{Pos}_2 = \mathrm{Pos}_2(\mathbf{C})$ *be the space of positive definite Hermitian matrices. Then the map*

$$g \longmapsto gg^*$$

is a bijection $G/K \xrightarrow{\approx} \mathrm{SPos}_2$, *so of* $UA \xrightarrow{\approx} \mathrm{SPos}_2$.

Proof. Let $g \in G$. Let e_1, e_2 be the standard (vertical) unit vectors of $\mathbf{C}^2$. Let $g^{(i)} = ge_i$. We orthonormalize $g^{(1)}$, $g^{(2)}$ by the standard Gram–Schmidt process, which means we can find an upper triangular matrix $B = (b_{ij})$ $(b_{21} = 0)$ such that if we let

$$e_1' = b_{11}g^{(1)} \text{ and } e_2' = b_{12}g^{(1)} + b_{22}g^{(2)},$$

then e_1', e_2' are orthonormal unit vectors, and we can choose both b_{11} and b_{22} greater than 0. (This corresponds to dividing by the norm.) Thus $B \in AU = UA$.

Let k be the matrix such that $ke_i = e_i'$ for $i = 1, 2$. Then k is unitary. A direct computation shows that $gB = k$. Since $\det(g) = 1$ and the diagonal components of B are positive, it follows that $\det(k) = 1$, so $G = KAU = UAK$. To show uniqueness, we use the map $g \longmapsto gg^*$. Note that $gg^* = I$ (identity) if and only if $g \in K$. Suppose that

$$u_1 a^2 u_1^* = u_2 b^2 u_2^* \text{ with } u_1,\ u_2 \in u \text{ and } a,\ b \in A.$$

Putting $u = u_2^{-1}u_1$, we obtain $ua^2 = b^2 u^{*^{-1}}$. Since u, u^* are triangular in opposite directions, they must be diagonal, and finally $a^2 = b^2$, so $a = b$ because the diagonal elements are positive. This proves uniqueness.

Corollary 1.1.2. *Every element $g \in G$ has a decomposition, called* **polar**,

$$g = k_1 b k_2 \text{ with } k_1,\ k_2 \in K, \text{ and } b \in A.$$

The element b is uniquely determined up to permutation of diagonal components. Let A^+ be the set of $a = \text{diag}\,(a_1, a_2) \in A$ such that $a_1 \geqq 1$. Then $G = KA^+K$, and the decomposition $g = k_1 b k_2$ determines b uniquely in A^+.

Proof. By a basic theorem of linear algebra, there is an orthonormal basis of $\mathbf{C}^2$ consisting of eigenvectors of the form $g\,{}^t\bar{g}$, so $g\,{}^t\bar{g} = k_1 b^2 k_1^{-1}$ with $b \in A$. By the bijection in Theorem 1.1.1, there exists $k_2 \in K$ such that $g = k_1 b k_2$, and b^2 is the matrix of eigenvalues of $g\,{}^t\bar{g}$ (roots of the characteristic polynomial), which gives the uniqueness up to permutation.

Given $g = uak \in G$, with its Iwasawa decomposition, we let

$$g_A = a \text{ or also } a_g = a$$

denote the A-projection, and similarly for g_U and g_K.

1.1.1 Characters

A **character** of A is a continuous homomorphism $\chi : A \to \mathbf{C}^\times$ into the multiplicative group of nonzero complex numbers. Additively, let $a = \text{Lie}(A)$ be the $\mathbf{R}$-vector space of diagonal matrices $H = \text{diag}\,(h_1.h_2)$ with trace 0, so $A = \exp\, a$. Let $\check{a}$ be the dual space. Let α be the (additive) character on a defined by $\alpha(H) = h_1 - h_2$. If $a = \exp(H) = \text{diag}(a_1, a_2)$, then we define the **Iwasawa coordinate** y, by

$$y_a = \chi_\alpha(a) = a^\alpha = a_1/a_2 = a_1^2.$$

We note that α is a basis of $\check{a}$. Furthermore, y gives a group isomorphism of A with the positive multiplicative group. For complex s, we get

$$y_a^s = a^{s\alpha} = \chi_{s\alpha}(a),$$

with a complex (multiplicative) character $\chi_{s\alpha}$. We often write y instead of y_a and

$$a = a_y \text{ or } a(y) = \begin{pmatrix} y^{1/2} & 0 \\ 0 & y^{1/2} \end{pmatrix}.$$

Let

D = diagonal subgroup of G, with elements

$$z = \begin{pmatrix} z_1 & 0 \\ 0 & z_1^{-1} \end{pmatrix}.$$

T = torus subgroup = subgroup of **D** consisting of elements

$$\varepsilon = \begin{pmatrix} \varepsilon_1 & 0 \\ 0 & \varepsilon_1^{-1} \end{pmatrix} \text{ with } |\varepsilon_1| = 1.$$

Thus we have a direct product decomposition

$$\mathbf{D} = A\mathbf{T} \text{ with } z = a_z\varepsilon \text{ and } \mathbf{T} = \mathbf{D} \cap K.$$

The character α is actually the restriction of a character on **D**, namely

$$z^\alpha = z_1/z_2 = z_1^2.$$

The kernel of α in **D** is the group $\pm I$. We define $z \in \mathbf{D}$ to be **regular** if its diagonal components are distinct, in other words if $z^\alpha \neq 1$. More generally, an arbitrary square matrix is called **regular** if it can be conjugated to a regular diagonal matrix. Regularity will become relevant for measures in Section 1.2.

1.1.2 K-bi-invariant Functions

Let f be a function on G, which we write $f \in \mathrm{Fu}(G)$. We say that f is ***K*-bi-invariant** if $f(k_1gk_2) = f(g)$ for all $k_1, k_2 \in K$ and $g \in G$. By the polar decomposition, if b is the A-polar component of g, then $f(g) = f(b)$ and f is determined by its values on A. Let

$$w = \begin{pmatrix} 0 & 1 \\ -1 & 0 \end{pmatrix}$$

be called the **Weyl element**, and let the group of order 2 that it generates mod $\pm I$ be called the **Weyl group** W. Then w acts on **D** and A by conjugation, and for $z \in \mathbf{D}$,

$$z^w = z^{-1}.$$

A K-bi-invariant function is necessarily even. Proof. By the polar decomposition, $g = k_1bk_2$, so if φ is K-bi-invariant, then

$$\varphi(b) = \varphi\left(wbw^{-1}\right) = \varphi\left(b^{-1}\right).$$

Thus the restriction map from G to A induces bijections

$$\mathrm{Fu}(K\backslash G/K) \xrightarrow{\approx} \mathrm{Fu}_{\mathrm{even}}(A) \text{ and } C(K\backslash G/K) \xrightarrow{\approx} C_{\mathrm{even}}(A).$$

Since the Iwasawa coordinate y gives an isomorphism of A with $\mathbf{R}^+$, the function $v = \log y$ gives an isomorphism of A with $\mathbf{R}$. We hereby obtain a linear isomorphism

$$\mathrm{Fu}(K\backslash G/K) \xrightarrow{\approx} \mathrm{Fu}_{\mathrm{even}}(\mathbf{R}).$$

Given a K-bi-invariant function $f \in \mathrm{Fu}(K\backslash G/K)$, the corresponding even function on $\mathbf{R}$ will be denoted by f_+, so that we have

$$f(g) = f(b) = f_+(v).$$

Characters will arise naturally in connection with conjugation. We denote the **conjugation action** of G on itself by $\mathbf{c}$, so by definition,

$$\mathbf{c}(g)g' = gg'g^{-1}.$$

This action induces what we also call the conjugation action on various functors, one of which is now discussed naively and directly.

Action on the Lie algebra. Let n be the Lie algebra of U, by definition the algebra of strictly upper triangular matrices, so of the form

$$\begin{pmatrix} 0 & x \\ 0 & 0 \end{pmatrix}, \; x \in \mathbf{C}.$$

Then n is 1-dimensional as a complex vector space, with $\mathbf{C}$-basis

$$E_{12} = \begin{pmatrix} 0 & 1 \\ 0 & 0 \end{pmatrix}.$$

For n as a vector space over $\mathbf{R}$, a natural basis consists of E_{12} and $\mathbf{i}E_{12}$. We have

$$U = I + n = \exp n,$$

where exp is the exponential given by the usual power series.

Let $a = \mathrm{diag}(a_1, a_1^{-1}) \in A$ as above. Then E_{12} and $\mathbf{i}E_{12}$ are eigenvectors for the conjugation action by elements of A. In fact,

$$aE_{12}a^{-1} = a^{\alpha}E_{12} \text{ and } a\mathbf{i}E_{12}a^{-1} = a^{\alpha}\mathbf{i}E_{12}.$$

Thus $n = n_{\alpha}$ is an eigenspace, with eigencharacter α, which has multiplicity 2 viewing n as a vector space of dimension 2 over $\mathbf{R}$. Note that ${}^{t}n$ is a $(-\alpha)$-eigenspace. We call α, $-\alpha$ the **regular characters**. The trace of the representation on n is 2α, and the half-trace is $\rho = \alpha = \rho_G$ (this is notation that becomes significant in the higher-dimensional case). More generally, for $z \in \mathbf{D}$,

$$zu(x)z^{-1} = u(z^{\alpha}x).$$

We let $\delta = \delta_G$ be the **Iwasawa character** on A, namely

$$\delta_G(a) = a^{2\alpha} = y^2.$$

The exponent is the trace of the representation, viewed as being on a 2-dimensional **R**-space. The absolute Jacobian of conjugation on n viewed as an **R**-space is therefore

$$|\delta(z)| = a_z^{2\alpha}.$$

Remark. Until Chapter 6, it will be most convenient to work with the above decompositions. Readers may, however, already look at Section 6.0, which gives another convenient model for G/K, and another interpretation for the Iwasawa character y. See also Section 6.2 for the corresponding Riemannian distance on $G/K \cong \mathbf{H}^3$.

1.2 Haar Measures

We start with some general lemmas of measure theory on groups.

Proposition 1.2.1. *Let P be a locally compact group with two closed subgroups A, U such that A normalizes U, and such that the product*

$$A \times U \to AU = P$$

is a topological isomorphism. Then for $f \in C_c(P)$ the functional

$$f \mapsto \int_U \int_A f(au)\,da\,du = \int_A \int_U f(au)\,du\,da$$

is a Haar (left-invariant) functional on P.

Proof. Left invariance by A is immediate. Let $u_1 \in U$. Then

$$\begin{aligned}\int_A \int_U f(u_1 au)du\,da &= \int_A \int_U f(aa^{-1}u_1 au)du\,da\\ &= \int_A \int_U (f \circ a)\,(u_1^a u)\,du\,da \quad \text{where } u_1^a = a^{-1}u_1 a\\ &= \int_A \int_U f(au)\,du\,da \text{ by the Haar measure property,}\end{aligned}$$

thus proving our assertion.

Let G be a locally compact group with Haar measure dg. We recall that the **modular function** $\Delta = \Delta_G$ on G is the function (actually continuous homomorphism into $\mathbf{R}^+$) such that for all $f \in C_c(G)$,

$$\int_G f(gg_1)\,dg = \Delta(g_1) \int_G f(g)\,dg.$$

Proposition 1.2.2. *Notation is as in Proposition 1.2.1. There exists a unique continuous homomorphism* $\delta : A \to \mathbf{R}^+$ *such that for* $f \in C_c(U)$,

$$\int_U f\left(a^{-1}ua\right) du = \delta(a) \int_U f(u)\, du,$$

or in other words, for $f \in C_c(P)$,

$$\int_U f(ua)\, du = \delta(a) \int_U f(au)\, du.$$

If U is unimodular, then δ *is the modular function on P, that is,*

$$\Delta_P(p) = \Delta_P(au) = \delta(a).$$

Proof. The first statement is immediate, because the map

$$u \mapsto u^a = a^{-1}ua$$

is a topological group automorphism of U, which preserves Haar measure up to a constant factor, by uniqueness of Haar measure. For the second statement, we first note that the functional

$$f \mapsto \int_U \int_A (au)\, da\, du$$

is right invariant under U. Let $a_1 \in A$. Then

$$\begin{aligned}\int\int f(aua_1)\, da\, du &= \int\int f\left(aa_1a_1^{-1}ua_1\right) da\, du \\ &= \delta(a_1) \int\int f(au)\, da\, du,\end{aligned}$$

which proves the formula.

Proposition 1.2.3. *Let G be a locally compact group with two closed subgroups, P, K such that*

$$P \times K \to PK = G$$

is a topological isomorphism (not group isomorphism). Assume that G, K are unimodular. Let dg, dp, dk be given Haar measures on G, P, K respectively. Then there is a constant c such that for all $f \in C_c(G)$,

$$\int_G f(g)\, dg = c \int_P \int_K f(pk)\, dp\, dk.$$

If in addition $P = AU$ *as in Proposition 1.2.1, with U unimodular, so we have the product decomposition*

$$U \times A \times K \to G,$$

then

$$\int_G f(g)\,dg = c \int_U \int_A \int_K f(uak)\,\delta(a)^{-1}\,du\,da\,dk.$$

Proof. The first assertion comes from a standard fact of homogeneous spaces, that there exists a left-invariant Haar measure on $G/K = P$. See for instance [Lan 99], Chapter XVI, Theorem 1.5.1. The second integral formula simply comes from plugging in Proposition 1.2.2.

In the concrete application to $G = \mathrm{SL}_2(\mathbf{C})$ and its Iwasawa decomposition $G = UAK$, we have

$$P = UA = AU,$$

and U is indeed normal in P. All four of G, K, U, A are unimodular, so all the hypotheses in the previous propositions are satisfied. The character $a \mapsto \delta(a)$ in Proposition 1.2.3 is the Iwasawa character, $\delta(a) = a^{2\alpha}$. We normalize the Haar measure by taking $c = 1$. Thus we define the **Iwasawa measure** to be the Haar measure such that for $f \in C_c(G)$ we have

$$\int_G f(g)\,dg = \int_K \int_U \int_A f(auk)\,da\,du\,dk = \int_K \int_A \int_U f(uak)\,a^{-2\alpha}\,da\,du\,dk,$$

where

$da = dy/y$ with $y = a^{\alpha}$;
$du =$ Euclidean measure $dx = dx_1 dx_2$ if $x = x_1 + \mathbf{i}x_2$, $x_1, x_2 \in \mathbf{R}$;
$dk =$ Haar measure on K giving K total measure 1.
Unless otherwise specified, all integral formulas are with respect to the Iwasawa measure.

The Iwasawa measure is canonically determined by the Iwasawa decomposition. We may use the notation

$$dg = d_{\mathrm{Iw}}g = d\mu_{\mathrm{Iw}}(g)$$

for the Iwasawa measure, when comparing it with other measures. Thus the basic Iwasawa measure integration formula can be written

$$\textbf{INT 1}: \quad \int_G f(g)\,d\mu_{\mathrm{Iw}}(g) = \int_G f(g)\,dg = \int_K \int_A \int_U f(uak)\,\delta(a)^{-1}\,du\,da\,dk.$$

Remark. The above normalization of the Haar measure fits perfectly the normalization that comes from the quaternionic model for G/K, i.e. the set of quaternions $z = x_1 + \mathbf{i}x_2 + \mathbf{j}y$, and the standard normalization of the measure on this homogeneous space for the action of $\mathrm{SL}_2(\mathbf{C})$ given by $(az+b)(cz+d)^{-1}$ with complex a, b, c, d. This measure is $dx_1 dx_2 dy/y^3$; cf. Section 6.2.

Next comes the **Jacobian** for the **commutator map**. For z diagonal regular in G, and $u \in U$, the commutator is $zuz^{-1}u^{-1}$. Define

$$J_{G,\mathrm{com}}(z) = |1 - z^{\alpha}|^{m(\alpha)} \text{ with } m(\alpha) = 2.$$

The exponent is the multiplicity $m(\alpha)$ of α. Then

$$\textbf{INT 2}: \quad \int_U f\left(z^{-1}uzu^{-1}\right) du = \frac{1}{J_{\text{com}}\left(z^{-1}\right)} \int_U f(u)\, du.$$

Proof. Matrix multiplication shows that

$$\begin{pmatrix} z_1^{-1} & 0 \\ 0 & z_1 \end{pmatrix} \begin{pmatrix} 1 & x \\ 0 & 1 \end{pmatrix} \begin{pmatrix} z_1 & 0 \\ 0 & z_1^{-1} \end{pmatrix} \begin{pmatrix} 1 & -x \\ 0 & 1 \end{pmatrix} = \begin{pmatrix} 1 & \left(z_1^{-2}-1\right)x \\ 0 & 1 \end{pmatrix},$$

whence the formula follows. Remember that the Jacobian is taken on $\mathbf{C}$ viewed as 2-dimensional over $\mathbf{R}$, so in the change of variables formula, the absolute value of the complex derivative gets squared.

From the product decomposition $\mathbf{D} = A\mathbf{T}$, we give the compact group $\mathbf{T}$ total Haar measure 1. Then we give $\mathbf{D}$ the product measure of $da = dy/y$ and this normalized measure on $\mathbf{T}$, which we write dw. Then $\mathbf{D}\backslash G$ has the **homogeneous space measure**, which makes the **group Fubini** theorem valid, that is, $\mu_{\mathbf{D}}\backslash G$ such that

$$\int_G f(g)\, dg = \int_{\mathbf{D}\backslash G} \int_{\mathbf{D}} f(w\dot{g})\, dw\, d\mu_{\mathbf{D}\backslash G}(\dot{g}).$$

Actually, under the correspondence $\mathbf{D}\backslash G \leftrightarrow U(\mathbf{T}\backslash K)$, for right K-invariant functions we have with respect to a decomposition $g = wuk$ ($w \in \mathbf{D}$, $u \in U$, $k \in K$),

$$d\mu_{\mathbf{D}\backslash G}(g) = du.$$

The next relation is the main one for this section.

Let $\varphi \in C_c(K\backslash G/K)$ and let z be a regular diagonal element. Then

$$g \mapsto \varphi\left(g^{-1}zg\right)$$

has compact support on $\mathbf{D}\backslash G$, and we have

$$\textbf{INT 3}: \quad \int_{\mathbf{D}\backslash G} \varphi\left(g^{-1}zg\right) dg = \frac{1}{J_{\text{com}}\left(z^{-1}\right)} \int_U \varphi(a_z u)\, du.$$

Proof. The statement about compact support is immediate from the matrix U-coordinate. Then

$$\begin{aligned} \int_{\mathbf{D}\backslash G} \varphi\left(g^{-1}zg\right) dg &= \int_U \varphi\left(u^{-1}zu\right) du = \int_U \varphi\left(zz^{-1}u^{-1}zu\right) du \\ &= \frac{1}{J_{\text{com}}\left(z^{-1}\right)} \int_U \varphi(zu)\, du \quad \text{by } \textbf{INT 2}, \end{aligned}$$

which proves the formula.

1.3 The Harish Transform and the Orbital Integral

We reformulate this last formula in terms of mappings that will subsequently yield the spherical transform of Harish-Chandra. Functions can be at first in $C_c(K\backslash G/K)$, but the formalism extends by continuity to larger spaces that include the Gaussians.

We **factorize** $J(z)$ for diagonal z as follows. Put

$$|\mathbf{D}|\left(z^{-1}\right) = |\mathbf{D}|\left(z\right) = \left|z^{\alpha/2} - z^{-\alpha/2}\right|^{m(\alpha)} = \left|z^{\alpha/2} - z^{-\alpha/2}\right|^{2}.$$

Then

$$J_{\mathrm{com}}\left(z^{-1}\right) = |\mathbf{D}|\left(z\right)\delta\left(a_z\right)^{-1/2} = |\mathbf{D}|\left(z\right)z_z^{-\alpha}. \tag{1.1}$$

Note that we can take $z^{\alpha/2} = z_1$, $z^{-\alpha/2} = z_1^{-1}$, but any square root will do since the absolute value signs will kill whatever sign we take.

We define the **orbital integral** by

$$\mathrm{Orb}_{\mathbf{D}\backslash G}\left(\varphi, z\right) = \int_{\mathbf{D}\backslash G} \varphi\left(g^{-1}zg\right)d\mu_{\mathbf{D}\backslash G}\left(g\right) \text{ with } \mu_{\mathbf{D}\backslash G} \text{ as in Section 1.2.}$$

We define the **Harish transform** of $\varphi \in C_c(K\backslash G/K)$ by

$$\left(\mathbf{H}\varphi\right)\left(z\right) = \delta\left(a_z\right)^{1/2}\int_U \varphi\left(a_z u\right)du = |\mathbf{D}|\left(z\right)\mathrm{Orb}_{\mathbf{D}\backslash G}\left(\varphi, z\right).$$

The equality on the right comes from **INT 3** and factorization (1.1). The first formula is applicable to all diagonal z, but the second, with the orbital integral, is applicable only to regular z. As usual in measure-integration theory, the equality extends to functions for which the integrals are absolutely convergent, by continuity. This will be important for us, and certain convergence properties will be dealt with at the appropriate time and place.

Remark. Harish-Chandra uses the notation F_f for what we call the Harish transform. He applies it to both expressions. See [Har 58], and further historical and notational comments in [JoL 01], p. 98.

We reproduce the basic formalism of the Harish transform, similar on $\mathrm{SL}_2(\mathbf{C})$ to that of $\mathrm{SL}_2(\mathbf{R})$; cf. [Lan 75/85].

Theorem 1.3.1. *For $\varphi \in C_c(K\backslash G/K)$, the Harish transform $\mathbf{H}\varphi$ is invariant under the Weyl group, i.e.,*

$$\left(\mathbf{H}\varphi\right)\left(a\right) = \mathbf{H}\varphi\left(a^{-1}\right).$$

Proof. By continuity, it suffices to prove the assertion when a is regular, so $|\mathbf{D}|(a) = |\mathbf{D}|(a^{-1})$. The relation is then clear from the fact that K-bi-invariance implies that φ is even and from the orbital integral formula. One can also argue directly on the other integral defining $\mathbf{H}$, using the Jacobian for $u \mapsto a^{-1}ua$.

The Harish transform is therefore a linear map

$$\mathbf{H} : C_c(K\backslash G/K) \longrightarrow C_c(A)^W,$$

where the upper index W means the space of functions invariant under W.

We consider next the behavior under products, notably the **convolution product** on G, defined for a specified Haar measure by

$$(h * f)(z) \int_G h\left(zg^{-1}\right) f(g)\, dg$$

on a class of pairs of functions for which the integral is absolutely convergent. For noncommutative G, the K-bi-invariance (and the evenness) allows a choice of writing variables on one side or another, with inverses or not, namely for ψ, f-measurable K-bi-invariant such that for all z the function $z \mapsto \psi(zg)f(g)$ is in L^1, the convolution $(\psi * f)(z)$ is defined by

$$\begin{aligned}\int_G \psi(zg) f(g)\, dg &= \int_G \psi\left(zg^{-1}\right) f\left(g^{-1}\right) dg = \int_G \psi(zg-1) f(g)\, dg \\ &= \int_G \psi\left(g^{-1}\right) f(gz)\, dg = \int_G \psi(g) f(gz)\, dg.\end{aligned}$$

Theorem 1.3.2. (Gelfand) *Let G be a locally compact unimodular group, and let K be a compact subgroup. Let τ be an antiautomorphism of G of order 2 such that given $x \in G$, there exist $k_1,\ k_2 \in K$ satisfying $x^\tau = k_1 x k_2$. Then the convolution is commutative (whenever the convolution integral is absolutely convergent). (Note: In our special case, τ = transpose conjugate.)*

Proof. The functions $\psi,\ f$ are τ-invariant, so is the Haar measure, and

$$\begin{aligned}(\psi * f)(z) &= \int \psi\left(zg^{-1}\right) f(g)\, dg \\ &= \int \psi\left({}^\tau g^{-1} \tau z\right) f\left({}^\tau g\right) dg && \text{[by } \tau\text{-invariance of } \psi, f] \\ &= \int \psi\left({}^\tau g^{-1}\right) f\left({}^\tau z\, {}^\tau g\right) dg && \text{[by } g \mapsto gz]. \\ &= \int \psi\left(g^{-1}\right) f\left(k_1 z k_2 {}^\tau g\right) dg && \text{[by } \tau\text{-invariance of } \psi].\end{aligned}$$

Let $g \mapsto {}^\tau g,\ g \mapsto k_2^{-1} g$, and use the K-bi-invariance of $\psi,\ f$ to get $f * \psi$, thereby concluding the proof.

Theorem 1.3.3. *For $\varphi,\ \psi \in C_c(K\backslash G/K)$,* we have

$$\mathbf{H}(\varphi * \psi) = \mathbf{H}\varphi * \mathbf{H}\psi,$$

i.e., on $C_c(K\backslash G/K)$ the Harish transform is an algebra homomorphism for the convolution product.

Proof. To shorten notation, let $a^{\rho} = \delta(a)^{1/2}$ so $\rho = \alpha$. Then

$$\mathbf{H}(\varphi * \psi)(a) = a^{\rho} \int_U (\varphi * \psi)(au)\,du = a^{\rho} \int_U \int_G \varphi(aug)\,\psi(g^{-1})\,dg\,du$$
$$= a^{\rho} \int_U \int_G \varphi(ag)\,\psi(g^{-1}u)\,dg\,du.$$

Let $g = bvk$ be the Iwasawa decomposition with $b \in A,\ v \in U,\ k \in K$. Then the last expression is

$$= a^{\rho} \int_U \int_A \int_U \varphi(abv)\,\psi(v^{-1}b^{-1}u)\,db\,dv\,du$$
$$= a^{\rho} \int_U \int_A \int_U \varphi(ab^{-1}v)\,\psi(v^{-1}bu)\,db\,du\,dv \qquad [\text{by } b \mapsto b^{-1}]$$
$$= a^{\rho} \int \int \int \varphi(ab^{-1}v)\,\psi(v^{-1}ub)\,\delta(b)^{-1}\,db\,dv\,du \qquad [\text{by } bu \mapsto ub].$$
$$= a^{\rho} \int \int \int \varphi(ab^{-1}v)\,\psi(vb)\,\delta(b)^{-1}\,db\,dv\,du \qquad [\text{by } u \mapsto uv].$$
$$= a^{\rho} \int \int \int \varphi(ab^{-1}v)\,\psi(bu)\,db\,dv\,du \qquad [\text{by } ub \mapsto bu].$$

But

$$\mathbf{H}\varphi * \mathbf{H}\psi(a) = \int_A \mathbf{H}\varphi(ab^{-1})\,\mathbf{H}\psi(b)\,db$$
$$= \int \int \int (ab^{-1})^{\rho}\,\varphi(ab^{-1}v)^{\rho}\,\psi(bu)\,dv\,du\,db,$$

which is the same as the last expression obtained above, thus proving the theorem.

Basic results concerning the Harish and spherical transforms are that they are invertible on various spaces, including C_c^{∞}, and Schwartz-type spaces, with L^1, L^2, L^p conditions. We introduce a new space, the space of Gaussians, defined below in Section 6.6. We shall then see in Theorem 1.6.1 that the Harish transform induces an isomorphism on the space of Gaussians, by means of an explicit formula. This Gaussian space is admirably suited for getting explicit formulas, especially leading to the construction of the heat kernel, carried out in Chapter 2. In [JoL 03b], it is shown that the Gauss space is dense in anything one wants.

1.4 The Mellin and Spherical Transforms

Let χ be a character on A (continuous homomorphism into $\mathbf{C}^{\times}$). Let $F \in C_c(A)$. We define the **Mellin transform** of F to be the function on the group of characters given by

$$\mathbf{M}F(\chi) = \int_A F(a)\chi(a)\,da.$$

We use the same Haar measure da on A as before, with the coordinate character y. Viewing y as a basis for the characters, we can write a character in the form

$$\chi(a) = y^s = a^{s\alpha}$$

with a complex variable s. Then the Mellin transform is written with $F_\alpha(y) = F(a)$:

$$(\mathbf{M}_\alpha F)(s) = (\mathbf{M}F_\alpha)(s) = \int_0^\infty F_\alpha(y)\,y^s\frac{dy}{y}.$$

This is the ordinary Mellin transform of elementary analysis. If F does not have compact support, the integral may converge only in a restricted domain of s, usually some half-plane where the transform is analytic.

We use the trivial relation that *if F is even* $(F(a) = F(a^{-1}))$, i.e., *invariant under the Weyl group, then* $\mathbf{M}F$ *is also even*. This is immediate from the invariance of Haar measure under $a \mapsto a^{-1}$. In other words,

$$\mathbf{M}F(\chi) = \mathbf{M}f(\chi^{-1}).$$

For $g \in G$ we let g_A be its Iwasawa projection on A. Then the function

$$g \mapsto \chi(g_A) \text{ also written } \chi(g)$$

is well defined, and is called the **Iwasawa lifting** of χ to G. As in Section 1.3 we let $\delta = \delta_G$ be the Iwasawa character, $\delta(a) = a^{2\alpha}$. We define the **spherical kernel** on the product of G with the character group to be the function given by

$$\Phi_\chi(g) = \Phi(\chi, g) = \int_K \left(\chi\delta^{1/2}\right)(kg)\,dk = \int_K \left(\chi\delta^{1/2}\right)((kg)_A)\,dk.$$

Let $a = \text{Lie}(A)$ be the real vector space of diagonal 2×2 matrices of trace 0,

$$H = \begin{pmatrix} h_1 & 0 \\ 0 & h_2 \end{pmatrix} \text{ with } h_2 = -h_1.$$

Let ζ denote a complex character of a (additive). Then χ may be written as χ_ζ, with ζ such that if $a = \exp(H)$, then

$$\chi_\zeta(a) = e^{\zeta(H)} = e^{\zeta(\log a)}.$$

Thus the Mellin transform can also be viewed as defined on the complex dual $\check{a}_{\mathbf{C}}$, and the spherical kernel may be written in terms of the additive variable ζ, in the form

$$\Phi_\zeta(g) = \Phi(\zeta, g), \quad \zeta \in \check{a}_{\mathbf{C}}.$$

The context should make clear which version we are using, hopefully the most convenient one in each instance. With this notation, we can write the spherical kernel in the form

$$\Phi(\zeta, g) = \int_K (kg)_A^{\zeta+\rho}\, dk.$$

We let $\mathbf{S}_{\mathrm{Iw}} = \mathbf{S}$ be the **integral (spherical) transform** defined by this kernel.

Theorem 1.4.1. *Let f be measurable and K-bi-invariant on G. Also suppose that the repeated integral*

$$\int_G \int_K \left(|\chi|\delta^{1/2}\right)((kg)_A)\, |f(g)|\, dk\, dg$$

is absolutely convergent. Then

$$\mathbf{S}f = \mathbf{MH}f,$$

that is, $\mathbf{S} = \mathbf{MH}$ *on the space of such functions.*

Proof. Straighforwardly,

$$\begin{aligned}
(\mathbf{S}f)(\chi) &= \int_G \Phi(\chi, g) f(g)\, dg && \text{[by definition]}\\
&= \int_K \int_G \left(\chi\delta^{1/2}\right)((kg)_A) f(g)\, dg\, dk && \text{[by Fubini]}\\
&= \int_K \int_G \left(\chi\delta^{1/2}\right)((g)_A) f(g)\, dg\, dk && \text{[by } K\text{-invariance of } f \text{ and } dg]\\
&= \int_A \int_U \chi(a)\delta^{1/2}(a) f(au)\, da\, du\\
&= \mathbf{M}(\mathbf{H}f)(\chi)
\end{aligned}$$

as desired.

Using the fact that **M**, **H** preserve invariance under the Weyl group, we get the following corollary:

Corollary 1.4.2. *The spherical kernel is invariant under W in the variable* χ, *that is,*

$$\Phi_\chi = \Phi_{\chi^{-1}}.$$

Proof. By the theorem, for every $f \in C_c^\infty(K\backslash G/K)$ we get $\mathbf{S}f(\chi) = \mathbf{S}f(\chi^{-1})$, so the integrals of Φ_χ and Φ_χ^{-1} against f give the same value. Hence $\Phi_\chi = \Phi_\chi^{-1}$.

Theorem 1.4.3. *The Mellin transform and the spherical transforms are multiplicative homomorphisms, that is, for* $\varphi,\ \psi \in C_c(K\backslash G/K)$, *and* $f,\ h \in C_c(A)$, *we have*

$$\mathbf{S}(\varphi * \psi) = \mathbf{S}(\varphi) S(\psi) \text{ and } \mathbf{M}(f * h) = \mathbf{M}(f)\mathbf{M}(h).$$

Proof. Directly from the definitions,

$$\mathbf{M}(f*h)(\chi) = \int_A (f*h)(a)\chi(a)\,da$$
$$= \int_A \int_A f(ab^{-1})\,h(b)\chi(a)\,db\,da.$$

we change the order of integration, and we let $a \mapsto ab$, to get

$$= \int_A \int_A f(a)h(b)\chi(a)\chi(b)\,db\,da$$
$$= \mathbf{M}f(\chi)\mathbf{M}h(\chi),$$

thus concluding the proof for **M**. We combine this with Theorem 1.3.2 to get the result for **S**.

Remark. The multiplicative property of Theorem 1.4.3 holds under conditions guaranteeing the absolute convergence of the integrals involved. In particular, they hold for the space of Gaussians defined in the next section.

The Mellin transform is just a Fourier transform with a change of variables, with its usual formalism. The Harish transform will be developed further as in the next section, which gives another expression for the orbital integral, allowing us to compute this integral explicitly for a more explicit space of test functions, the Gaussian functions, which are introduced in Section 1.6. Using $\mathbf{S} = \mathbf{MH}$, we shall then determine another integral kernel for the spherical transform in Section 1.7, as well as for its inverse transform.

In terms of the canonical basis $\alpha = \rho$ (complex case), if $\chi = \chi_{s\alpha}$, we may write

$$(\mathbf{S}f)(\chi_{s\alpha}) = (\mathbf{S}_{\alpha}f)(s).$$

On the other hand, if we use the notation $\chi = \chi_{\zeta}$ with $\zeta \in \check{a}_{\mathbf{C}}$, then we may also write

$$(\mathbf{S}f)\left(\chi_{\zeta}\right) = (\mathbf{S}f)(\zeta).$$

1.5 Computation of the Orbital Integral

Let $z \in G$ be regular, that is, z is conjugate to a diagonal matrix with distinct diagonal elements (the eigenvalues),

$$z \sim \tilde{\eta} = \begin{pmatrix} \eta & 0 \\ 0 & \eta^{-1} \end{pmatrix} \quad \text{with } \eta \in \mathbf{C}, \eta \neq 0.$$

Without loss of generality, we may suppose that $|\eta| \geqq 1$. If $|\eta| > 1$ then η is uniquely determined among the two eigenvalues η, η^{-1} of z.

From the conjugation action on matrices, one sees that the centralizer in G of a regular diagonal element is the set of diagonal matrices. In particular, the isotropy group of $\tilde{\eta}$ under the conjugation action is the diagonal group $\mathbf{D}$. Hence the isotropy group G_z is conjugate to $\mathbf{D}$ by the same conjugation that diagonalizes z to $\tilde{\eta}$. We write $\eta = \eta(z)$ and $\tilde{\eta} = \tilde{\eta}(z)$.

Since conjugation preserves Haar measure, we have an equality or orbital integrals

$$\int_{G_z \backslash G} \varphi\left(g^{-1} z g\right) dg = \int_{\mathbf{D} \backslash G} \varphi\left(g^{-1} \tilde{\eta} g\right) dg. \tag{1.2}$$

Thus the Harish transform can be computed for z in diagonal form, and we use indiscriminately the identity

$$(\mathbf{H}\varphi)(z) = |\mathbf{D}|(\tilde{\eta}(z))\mathrm{Orb}_{G_Z \backslash G}(\varphi, z) = \mathbf{D}(\eta)\mathrm{Orb}_{\mathbf{D}/G}(\varphi, \tilde{\eta}).$$

We now go on with a theorem that allows for a more explicit determination of the orbital integral, and therefore of the Harish transform. As in Section 1.2, we normalize Haar measure so that $\mathbf{T}$ as well as K has measure 1. On $\mathbf{D}$ we put the product measure of $da = dy/y$ and this normalized measure on $\mathbf{T}$, which we write dw (w variable in $\mathbf{D}$). Then for $f \in L^1(G/K)$, we have

$$\int_G f(g)\, dg = \int_U \int_{\mathbf{D}} f(wu)\, dw\, du. \tag{1.3}$$

The dg on the left is of course the Iwasawa measure of Section 1.2. From the modified Iwasawa decomposition $G = \mathbf{D}UK$, we get

$$\begin{aligned} \mathrm{Orb}_{\mathbf{D}\backslash G}(\varphi, \tilde{\eta}) &= \int_U \varphi\left(u^{-1} \tilde{\eta} u\right) du = \int_{\mathbf{C}} \varphi(u(-x)\tilde{\eta} u(x))\, dx \\ &= \int_{\mathbf{C}} \varphi\left(M_\eta(x)\right) dx, \end{aligned} \tag{1.4}$$

where

$$M_\eta(x) = u(x)^{-1} \tilde{\eta} u(x) = \begin{pmatrix} \eta & x(\eta - \eta^{-1}) \\ 0 & \eta^{-1} \end{pmatrix}.$$

The function φ is given in terms of polar coordinates, so we have to express the polar A-coordinate of $M_\eta(x)$ in terms of x. We shall deal with the (x, y) coordinates, polar A-coordinates, and ordinary polar coordinates in $\mathbf{C}$. The final expression will be given most simply in terms of the y-coordinate. The end result is the following theorem:

Theorem 1.5.1. *Let the Haar measure on G be the Iwasawa measure as in Section 1.2. Let φ be measurable on G, K-bi-invariant (so even), and such that* *the orbital integral* $\mathrm{Orb}_{\mathbf{D}\backslash G}(\varphi, \tilde{\eta})$ *is absolutely convergent for regular* $\tilde{\eta}$. *For* $b \in A$, $y = b^{\alpha} \geqq 1$, *write*

$$\varphi(b) = \varphi_{\alpha}(y(b)) = \varphi_{\alpha}(y).$$

Then for $|\eta| \geqq 1$ *and* $\tilde{\eta}$ *regular, we have*

$$\mathrm{Orb}_{\mathbf{D}\backslash G}(\varphi, \tilde{\eta}) = \frac{2\pi}{|\eta - \eta^{-1}|^2} \int_{|\eta|^2}^{\infty} \varphi_{\alpha}(y) \frac{y - y^{-1}}{2} \frac{dy}{y}.$$

Or in terms of the additive variable $v = \log y$, *using the notation*

$$\varphi(b) = \varphi_{\alpha,+}(\log y) = \varphi_{\alpha,+}(v),$$

the formula becomes

$$\begin{aligned}\mathrm{Orb}_{\mathbf{D}\backslash G}(\varphi, \tilde{\eta}) &= \frac{2\pi}{|\eta - \eta^{-1}|^2} \int_{|\eta|^2}^{\infty} \varphi_{\alpha,+}(\log y) \frac{y - y^{-1}}{2} \frac{dy}{y} \\ &= \frac{2\pi}{|\eta - \eta^{-1}|^2} \int_{2\log|\eta|}^{\infty} \varphi_{\alpha,+}(v) \sinh(v)\, dv.\end{aligned}$$

Proof. The proof will be mostly in computing the Jacobians for the various coordinates, to express the orbital integrand in a simple form, and we shall also determine the limits of integration in terms of the y-variable.

To go back and forth between the variable x in $M_{\eta}(x)$ and the A-variable, we use the quadratic map, namely,

$$\text{If } M_{\eta}(x) = k_1 b k_2 \text{ (with } k_1, k_2 \in K, b \in A\text{), then } M_{\eta}(x)\, {}^t\overline{M_{\eta}(x)} = k_1 b^2 k_1^{-1}. \quad (1.5)$$

For the record,

$$M_{\eta}(x)^t\, \overline{M_{\eta}(x)} = \begin{pmatrix} |\eta|^2 + |x|^2 |\eta - \eta^{-1}|^2 & x\eta^{-1}(\eta - \eta^{-1}) \\ \bar{x}\bar{\eta}^{-1}(\bar{\eta} - \bar{\eta}^{-1}) & |\eta|^{-2} \end{pmatrix}. \quad (1.6)$$

So b^2 can be computed from the eigenvalues of $M_{\eta}\, {}^t\overline{M_{\eta}}$, namely the roots of the **characteristic polynomial**

$$T^2 - \tau(r) T + 1 = 0,$$

where $r = |x|$, and $\tau(r)$ is the trace of the matrix $M_{\eta}\, {}^t\overline{M_{\eta}}$:

$$\tau(r) = |\eta|^2 + |\eta|^{-2} + r^2 |\eta - \eta^{-1}|^2. \quad (1.7)$$

We have $r^2 = x\bar{x}$ and

$$dx = dx_1\, dx_2 = r\,dr\,d\theta.$$

With $b = \mathrm{diag}(b_1,\ b_2),\ b_1 > 1$, we can solve for b as a function of r (and so as a function of x). It is convenient to give a symbol for the **discriminant of the characteristic polynomial**

$$\Delta = \tau(r)^2 - 4.$$

Then the roots of the characteristic polynomial are

$$y(b) = b_1^2 = \frac{\tau(r) + \Delta^{1/2}}{2} \text{ and } b_2^2 = \frac{\tau(r) - \Delta^{1/2}}{2}, \tag{1.8}$$

or also $\tau + \Delta^{1/2} = 2y$. Since $b_2^2 = b_1^{-2}$ we get from (1.8) the value of Δ in terms of y,

$$\Delta^{1/2} = y - y^{-1}. \tag{1.9}$$

Note that y is a variable on the positive multiplicative group. In terms of the multiplicative coordinate $y = y(b) = b^{\alpha}$, we write

$$\varphi(b) = \varphi_{\alpha}(y). \tag{1.10}$$

We change variables in the integral from (1.4), $dx_1\ dx_2 = r\ dr\ d\theta$, to get

$$\int_{\mathbf{C}} \varphi(M_{\eta}(x))\,dx = 2\pi \int_0^{\infty} \varphi(b(r))\,r\,dr. \tag{1.11}$$

We have to write down the Jacobian between the variables y and r. By (1.7), we get

$$d\tau = |\eta - \eta^{-1}|^2\, 2r\,dr.$$

By (1.8),

$$dy = \frac{1}{2}\left(\frac{\Delta^{1/2} + \tau}{\Delta^{1/2}}\right) d\tau = \frac{y}{\Delta^{1/2}}\, d\tau.$$

Hence

$$\frac{dy}{y} = \frac{|\eta - \eta^{-1}|^2}{\Delta^{1/2}} 2r\,dr, \text{ or using (8), } r\,dr = \frac{y - y^{-1}}{2} \frac{1}{|\eta - \eta^{-1}|^2} \frac{dy}{y}. \tag{1.12}$$

Putting (1.4) and (1.10), (1.11), (1.12) together yields precisely the integrand in Theorem 1.3.2, so there remains only to determine the limits of integration.

The polar r ranges over $(0, \infty)$. The expression (1.7) for $\tau(r)$ shows that if we put

$$\ell = |\eta|^2 + |\eta|^{-2},$$

then

$$\tau \in (\ell, \infty)\,,\; \Delta \in \left(\ell^2 - 4, \infty\right), \ell^2 - 4 = \left(|\eta|^2 - |\eta|^{-2}\right)^2,\; \ell + \sqrt{\ell^2 - 4} = 2\,|\eta|^2\,. \tag{1.13}$$

From (1.8), we get

$$y = \frac{\tau + \Delta^{1/2}}{2} \in \left(\frac{\ell + \sqrt{\ell^2 - 4}}{2}, \infty\right) = \left(|\eta|^2, \infty\right), \tag{1.14}$$

which determines the limits of integration, and concludes the proof of Theorem 1.5.1.

Corollary 1.5.2. *For $a \in A^+$ (i.e. $a^\alpha > 1$, so a regular), we have*

$$(\mathbf{H}\varphi)(a) = 2\pi \int_{a^\alpha}^{\infty} \varphi_\alpha(y) \frac{y - y^{-1}}{2} \frac{dy}{y}.$$

Remark. Readers will note the factor 2π that came in the theorem and its corollary. The origin of this factor is in having taken the Euclidean measure on **C**, which in ordinary polar coordinates gives the circle measure 2π. In Lie theory, the natural normalization of measures is to give all compact groups (except points) measure 1. Hence the Iwasawa measure can be normalized by dividing by 2π. This normalization will be seen to be the one that also fits polar integration, in terms of the polar decomposition of G. This will be carried out in Section 1.7.

The computation of Theorem 1.5.1 will be completed more explicitly in the next section, Theorem 1.6.1, for Gaussians.

1.6 Gaussians on G and Their Spherical Transform

Having given a formula for the orbital integral, and therefore for the Harish transform, we want a space of test functions to which we can apply this formula in a simple way. There exist a number of such spaces in current use, such as the Schwartz space, or functions with compact support. The latter really don't occur that much in real-life situations, and in the general theory one needs some analysis of the continuity properties of the various transforms to extend them from the space of functions with compact support to more general functions. For our purposes, we define the space of Gaussians on G so that under the spherical transform, they correspond to ordinary Gaussian functions on **R**. Here goes.

To motivate the forthcoming normalization, we recall that if v is a variable on **R**, then the function

$$v \mapsto e^{-v^2/2}$$

is self-dual for the Fourier transform with respect to the Haar measure $dv/\sqrt{2\pi}$. This measure is also the measure that causes Fourier inversion to be true without any extraneous constant factor. We say that this measure is **Fourier normalized**.

In addition, we shall meet the factor

$$\frac{v}{\sinh v} = \frac{2v}{e^v - e^{-v}} = \frac{2\log y}{y - y^{-1}}.$$

It is used to create cancellations whose effect is that whatever one wants to prove "works" by using the above factor. The main point is that there is such a factor having this desirable effect.

By a **basic G-gaussian** (**basic Gaussian** for short) we mean a function on G that is K-bi-invariant, and with some constant $c > 0$, for all $b \in A$ has the formula

$$\begin{aligned} \varphi_c(b) &= e^{-(\alpha(\log b))^2/2c} \frac{\alpha(\log b)}{\sinh \alpha(\log b)} \\ &= e^{-(\log y)^2/2c} \frac{\log y}{(y - y^{-1})/2} \qquad \text{with } y = y(b) \qquad (1.15) \\ &= e^{-v^2/2c} \frac{v}{\sinh v} \qquad \text{with } v = \log y. \end{aligned}$$

We can then deal with the vector space **Gauss**(G) generated by the basic Gaussian functions as a natural space of test functions, which is especially important because in the long run this space is adapted to the use of the heat Gaussian. However, the heat differential equation will be irrelevant for a while, so we may as well deal with any Gaussian.

We now want to determine explicitly both the orbital integral with φ_c and its spherical transform. Since $\mathbf{S} = \mathbf{MH}$, we start with the orbital integral, which will also give us $\mathbf{H}\varphi_c$. As usual, the Haar measure on G is the Iwasawa measure.

Theorem 1.6.1. *For the orbital integral, with the Gaussian φ_c, and regular diagonal z, we have the value*

$$\mathrm{Orb}_{\mathbf{D}\backslash G}(\varphi_c, z) = \frac{2\pi c}{|z^{\alpha/2} - z^{-\alpha/2}|^2} e^{-(\log|z^\alpha|)^2/2c}.$$

For $a \in \mathrm{A}$ regular, $y = y(a)$, the Harish transform is given by

$$(\mathbf{H}\varphi_c)(a) = 2\pi c e^{-(\alpha(\log a))^2/2c} = 2\pi c e^{-(\log y)^2/2c}.$$

Proof. We write $z = \tilde{\eta}$ and take $\varphi = \varphi_c$ in Theorem 1.5.1. The factor $(\sinh v)^{-1}$ was used so that there would be a cancellation that allows us to evaluate the orbital integral exactly, with the stated answer because of the v-factor. We then use the definition of the Harish transform as a product of the orbital integral times the $|\mathbf{D}|(a)$ factor. This factor cancels, giving the stated answer.

Thus we see that under the Harish transform, the G-Gaussians go to ordinary Gaussians on A, with the constant appropriately placed. The factor 2π came originally from having taken the ordinary Euclidean measure on $\mathbf{C}$ ($\cong U$), so polar coordinates for which the circle has measure 2π, which is standard, but is given priority over the counterconvention to give compact groups measure 1.

Next comes the Mellin transform. We use the y-variable, $\alpha(\log a) = \log y$.

Theorem 1.6.2. *Write a character on a in the form $\zeta = s\alpha$ with complex s. Then the spherical transform is*

$$(\mathbf{S}\varphi_c)(s\alpha) = (\mathbf{MH}\varphi_c)(s\alpha) = (2\pi c)^{3/2}\, e^{s^2c/2}.$$

On the imaginary axis $s = \mathbf{i}r$; this gives

$$(\mathbf{S}\varphi_c)(\mathbf{i}r\alpha) = (\mathbf{MH}\varphi_C)(\mathbf{i}r\alpha) = (2\pi c)^{3/2}\, e^{-r^2c/2}.$$

Proof. By Theorem 1.6.1 and $\mathbf{S} = \mathbf{MH}$ we have

$$\frac{1}{2\pi c}(\mathbf{S}_\alpha\varphi_c)(s) = \int_0^\infty e^{-(\log y)^2/2c} y^s \frac{dy}{y} = \int_{-\infty}^{\infty} e^{-v^2/2c} e^{vs} dv.$$

Because of the rapid decay of Gaussians, the Mellin transform is entire in s, so it suffices to determine the integral when $s = \mathbf{i}r$ (r real) is pure imaginary. One then uses the self-duality of the function $e^{-v^2/2}$ with respect to $dv/\sqrt{2\pi}$, and a change of variables to get rid of c, to conclude the proof.

We let the **Gauss space** on $\mathbf{R}$, Gauss($\mathbf{R}$), be the vector space generated by the Gaussian functions $r \mapsto e^{-r^2c/2}$, with all constants $c > 0$. Sometimes we deal with the imaginary axis $\mathbf{iR}$, and Gauss($\mathbf{iR}$) is the space generated by the functions having the above value at $\mathbf{i}r$. We let Gauss(G) be the G-**Gauss space**, i.e., the vector space generated by the basic G-Gaussians.

Corollary 1.6.3. *The spherical transform gives a linear isomorphism*

$$\text{Gauss}(G) \to \text{Gauss}(\mathbf{R}).$$

Proof. Immediate from the linear independence of the exponential functions with different constants c.

We tabulate one more useful integral.

Corollary 1.6.4. *With the Iwasawa measure, the total integral of φ_c is*

$$\int_G \varphi_c(g)\, d\mu_{\text{Iw}}(g) = (2\pi c)^{3/2}\, e^{c/2}.$$

Proof. Take $s = -1$ in Theorem 1.6.2, and recall that $\alpha = \rho$. The spherical kernel is given by the integral

$$\Phi(\zeta, g) = \int_K (kg)_A^{\zeta+\rho}\, dk.$$

At $\zeta = -\rho$, the spherical kernel has value 1, and the stated answer falls out of Theorem 1.6.2.

Note the appearance of a factor 2π in the last three formulas concerning the Harish transform, the spherical transform, and Haar measure. As on the real line when $d\theta/\sqrt{2\pi}$ is a useful normalization giving the circle measure 1, it will be useful to divide by 2π, as will be carried out in the next section. The world is made up so that this normalization is also the normalization that makes Fourier inversion come out without an extra constant.

1.6.1 The Polar Height

Let $g \in G$, with polar decomposition

$$g = k_1 b k_2,\ k_1,\ k_2 \in K \text{ and } b \in A.$$

The element $b \in A$ is determined only up to a permutation of its diagonal components. We call $y = y_{\mathbf{p}}(g) = b^{\alpha}$ or y^{-1} a **polar coordinate** (or polar A-coordinate) of g. We defined Gaussians using this coordinate. Note that $(\log\, y)^2$ is uniquely determined by g, and is real analytic on G. We now define the **polar height** σ to be the K-bi-invariant function such that $\sigma \geqq 0$ and

$$\sigma^2(g) = \sigma^2(b) = (\log y)^2 = v^2.$$

Then the Gaussian φ_c can be written in the form

$$\varphi_c = e^{-\sigma^2/2c}\frac{\sigma}{\sinh\sigma} = e^{-(\log y)^2/2c}\frac{\log y}{\sinh(\log y)}. \tag{1.16}$$

Note that $v/\sinh\, v$ is a power series in $v^2 = (\log\, y)^2$, so real analytic on G. Properties of the polar height will be given in Section 1.8. We first go straight for spherical inversion.

We relate the σ definition to the norm associated with a positive definite scalar product on a, in the present case the **trace form**. For $H \in a$, we define the symmetric bilinear form B by

$$B(H,H) = \mathrm{tr}(HH) = |H|^2 \text{ for abbreviation.}$$

For $b = \text{diag}(b_1, b_2) \in A$ we have $\log b = \text{diag}(\log b_1, \log b_2)$, and

$$|\log b|^2 = (\log b_1)^2 + (\log b_2)^2 = 2(\log b_1)^2 = (\log y)^2/2,$$

so

$$\sigma^2(b) = 2|\log b|^2 = (\log y)^2.$$

Remark. In (1.16) there is an advantage to have the formula written entirely in terms of σ, which was defined in terms of the group coordinate y. This advantage does not persist in higher rank, in which case one has to use a G-invariant bilinear form B, suitably scaled, to define σ. Even here, we have the natural scaling factor 2. We discuss scaling the form systematically in Sections 2.4 and 2.5.

In higher rank, one also faces a finite set of characters α, occurring in the regular representation of a on n, so the formula for a Gaussian involves a finite product

$$\varphi_c(b) = e^{-|\log b|^2/2c} \prod_{\alpha} \frac{\log b^{\alpha}}{\sinh \log b^{\alpha}}.$$

The trace form on a can be extended in two natural ways to the complexification, namely a symmetric way and a Hermitian way. The symmetric way will be relevant in the next section, so we add some comments concerning it.

The trace form induces a corresponding positive definite scalar product on the dual space $\check{a}$, because it induces a natural isomorphism of a with $\check{a}$. Specifically, if $\lambda \in \check{a}$, then there is an element $H_\lambda \in a$ such that for all $H \in a$,

$$B(H_\lambda, H) = \lambda(H).$$

Then by definition, for $\lambda, \xi \in \check{a}$, we have

$$\check{B}(\xi, \lambda) = B(H_\xi, H_\lambda),$$

so for example $\check{B}(\alpha, \alpha) = 2$.

We can then extend the form from $\check{a}$ to a $\mathbf{C}$-bilinear form $\check{B}_{\mathbf{C}}$ on its complexification, i.e., to characters $\zeta \in \check{a}_{\mathbf{C}}$ that can be written in the form $\zeta = \xi + \mathbf{i}\lambda$, with $\xi, \lambda \in \check{a}$. Of course, we call ξ and λ the **real** and **imaginary parts** of ζ respectively. Then by definition

$$\check{B}_{\mathbf{C}}(\zeta, \zeta) = \check{B}(\zeta, \zeta) + 2\mathbf{i}\check{B}(\xi, \lambda) - \check{B}(\lambda, \lambda).$$

In particular, writing $\zeta = s\alpha$, $s \in \mathbf{C}$, we have

$$\check{B}_{\mathbf{C}}(\mathbf{i}\lambda, \mathbf{i}\lambda) = -\check{B}(\lambda, \lambda) \text{ and } \check{B}_{\mathbf{C}}(s\alpha, s\alpha) = 2s^2.$$

Thus the form $\check{B}_{\mathbf{C}}$ is $\mathbf{C}$-bilinear, not Hermitian. Its restriction to the imaginary axis $\mathbf{i}\check{a}$, however, is negative definite, corresponding to minus the given form on the imaginary part from $\check{a}$, itself corresponding to minus the original form on a.

1.7 The Polar Haar Measure and Inversion

Commonly, spaces of test functions range over functions with compact support (continuous, $\mathbf{C}^\infty$), and Schwartz spaces, possibly with L^1, L^2, L^p conditions. We shall use the Gauss space in a natural way as we develop the polar Haar measure.

In Section 1.1 we already noted the polar decomposition $G = KAK$. A K-bi-invariant function is thus determined by its values on the polar A-component b when $g = k_1 b k_2$. We want an integration formula in terms of this A-component. For this, we define the **polar Jacobian**

$$J_{\mathbf{p}}(b) = \left(\frac{b^\alpha - b^{-\alpha}}{2}\right)^{m(\alpha)} = \left(\frac{y - y^{-1}}{2}\right)^2$$

with $y = y_{\mathbf{p}}(b) = b^\alpha$ and $m(\alpha) = 2$. In Section 1.2 we define the Iwasawa measure $d\mu_{\mathrm{Iw}}$. Now we define the **polar measure** $d\mu_{\mathbf{p}}$ by

$$d\mu_{\mathbf{p}}(b) = J_{\mathbf{p}}(b)\,db = \left(\frac{e^v - e^{-v}}{2}\right)^2 dv = (\sinh v)^2\,dv$$

in terms of the v-variable, $v = \log\, y$. It is a Jacobian computation due to Harish-Chandra in general that the functional

$$\varphi \mapsto \int_K \int_K \int_A \varphi(k_1 b k_2)\, J_{\mathbf{p}}(b)\, db\, dk_1\, dk_2$$

is a Haar functional on G; cf. [Hel 84], Chapter I, Theorem 5.8; [GaV 88], Proposition 2.4.11; and [JoL 01], Chapter VI, Theorem 1.5. For K-bi-invariant functions, one does not need the two K integrations in the above formula. The total dk-measure of K is assumed to be normalized to 1. The question arises how the polar Haar measure is related to the Iwasawa measure. The answer is the following theorem:

Theorem 1.7.1. *For $\varphi \in$ Gauss (G), we have*

$$\frac{1}{2\pi}\int_G \varphi(g)\, d\mu_{\mathrm{Iw}}(g) = \int_A \varphi(b)\, \mathrm{J}_{\mathbf{p}}(b)\, db = \int_A \varphi d\mu_{\mathbf{p}}.$$

Proof. This is a routine verification at the level of elementary calculus, using the basic Gaussian functions φ_c. We want to show that the integral of φ_c with respect to the polar measure is equal to $1/2\pi$ times the value found in Corollary 6.4, namely

$$(2\pi)^{1/2} c^{3/2} e^{c/2}.$$

In terms of the variable v, we see that one factor sinh v cancels in the Jacobian, and the polar integral over A is thus equal to

$$\int_A \varphi_C(b)\, d\mu_{\mathbf{p}}(b) = \int_{-\infty}^{\infty} e^{-v^2/2c} v \left(\frac{e^v - e^{-v}}{2}\right) dv = \int_{-\infty}^{\infty} e^{-v^2/2c} e^v v\, dv.$$

We use the **calculus identity**

$$\frac{1}{s}\int_{-\infty}^{\infty} e^{sx}e^{-x^2/2c}x\,dx = (2\pi)^{1/2}c^{3/2}e^{s^2c/2}$$

$$= \frac{1}{2\pi}(\mathbf{S}_\alpha\varphi_c)(s) \text{ by Theorem 1.6.2.} \tag{1.17}$$

We integrate by parts, with $u = e^{sx}$, $du = se^{sx}\,dx$. The boundary term is 0, and the extra factor x disappears, while at the same time s comes out as a factor to cancel the $1/s$. Then (1) drops out, and putting $s = 1$, so does the equality stated in Theorem 1.7.1.

Corollary 1.7.2. *For all* $f \in C_c(K\backslash G/K)$ *one has*

$$\frac{1}{2\pi}\int_G f(g)\,d\mu_{\mathrm{Iw}}(g) = \int_A f(b)\,d\mu_{\mathbf{p}}(b).$$

Proof. Since the formula holds with constant factor 1 on the Gauss space, it holds with constant factor 1 for any test function for which the integrals are absolutely convergent.

Having two natural measures, the Iwasawa measure and the polar measure, we get two normalizations of the spherical transform, which we denote by

$$\mathbf{S}_{\mathrm{Iw}} \text{ and } \mathbf{S}_{\mathbf{p}}, \text{ so that } \mathbf{S}_{\mathbf{p}} = \frac{1}{2\pi}\mathbf{S}_{\mathrm{Iw}}.$$

The $\mathbf{S}$ of preceding sections is $\mathbf{S}_{\mathrm{Iw}}$. We may call $\mathbf{S}_{\mathbf{p}}$ the **polar normalized spherical transform**, or simply the **polar spherical transform**. We shall exhibit an integral kernel for $\mathbf{S}_{\mathbf{p}}$ quite different from the one used in the definition of spherical transform. It was the K-invariantization of a character; it will now be related to polar coordinates. We let $\check{B}_{\mathbf{C}}$ be the dual of the $\mathbf{C}$-bilinear trace form, so $\check{B}_{\mathbf{C}}(\alpha,\alpha) = 2$. Let

$$\Phi_{\mathbf{p}}(\zeta,a) = \frac{\check{B}(\alpha,\rho)}{\check{B}_{\mathbf{C}}(\alpha,\zeta)}\frac{a^{\zeta}-a^{-\zeta}}{a^{\rho}-a^{-\rho}}$$

Writing $\zeta = s\alpha$ with $s \in \mathbf{C}$, and recalling that $\alpha = \rho$ (complex case), we write this formula in the form

$$\Phi_{\mathbf{p}}(s\alpha,a) = \Phi_{\mathbf{p},\alpha}(s,a) = \frac{1}{s}\frac{y^s-y^{-s}}{y-y^{-1}} = \frac{1}{s}\frac{e^{sv}-e^{-sv}}{e^v-e^{-v}} = \frac{1}{s}\frac{\sinh sv}{\sinh v}.$$

Theorem 1.7.3. *For all basic Gaussians, taking the polar measure convolution on A, we have*

$$(\Phi_{\mathbf{p}} *_{A,\mathbf{p}} \varphi_c)(s\alpha) = \mathbf{S}_{\mathbf{p}}\varphi_c(s\alpha),$$

so $\Phi_{\mathbf{p}}$ *is an integral kernel for* $\mathbf{S}_{\mathbf{p}}$ *(with polar measure).*

Proof. The computation is at the same level of calculus as in the proof of Theorem 1.6.2. The convolution integral on the left by definition is equal to

$$\int_A \Phi_{\mathbf{p}}(s\alpha, a)\varphi_c(a)\,d\mu_{\mathbf{p}}(a) = \frac{1}{s}\int_{-\infty}^{\infty} \frac{e^{sv}-e^{-sv}}{e^v-e^{-v}}\,e^{-v^2/2c}\,\frac{2v}{e^v-e^{-v}}\left(\frac{e^v-e^{-v}}{2}\right)^2 dv$$

$$= \frac{1}{s}\int_{-\infty}^{\infty}\left(\frac{e^{sv}-e^{-sv}}{2}\right)e^{-v^2/2c}y\,dv$$

$$= \frac{1}{s}\int_{-\infty}^{\infty} e^{sv}e^{-v^2/2c}v\,dv.$$

We apply the calculus identity to conclude the proof.

Note. The spherical kernel in polar coordinates on $SL_n(\mathbf{C})$ is originally due to Gelfand–Naimark [GeN 50/57], Section 20, 3, p. 101, of the German translation.

We observe that the polar spherical kernel has a product structure. The first factor has a name, the **Harish-Chandra c-function**, defined for our purposes by

$$c(\zeta) = \frac{\check{B}(\alpha,\rho)}{\check{B}_{\mathbf{c}}(\alpha,\zeta)} = \frac{1}{s}\ \text{if}\ \zeta = s\alpha.$$

The second factor comes from a skew-symmetrization procedure, namely

$$\frac{a^{\zeta}-a^{-\zeta}}{a^{\rho}-a^{-\rho}}.$$

We know from Corollary 6.3 that the spherical transform is a linear isomorphism between the Gauss spaces. We want to exhibit the measure on $\check{a}$ with respect to which the transpose of the spherical kernel induces the inverse integral transform. We do this as follows.

As to the Haar measure on $\check{a}$, we have the Haar measure $da = dy/y = dv$ on $\check{a} \cong \mathbf{R}$. We let $d\mu_{\text{Fou}}(\lambda)$ be the Haar measure on $\check{a}$ that makes Fourier inversion come out without any extra constant factor. The **Fourier transform F** is defined by

$$(\mathbf{F}f)(\lambda) = \int_a f(H)^{-\mathbf{i}\lambda(H)}\,dH.$$

The isomorphism of a with $\mathbf{R}$ is taken via the coordinate v. Thus if $\lambda = r\alpha$ on a, with $r \in \mathbf{R}$, then

$$(\mathbf{F}f)(\lambda) = (\mathbf{F}_{\alpha,+}f)(r) = \int_{\mathbf{R}} f_{\alpha,+}(v)\,e^{-\mathbf{i}rv}\,dv.$$

The **Fourier normalized measure** is then defined to be

$$d\mu_{\text{Fou}}(\lambda) = \frac{dr}{2\pi}.$$

We define the **Harish-Chandra measure** on $\breve{a}$ by

$$d\mu_{HC}(\lambda) = |c(\mathbf{i}\lambda)|^{-2} d\mu_{Fou}(\lambda) = \frac{r^2 dr}{2\pi}.$$

We let ${}^t\Phi_{\mathbf{p},\mathrm{HC}}$ be the integral operator having the kernel function $\Phi_{\mathbf{p}}(\mathbf{i}\lambda, a)$ with respect to the Harish-Chandra measure, so for an even function h such that the integral is absolutely convergent, by definition

$$\left({}^t\Phi_{\mathbf{p},HC}h\right)(a) = \int_{\breve{a}} h(\mathbf{i}\lambda)\Phi_{\mathbf{p}}(\mathbf{i}\lambda, a)\, d\mu_{HC}(\lambda)$$

$$= \int_{\mathbf{R}} h_\alpha(\mathbf{i}r)(-\mathbf{i}r)\frac{e^{\mathbf{i}rv} - e^{-\mathbf{i}rv}}{e^v - e^{-v}}\frac{dr}{2\pi} \qquad [\text{with } v = \alpha(\log a)]. \quad (1.18)$$

Theorem 1.7.4. *The integral operator ${}^t\Phi_{\mathbf{p},\mathrm{HC}}$ is the inverse of $\mathbf{S}_{\mathbf{p}}$ on the Gauss spaces*

$$\mathbf{S}_{\mathbf{p}} : \mathrm{Gauss}(G) \longrightarrow \mathrm{Gauss}(\mathbf{iR}).$$

The map $\mathbf{S}_{\mathbf{p}}$ is an L^2-isometry with respect to the polar measure on G and the Harish-Chandra measure on $\breve{a}$, the integral scalar product taken on $\breve{a}$.

Proof. As to the first statement, we let $h = \mathbf{S}_{\mathbf{p}}\varphi_c$, that is,

$$h_\alpha(\mathbf{i}r) = (2\pi)^{1/2} c^{3/2} e^{-r^2 c/2}.$$

We plug this in formula (2). By a change of variables $v \mapsto -v$, we can omit the term with $-e^{-\mathbf{i}rv}$ provided we multiply the integral by 2, which combined with $(e^v - e^{-v})^{-1}$ gives $1/\sinh v$. Then the integral can be evaluated, say by plugging into the calculus identity (1), showing that ${}^t\Phi_{\mathbf{p},\mathrm{HC}}\mathbf{S}_{\mathbf{p}}\varphi_c = \varphi_c$.

As to the second statement, we first make explicit the L^2-scalar product on G, for the polar measure. For two basic Gaussians φ_c, $\varphi_{c'}$, we get, integrating over $A \cong (-\infty, \infty)$;

$$< \varphi_c, \varphi_{c'} >_{\mathbf{p}} = \int_{-\infty}^{\infty} e^{-v^2/2c}\frac{2v}{e^v - e^{-v}} e^{-v^2/2c'}\frac{2v}{e^v - e^{-v}}\left(\frac{e^v - e^{-v}}{2}\right)^2 dv$$

$$= \int_{-\infty}^{\infty} e^{-v^2(c+c')/2cc'} v^2\, dv$$

$$= (cc')^{3/2}\int_{-\infty}^{\infty} e^{-v^2(c+c')/2} v^2\, dv \qquad [\text{by } v \mapsto v(cc')^{1/2}].$$

On the other hand,

$$\langle \mathbf{S_p}\varphi_c, S_{\mathbf{p}}\varphi_{c'}\rangle_{\mathrm{HC}} = \int_{-\infty}^{\infty} (2\pi)^{1/2} c^{3/2} e^{-r^2 c/2} (2\pi)^{1/2} c'^{3/2} e^{-r^2 c'/2} r^2 \frac{dr}{2\pi}$$

$$= (cc')^{3/2} \int_{-\infty}^{\infty} e^{-r^2(c+c')/2} r^2 \, dr.$$

This proves the theorem.

Remark. In [JoL 03b], it is shown how the above isometry extends by continuity to all of $L^2(K\backslash G/K)$ because of the denseness of the Gauss space.

We have now concluded the basic analysis of spherical inversion. The next chapter will give a major application.

1.8 Point-Pair Invariants, the Polar Height, and the Polar Distance

Having finished spherical inversion, we give more properties of the height, notably its being a distance function. This property becomes especially significant in the higher-rank case.

Let φ be a K-bi-invariant function on G. Following Selberg, we define its associated **point-pair invariant** $\mathbf{K}_\varphi$ to be the function defined by

$$\mathbf{K}_\varphi(z,w) = \varphi\left(z^{-1}w\right).$$

Immediately from the K-bi-invariance of φ, we conclude that $\mathbf{K}_\varphi$ *is defined on* $G/K \times G/K$.

The function $\mathbf{K}_\varphi$ *is symmetric*, that is,

$$\mathbf{K}_\varphi(z,w) = \mathbf{K}_\varphi(w,z),$$

because φ is even. (In higher dimensions, one has to assume the evenness condition in addition to the K-bi-invariance.)

The point-pair invariant $\mathbf{K}_\varphi$ *is G-invariant*, i.e.,

$$\mathbf{K}_\varphi(gz,gw) = \mathbf{K}_\varphi(z,w),$$

immediately from the definition.

A K-bi-invariant function on G is determined by its values on A because of the polar decomposition $G = KAK$. If $g = k_1 b k_2$, then b is uniquely determined up to a permutation of its diagonal elements. Hence a function on A that is invariant under permutations extends to a K-bi-invariant function on G.

We now come to the polar height, defined in Section 1.6 for $b \in A$ by $\sigma \geqq 0$,

$$\sigma^2(b) = (\log y_b)^2 = (\log b^\alpha)^2,$$

and K-bi-invariant. We first compare the polar height and the height of the A-Iwasawa component. The polar height is bigger.

Proposition 1.8.1. *Let $g = uak = k_1 b k_2$ be the Iwasawa, respectively polar, decomposition. Then*

$$\sigma(a) \leqq \sigma(b), \text{ or alternatively, } |\log a| \leqq |\log b|.$$

In general reductive Lie group theory, this result is due to Harish-Chandra. For bibliographical references about a much larger context involving Horn, Lenard, Thompson, Kostant, see [JoL 01a], Chapter I, Theorem 1.6.2, and the accompanying comments.

Proof. Let e_1, e_2 be the unit column vectors of $\mathbf{C}^2$. Let $a = \text{diag}(a_1, a_2)$ and $b = \text{diag}(b_1, b_2)$. Without loss of generality, we may assume a_1, $b_1 \geqq 1$. It suffices to prove that $a_1 \leqq b_1$. We have $ua = av$ with $v \in U$. Write $av = k_1 bk'$. Let $|\cdot|$ be the Euclidean norm on $\mathbf{C}^2$. We have

$$a_1^2 = |ave_1|^2 = |k_1 bk' e_1|^2.$$

Since k' is unitary, there are numbers c_1, c_2 such that $k'e_1 = c_1 a_1 + c_2 a_2$, and $|c_1|^2 + |c_2|^2 = 1$. Then

$$|bk'e_1|^2 = |b_1 c_1|^2 + |b_2 c_2|^2 = b_1^2 |c_1|^2 + b_2^2 |c_2|^2.$$

It suffices to show that for $y \geqq 1$ $(y = b_1^2)$ we have

$$y|c_1|^2 + y^{-1}|c_2|^2 \leqq y.$$

Clearing denominators, the inequality falls out.

The following property is deeper, and is due to Cartan–Harish-Chandra.

Theorem 1.8.2. Triangle Inequality. *For g, $g' \in G$ we have*

$$\sigma(gg') \leqq \sigma(g) + \sigma(g').$$

See [Lan 99], on *Bruhat–Tits spaces*, which provides an elementary self-contained treatment sufficient for our purposes. See also [JoL 01a], Chapter 10.

Proposition 1.8.3. We then define the **polar distance**

$$d_{\mathbf{p}}(z, w) = \sigma(z^{-1}w),$$

which is the associated point-pair invariant. As we saw, $d_{\mathbf{p}}$ is defined on G/K.

To make the link with [Lan 99], recall the isomorphism $G/K \cong \mathrm{Pos}_n(\mathbf{C})$ (positive definite Hermitian matrices) by $g \mapsto gg^*$ (with $g^* = \bar{g}$). Let

$$y_{\mathbf{p}}(g) = b^{\alpha} = b_1/b_2 = b_1^2$$

be the polar y-coordinate of g. Then

$$\sigma(g) = |\log y_{\mathbf{p}}(g)| = d_{\mathbf{p}}\left(gK, e_{G/K}\right),$$

where $e_{G/K}$ is the unit coset of G/K.

Theorem 1.8.4. *The function $d_{\mathbf{p}}$ is a distance function on G/K.*

Proof. This is an immediate consequence of the triangle inequality in Theorem 1.8.2, taking its usual form in terms of $d_{\mathbf{p}}$. A full treatment on $\mathrm{Pos}_n(\mathbf{R})$ is in [Lan 99], mentioned above. The analogous treatment for $\mathrm{Pos}_n(\mathbf{C})$ or $\mathrm{SPos}_n(\mathbf{C})$ runs exactly the same way. Thus we get a natural G-invariant distance on G/K, which is thereby a Riemannian manifold, with G acting by translation as a group of isometries.

See also [JoL 01], Section 10.7, leading into the heat kernel.

The polar height will be seen in the next chapter as the function that comes most basically into the definition of a normalized Gaussian, the heat Gaussian, giving rise to its point-pair invariant, the heat kernel. See especially the items on scaling in Section 2.5, culminating with Corollary 5.6.

Chapter 2
The Heat Gaussian and Heat Kernel

The Gaussian space is very useful when computing certain explicit simple formulas which come up in spherical inversion theory. One also wants to know that an identity for all Gaussian functions implies a similar identity for more general test functions usually occurring in analysis. This section provides a background for the systematic approach in [JoL 03b].

For some purposes, one wants to normalize the Gaussian functions more precisely. The ultimate normalization is that of the heat kernel. The structure of the heat kernel was discovered by Gangolli in his fundamental paper [Gan 68], including the fact that it is the inverse image of a normalized Gaussian on Euclidean space under the spherical transform. Our Theorems 2.5.1 and 1.6.1 of Chapter 1 will allow us to get Gangolli's theorem by our simple explicit handling of the spherical transform. One then gets a formula for the heat kernel in terms of a Gaussian, which exhibits a quadratic exponential decay.

The existence of such a formula in closed form holds only on complex groups G. On real groups, it is given in integral form. The general formalism and structure of this integral and its connection with spherical inversion is explained in [JLS 03], as a special case of totally geodesic embeddings of symmetric spaces into each other. This general framework also provides a more extensive context for the Flensted–Jensen transform, going from the complex to the real case [FlJ 78].

2.1 Dirac Families of Gaussians

We shall use the **height function** $\sigma_G = \sigma$ defined on an element $g = k_1 b k_2$ (polar decomposition) by

$$\sigma^2(g) = (\log b^{\alpha})^2 = (\log y_{\mathbf{p}})^2$$

cf. Section 1.6. Recall that σ is K-bi-invariant.

A **Weierstrass–Dirac family** $\{\psi_c\}$ for $c \to 0$, with respect to a Haar measure dg, is a family of continuous K-bi-invariant functions on G satisfying the following properties:

WD 1. We have $\psi_c \geqq 0$ for all $c > 0$.
WD 2. $\int_G \psi_c(g)dg = 1$.
WD 3. Given $\delta > 0$, $\lim_{c\to 0} \int_{\sigma(g)>\delta} \psi_c(g)dg = 0$.

Proposition 2.1.1. *Let φ_c be the Gaussian of Section 1.6. Let $d\mu(g) = dg$ be a given Haar measure. Let $I_\mu(\varphi_c)$ be the total integral of φ_c over G, and let*

$$\psi_{c,\mu} = \varphi_c / I_\mu(\varphi_c).$$

Then $\{\psi_{c,\mu}\}$ is a WD family with respect to μ.

Proof. The first two conditions are obvious from the definitions. As to **WD 3**, it suffices to prove it for one choice of Haar measure, say the polar measure. By Chapter 1, Corollary 1.6.4, we know the explicit value

$$\psi_{c,\mathbf{p}}(g) = \frac{2\pi e^{-c/2}}{(2\pi c)^{3/2}}\varphi_c(g) = \frac{2\pi e^{-c/2}}{(2\pi c)^{3/2}} e^{-\sigma^2(g)/2c}\frac{\sigma}{\sinh\sigma}. \tag{2.1}$$

The third condition follows at once from the quadratic exponential decay of the function, and the fact that the polar measure grows only linearly exponentially. (We formalize these notions below.)

Note that $\sigma/(\sinh\sigma)$ is defined naturally in terms of the group coordinate y,

$$\frac{\sigma}{\sinh\sigma} = \frac{\log y}{\sinh(\log y)} \text{ and } y = b^\alpha.$$

That one can use this same function for the $\sigma^2(g)$ in the (quadratic) exponent is an accident of the low dimension, which does not persist in higher dimension (or higher rank), but we take advantage of it here, with a warning.

The WD family with respect to the polar Haar measure will be called the **Gaussian polar WD family**.

2.1.1 Scaling

Different Haar measures differ by a constant factor, and give rise to a corresponding WD family. The Iwasawa measure gives rise to the Iwasawa Dirac family, obtained from (2.1) by leaving out the scaling factor 2π.

In Section 1.4 we defined the spherical transform with respect to the Iwasawa measure. Since we shall scale measures, we must now index this transform to indicate the dependence on the measure, and thus write $\mathbf{S}_{\mathrm{Iw}}$. For the polar measure,

$\mu_{\mathbf{p}} = (2\pi)^{-1}\mu_{\mathrm{Iw}}$. More generally, if μ, μ' are two Haar measures and $\mu' = c'\mu$, we have

$$\mathbf{S}_{\mu'} = c'\mathbf{S}_{\mu}.$$

This definition is consistent, and gives the relation

$$\psi_{c,c'\mu} = c'^{-1}\psi_{c,\mu}. \tag{2.2}$$

For any Haar measures μ, μ', if $\zeta = s\alpha$, and B as before is the trace form on a, we have

$$\mathbf{S}_{\mu'}\psi_{c,\mu'}(\zeta) = \mathbf{S}_{\mu}\psi_{c,\mu}(\zeta) = e^{-c/2}e^{\check{B}}(\zeta,\zeta)\,c/4 = e^{-c/2}e^{s^2c/2}. \tag{2.3}$$

In particular, $\mathbf{S}_{\mu}(\psi_{c,\mu})$ *is independent of the choice of Haar measure.*

For example, we can say that formula (2.1) for $\psi_{c,\mathbf{p}}$ is the unique G-Gaussian having the polar spherical image equal to the expression written in (2.3), namely

$$e^{-c/2}e^{s^2c/2}.$$

To fit the final normalization used afterward, we let $c = 4t$ and define the **μ-heat Gaussian** (justifiably by Theorem 2.5.1 below) to be

$$\mathbf{g}_{t,\mu}(g) = \psi_{4t,\mu}(g).$$

We also define the function

$$E_t(s\alpha) = e^{2(s^2-1)t},$$

which will be shown to be an eigencharacter (Proposition 2.4.4 and Chapter 3, Theorem 3.4.4). For the moment, repeating (2.3) with this final notation, we have the following:

Theorem 2.1.2. *The function $\mathbf{g}_{t,\mu}$ is the unique Gaussian satisfying*

$$\mathbf{S}_{\mu}(\mathbf{g}_{t,\mu}) = E_t.$$

These Gaussians form a WD family for the measure μ.

Note that if $\mu' = c'\mu$, then the convolutions scale by the same factor:

$$f_1 *_{\mu'} f_2 = c' f_1 *_{\mu} f_2.$$

We write down explicitly the heat Gaussian with the Iwasawa measure μ_{Iw}:

$$\mathbf{g}_{t,\mathrm{Iw}}(g) = (8\pi t)^{-3/2}e^{-2t}e^{-\sigma^2(g)/8t}\frac{\sigma}{\sinh\sigma}, \tag{2.4$_{\mathrm{Iw}}$}$$

so

$$\mathbf{g}_{t,\mathbf{p}}(g) = 2\pi\mathbf{g}_t,\mathrm{Iw}(g), \tag{2.4$_{\mathbf{p}}$}$$

and in general, if $\mu' = c'\mu$, then

$$\mathbf{g}_{t,c'\mu} = c'^{-1}\mathbf{g}_{t,\mu}.$$

Further comments on scaling will be made in Sections 2.4 and 2.5 concerning the Casimir operator and the heat equation when we scale the scalar product (trace form). Then the factor e^{-2t} in (2.4$_{\mathrm{Iw}}$) will be written properly in full in terms of B, namely

$$e^{-2t} = e^{-\check{B}}(\rho,\rho)t, \text{ with } \rho = \alpha.$$

By scaling the form B, we shall see that we can get rid of a factor 2.

In light of Theorem 2.1.2, we often omit the subscripts **p**, Iw, or μ in general, and write simply $\mathbf{g}_t$ and $\mathbf{S}(\mathbf{g}_t)$. Writing $\mathbf{g}_t$ by itself, it is understood that a Haar measure has been fixed. We define the **heat kernel** in each case to be the point-pair invariant $\mathbf{K}_{\mathbf{g}_t} = \mathbf{K}_{\mathbf{g}_{t,\mu}}$, abbreviated $\mathbf{K}_t$, so

$$\mathbf{K}_t(z,w) = \mathbf{g}_t\left(z^{-1}w\right).$$

From the general properties of Section 1.8, we know that *the heat kernel is defined on $G/K \times G/K$. It is symmetric and G-invariant.*

The same remarks as above apply to the heat kernel. For instance, we have that

$$\mathbf{K}_{t,\mu} *_{\mu} f \textit{ is independent of } \mu \tag{2.5}$$

for any function f and any Haar measure μ, because the scaling factor cancels.

The next section will give essentially general consequences of the Weierstrass–Dirac property in the context of convolutions. The justification for the terminology "heat kernel" will follow from Section 2.5, where it will be shown that $\mathbf{g}_t$ and $\mathbf{K}_t$ aside from being WD families also satisfy the heat equation, independent of the Haar measure.

2.1.2 Decay Property

Next we formalize the decay property of Gaussians, and compare it with a growth property for use when we take convolutions in the next section.

We define a function f on G to have **quadratic exponential decay** (QED) **with respect to** σ if there exists $C > 0$ such that

$$f(g) = O\left(e^{-C\sigma^2(g)}\right) \text{ for } \sigma(g) \to \infty.$$

To make the reference to σ explicit in the notation, one writes QED_σ.

We usually omit the reference to σ if no ambiguity can occur. We say that a function has (at most) **linear exponential growth** LEG_σ, or simply LEG, if there exists $C' > 0$ such that

$$f(g) = O\left(e^{C'\sigma(g)}\right) \text{ for } \sigma(g) \to \infty.$$

Note that a bounded function is in LEG. Functions will be assumed measurable.

Lemma 2.1.3. *For each $t > 0$, the heat Gaussian has quadratic exponential decay. Uniformly for z in a compact set and $0 < t \leqq t_0 < \infty$, the heat kernel as a function $w \longmapsto \mathbf{K}_t(z,w)$ has quadratic exponential decay.*

Proof. This is obvious from the formula and the triangle inequality for the σ-function (Chapter 1 Theorem 8.2).

We have the following properties under multiplication: QED_σ *is closed under multiplication.* LEG_σ *is closed under multiplication. Furthermore,* $\mathrm{QED}_\sigma \cdot \mathrm{LEG}_\sigma \subset \mathrm{QED}_\sigma$. In other words:

If f_1 has quadratic exponential decay and f_2 has linear exponential growth, then the product $f_1 f_2$ has quadratic exponential decay.

This is obvious, but very practical. Generally speaking, we deal with integrals of products where to ensure absolute convergence, one function may grow, but the other function has to decay (much) faster. There are many variations of precise conditions making certain conclusions valid.

For $\mathrm{QED}_\sigma^\infty$ and $\mathrm{LEG}_\sigma^\infty$, see Section 3.1.

Remarks. Gårding [Gar 60] (following a paper of Gaffney [Gaf 59]) used the space of exponential growth, and the space of arbitrarily rapid exponential decay, in connection with the heat kernel on a Riemannian manifold, and convolution. We exploit the quadratic exponential decay, which deserves being extracted as a significant measure of decay in its own right, in large part because the heat Gaussian or heat kernel exhibits this sort of decay.

We used a coordinate from group theory to define the growth or decay. In Section 4.1 we define the Hermitian norm, and see that the present condition of exponential linear growth amounts to polynomial growth with respect to this norm. This is what Harish-Chandra calls a "slowly increasing function." See [Har 68], pp. 12 and 13.

2.2 Convolution, Semigroup, and Approximations Properties

We start with the **semigroup property**. In Section 2.1, we gave some formulas that amount to seeing how the heat Gaussian and kernel change under scaling the Haar measure (i.e., multiplying it by a positive constant), in the two cases that are relevant to us, the Iwasawa measure and the polar measure. The choice of such a measure also determines the convolution operation. The next theorems are valid with these compatible choices.

Theorem 2.2.1. *The heat Gaussian and kernel satisfy the semigroup property:*

$$\mathbf{g}_t * \mathbf{g}_{t'} = \mathbf{g}_{t+t'} \text{ and } \mathbf{K}_t * \mathbf{K}_{t'} = \mathbf{K}_{t+t'}.$$

The space of Gaussians Gauss (G) is closed under the convolution product.

Proof. The only difficulty in light of Chapter 1, Theorem 4.3, is that we don't know that $\mathbf{g}_t * \mathbf{g}_{t'}$ is a Gaussian. A priori it might lie in a bigger space. One might actually compute the convolution integral, but we argue another way, avoiding another computation, say for the polar heat kernel. We know from Chapter 1, Theorem 1.7.4, that $\mathbf{S_p}$ is an L^2-isometry on the space of Gaussians, and in particular, it is injective on L^2. The convolution product $\mathbf{g}_t * \mathbf{g}_{t'}$ is in L^2 by elementary integration theory (recalled at the end of this section), so Chapter 1, Theorem 1.4.3, and the injectivity of $\mathbf{S_p}$ apply to conclude the proof of the convolution formula. Since the heat Gaussians range over all Gaussians up to a constant factor, the last statement is also proved.

Notational remark. It is sometimes convenient to use the notation $\mathbf{K}_t *$ for the convolution operator determined by $\mathbf{K}_t$.

Theorem 2.2.2. *Let f be measurable and such that the convolution integral $\mathbf{K}_{t+t'} * f$ is absolutely convergent. Then we have the associativity formula*

$$(\mathbf{K}_t * \mathbf{K}_{t'}) * f = \mathbf{K}_t * (\mathbf{K}_{t'} * f).$$

The convergence condition is satisfied, for instance, if f is bounded measurable, or even better in $\mathbf{LEG}_\sigma$, or in L^1, or in L^2.

Proof. This is merely Fubini's theorem.

Theorem 2.2.3. *Let f be in* LEG_σ, *or in L^1 or in L^2. Then for every $t > 0$ the convolution integral*

$$(\mathbf{K}_t * f)(z) = \int_G \mathbf{K}_t(z,w) f(w) dw$$

is absolutely convergent, uniformly for z in a compact set. The function $\mathbf{K}_t * f$ is continuous.

Proof. By Lemma 2.1.3, for z in a compact set, the function $w \longmapsto \mathbf{K}_t(z,w)$ has quadratic exponential decay uniformly in z, so the assertion is immediate either from the bound on $\mathbf{K}_t$ or from the L^1 property of f, or the Schwarz inequality if f is in L^2. The continuity of $\mathbf{K}_t * f$ follows. (Split up the domain of integration for z in some fixed ball such that the integral of $\mathbf{K}_t(z,w)dw$ outside a larger ball is less than ε.)

If the function f in Theorem 2.2.3 isn't bounded, there is still control over it. In the present case of Gaussian convolution, one can deal with test functions having a nontrivial rate of growth such as characters, which have linear exponential growth, as we shall see in Section 3.4.

2.2.1 Approximation Properties

Next come the properties pertaining to the Weierstrass–Dirac conditions. We start by repeating Proposition 2.1.1 in the present context.

Theorem 2.2.4. *The heat kernel satisfies the conditions defining a WD family of kernels.*
Explicitly, for a fixed choice of Haar measure μ, and the corresponding heat Gaussian, we have:

WD 1. $\mathbf{K}_t \geqq 0$ (actually > 0) for all t.
WD 2. For all z, we have

$$\int_G \mathbf{K}_t(z,w)\,d_\mu(w) = 1.$$

WD 3. Given $\delta > 0$, for all z,

$$\lim_{t \to 0} \int_{d(z,w)>\delta} \mathbf{K}_t(z,w)\,d\mu(w) = 0.$$

These are the conditions that define a WD family on a Riemannian manifold when there is no group that allows one to pass from the one-variable function $\mathbf{g}_t$ to the two-variable function $\mathbf{K}_t$. Thus these properties can be viewed as applying to the space G/K as a Riemannian manifold. Convolution is then taken with functions on G/K, under appropriate absolute convergence of the convolution integral.

One formulation of an approximation theorem associated with Weierstrass–Dirac families first manifested itself in Weierstrass [Wei 85]. For textbook treatments, see for instance [Lan 93] Chapter 8 or [Lan 99] (on approximation theorems). The most basic approximation theorem valid for all WD families is the following:

Theorem 2.2.5. *Let f be a bounded measurable function, and $\{\mathbf{K}_t\}$ a WD family. Then for $t \to 0$, $\mathbf{K}_t * f$ converges to f uniformly on every set where f is uniformly continuous.*

However, with the quadratic exponential decay, we can do better than that. The pattern of proof is the same as that of the standard case just quoted, but we have to use the faster decay of the heat kernel.

Theorem 2.2.6. *Let f be continuous on G/K with linear exponential growth. Then $\mathbf{K}_t * f \to f$ as $t \to 0$, uniformly on every compact set Ω.*

Proof. Let $\delta > 0$ be such that if $\sigma(z^{-1}w) < \delta$ then $|f(w) - f(z)| < \varepsilon$ for $z \in \Omega$. Then

$$\begin{aligned} |(\mathbf{K}_t * f)(z) - f(z)| &\leqq \int_G \mathbf{K}_t(z,w)\,|f(w) - f(z)|\,d_{\mu_{\mathbf{p}}}(w) \\ &= \int_{\sigma(z^{-1}w)\leqq\delta} + \int_{\sigma(z^{-1}w)\geqq\delta} \mathbf{K}_t(z,w)\,|f(w) - f(z)|\,d\mu(w) \\ \text{[by WD 2]} &\leqq \varepsilon + C'' \int_{\sigma(z^{-1}w)\geqq\delta} \mathbf{K}_t(z,w)\,e^{C'\sigma(w)}\,d\mu(w). \end{aligned}$$

The product inside the integral has of course quadratic exponential decay for each t, but from the formula, this decay actually makes the product decrease to 0 as $t \to 0$, so that one can apply the dominated convergence theorem. Hence the integral on the right tends to 0 as $t \to 0$, thus concluding the proof.

We want the analogous result for approximation in L^1 and L^2. We define **convolution** in a form to fit our point-pair invariant, namely

$$(h\#f)(z) = \int_G h\left(z^{-1}w\right) f(w)\,dw.$$

We need standard basic inequalities from integration theory on a unimodular locally compact group as follows.

Proposition 2.2.7. (Hölder inequality). *Let $\varphi \in L^p$ and $\psi \in L^q$ for $1/p + 1/q = 1$, on some measured space X. Then $|\varphi||\psi| \in L^1$, and*

$$\int_X |\varphi||\psi| d_\mu \leqq ||\varphi||_p ||\psi||_q.$$

Proposition 2.2.8. *Let $h \in L^1(G)$ and $f \in L^p(G)$ with $p \geqq 1$. Then*

$$||h\#f||_p \leqq ||h||_1 ||f||_p.$$

Readers will find proofs in standard texts on integration and measure theory, e.g., [Lan 93], Chapter 8, Theorem 8.5.1, and Chapter 8, Theorem 8.1.2. The proofs are usually given on Euclidean space, but they go over to the general group setting.

We reproduce the proof for the convenience of the reader, and because it will serve as a model for the Γ-periodization in Chapter 8.

Proof. Proof of Proposition 2.2.9. We start with the case $p = 1$, so let φ and ψ be in L^1. Then by definition,

$$||\varphi\#\psi||_1 = \int_G \left| \int_G \varphi\left(z^{-1}w\right) \psi(w)\,dw \right| dz.$$

Interchanging the order of integration, we have absolute convergence. Hence we can apply Fubini to see that the defining integral converges absolutely. Then we note that the interchanged integral $dz\,dw$ of the absolute values of the functions splits into a product, which yields the desired inequality

$$||\varphi\#\psi||_1 \leqq ||\varphi||_1\,||\psi||_1\,.$$

Next suppose $p > 1$, and let q be such that $1/p + 1/q = 1$. The functions h and f^p are in L^1, so the function

$$w \mapsto \left|h\left(z^{-1}w\right)\right|^{1/p} |f(w)|$$

is in L^p by the (1, 1) case of the proposition treated first. Furthermore,

$$w \mapsto \left|h\left(z^{-1}w\right)\right|^{1/q}$$

is in L^q. By Proposition 2.2.7 we obtain

$$\begin{aligned}\int \left|h\left(z^{-1}w\right)\right| |f(w)|\,dw &= \int \left|h\left(z^{-1}w\right)\right|^{1/p} |f(w)| \left|h\left(z^{-1}w\right)\right|^{1/q} dw \\ &\leqq \left[\int \left|h\left(z^{-1}w\right)\right| |f(w)|^p\,dw\right]^{1/p} \left[\int \left|h\left(z^{-1}w\right)\right| dw\right]^{1/q}.\end{aligned}$$

Raising both sides to the p power and using Haar measure invariance $w \mapsto zw$, we get

$$|(h\#f)(z)|^p \leqq \int \left|h\left(z^{-1}w\right)\right| |f(w)|^p\,dw \cdot ||h||_1^{p/q}.$$

So integrating dz and using Fubini gives

$$||h\#f||_p^p \leqq ||h||_1\,||f||_p^p\,||h||_1^{p/q} = ||h||_1^p\,||f||_p^q$$

because $1 + p/q = p$. This concludes the proof.

Theorem 2.2.9. *Let $f \in L^p(G/K)$ with $p \geqq 1$. Then $\mathbf{K}_t * f$ is L^p-convergent to f for $t \to 0$.*

Proof. Since $C_c(G/K)$ is L^p-dense in L^p, it suffices to prove the theorem for $f \in C_c(G/K)$, after picking $\varphi \in C_c(G/K)$ L^p-close to f, writing

$$\mathbf{K}_t * f - f = \mathbf{K}_t * (f - \varphi) + \mathbf{K}_t * \varphi - \varphi + \varphi - f,$$

and using Proposition 2.2.8. So let $f \in C_c(G/K)$. Let V be a compact neighborhood on G/K of Supp $(f) \cup$ origin, and also assume V even (i.e., if $z \in V$ then $z^{-1} \in V$). Let VV be the set of products zw with $z, w \in V$. Then VV is a compact, even neighborhood of supp$(f) \cup$ origin. We have

$$||\mathbf{K}_t * f) - f||_1 \leqq \int_G \left| \int_V \mathbf{K}_t(z,w) f(w)\,dw - f(z) \right| dz.$$

We split the integral over G into a sum of integrals over the complement of VV and VV,

$$\int_G \ldots dw = \int_{VV} + \int_{G-VV} \cdots dw.$$

On VV the expression under the integral approaches 0 as $t \to 0$ by Theorem 2.2.5. On the complement G-VV we use Fubini, and the boundedness of f, the fact that $f(w) = 0$ if $w \notin VV$, and **WD 3** to conclude the proof.

Remark. We really don't care about $p \neq 1$, 2, but to include those two cases, it's shorter to use the single letter p.

For further approximation properties of Gaussians, see [JoL 03]. Gaussians can be used to approximate more general functions. One can verify formulas first for Gaussians, and then by continuity extend the formulas to more general spaces. Even more importantly, Gaussians can be used as the natural functions giving rise to explicit theta inversion formulas in the context of semisimple Lie groups.

2.3 Complexifying t and the Null Space of Heat Convolution

The approximation properties of convolution come from letting $t \to 0$. For some other applications, one has to let t go to the right rather than the left. We are thus led to complexify t as follows.

The formula for the Iwasawa $\mathbf{g}_t$ from Section 2.1 is

$$\mathbf{g}_t = \frac{1}{(8\pi t)^{3/2}} e^{-2t} e^{-\sigma^2/8t} \frac{\sigma}{\sinh\sigma}.$$

The factor $\sigma/\sinh\sigma$ has linear exponential growth. We may replace t wherever it occurs by a complex variable

$$\tau = \mathrm{t} + \mathbf{i}r,$$

having real part $t = \mathrm{Re}(\tau) > 0$, so τ ranges over the right half-plane. Then

$$\mathbf{g}_\tau(z) = \frac{1}{(8\pi\tau)^{3/2}} e^{-2\tau} e^{-\sigma^2(z)/8\tau} \frac{\sigma(z)}{\sinh\sigma(z)}. \tag{2.6}$$

In the right half-plane, the term $\tau^{3/2}$ has the natural value with $-\pi < \theta < \pi$. Then

$$\frac{\sigma^2(z)}{8\tau} = \frac{\sigma^2(z)(t - \mathbf{i}r)}{8(t^2 + r^2)}$$

Hence

$$|\mathbf{g}_\tau(z)| = \frac{1}{(8\pi|\tau|)^{3/2}} e^{-2t} e^{-\sigma^2(z)t/8(t^2+r^2)} \frac{\sigma(z)}{\sinh\sigma(z)}. \tag{2.7}$$

We can then define $\mathbf{K}_\tau(z, w) = \mathbf{g}_\tau(z^{-1}w)$ as before, with complexified subscript.

Proposition 2.3.1. *The function* $\mathbf{g}_\tau(z)$ *is holomorphic in* τ *for* $\mathrm{Re}(\tau) > 0$.

Proof. This is immediate from (2.7).

Of course, Proposition 2.3.1 applies to the heat Gaussian associated with any scaling of Haar measure and of the trace form; cf. Theorem 2.5.5.

Theorem 2.3.2. *Let f be such that the convolution integral*

$$\int_G \mathbf{K}_t(z,w) f(w)\, dw$$

is absolutely convergent, uniformly for t, z in compact sets. Suppose that there is one value $t_0 > 0$ *such that*

$$\mathbf{K}_{t_0} * f = 0.$$

Then $\mathbf{K}_t * f = 0$ *for all t. In particular,* $f = 0$ *wherever f can be approximated by the convolution* $\mathbf{K}_t * f$ *with* $t \to 0$.

Proof. By Theorem 2.2.2 we know that for all $t > 0$,

$$\mathbf{K}_{t_0+t} * f = 0.$$

By Proposition 2.3.1 we know that $(\mathbf{K}_\tau * f)(z)$ is holomorphic in τ on the right half-plane. For fixed $z \in G$, the function $\tau \mapsto (\mathbf{K}_\tau * f)(z)$ is 0 on a line segment, hence it is identically 0 for all τ. This proves the theorem.

The final conclusion depending on approximation applies to the cases in which we have proved the approximation, namely:

- Pointwise approximation as in Theorems 2.2.5 and 2.2.6;
- L^p approximation (especially $p = 1$ and 2) as in Theorem 2.2.9.

For example, if f is continuous and bounded, or continuous and in L^1 or L^2, then the conclusion is simply $f = 0$ pointwise, i.e., $f(z) = 0$ for all z.

The above theorem is the first concerning convolution eigenvalues, in the present case eigenvalue 0. In Chapter 3, we shall consider further theorems concerning eigenvalues. There will be a further distinction, with eigenvalue 1, meaning that f is a fixed point of heat convolution, and eigenvalue $\neq 1$. See Chapter 3, Proposition 2.2.7, and Section 3.4.

2.4 The Casimir Operator

This section gives a major application of spherical inversion. We show how one Gaussian can be normalized in a significant way. We have to discuss invariant differential operators, because one of the two normalization conditions will be a differential equation. We recall briefly such operators in the special case at hand. For more details, see for instance [JoL 01], Chapter 2. The whole section specializes Gangolli's approach [Gan 68].

Let g be the Lie algebra of $G = \mathrm{SL}_2(\mathbf{C})$. Thus g consists of the 2×2 complex matrices with trace 0. We view g as a 6-dimensional vector space over $\mathbf{R}$. Then g has subspaces:

$n =$ strictly upper triangular complex matrices, so $U = \exp(n)$, and $n \subset \mathbf{C}$;
$\bar{n} = {}^t n =$ transpose of $n =$ strictly lower triangular matrices;
$a = \mathrm{Lie}(A) =$ real diagonal matrices with trace 0;
$k = \mathbf{R}$-space of skew-Hermitian matrices with trace 0, $k = \mathrm{Lie}(K)$.
$p = \mathbf{R}$-space of Hermitian matrices with trace 0;

All of the above are Lie subalgebras except p. Note that $\mathbf{i}a$ is also a (real) Lie subalgebra, and

$$\mathbf{i}a = \mathrm{Lie}(\mathbf{T}), \text{ where } \mathbf{T} = \text{circle} = \exp(\mathbf{i}a).$$

The bracket product $(Z,W) \mapsto [Z,W]$ is $\mathbf{C}$-bilinear, and gives rise to the regular representation of g on itself, with left action. Because of the low dimension, we have $n = g_\alpha$ equal to the α-eigenspace of the regular action of a on g. Note that $\bar{n} = g_{-\alpha}$. The 0-eigenspace of the bracket action of a on g is $a + \mathbf{i}a$.

An $\mathbf{R}$-basis of a is given by the matrix

$$H_\alpha \begin{pmatrix} 1 & 0 \\ 0 & -1 \end{pmatrix}.$$

We let $\mathbf{i}H_\alpha = H_\alpha^{(\mathbf{i})}$ be the corresponding basis of $\mathbf{i}a$, so H_α, $H_\alpha^{(\mathbf{i})}$ form a basis of $a + \mathbf{i}a$.

The matrix

$$E_\alpha = E_{12} = \begin{pmatrix} 0 & 1 \\ 0 & 0 \end{pmatrix}$$

is a $\mathbf{C}$-basis of n, and E_α, $\mathbf{i}E_\alpha = E_\alpha^{(\mathbf{i})}$ form an $\mathbf{R}$-basis of n. Similarly, $E_{-\alpha} = E_{21}$ and $E_{-\alpha}^{(\mathbf{i})} = \mathbf{i}E_{-\alpha}$ form an $\mathbf{R}$-basis of $\bar{n}$. With the above notation, we have the direct sum decomposition

$$g = a + \mathbf{i}a + g_\alpha + g_{-\alpha} = a + \mathbf{i}a + n + \bar{n}\,. \tag{2.1}$$

On g we take the **real trace form** B, defined by

$$B(Z,W) = \mathrm{Re}\ \mathrm{tr}(ZW).$$

This form is positive definite on a, and G-invariant for the conjugation action of G on g. Note that $(a + \mathbf{i}a)$ is orthogonal to $(g_\alpha + g_{-\alpha})$, and a is orthogonal to $\mathbf{i}a$. However, the natural basis

$$\left\{H_\alpha, H_\alpha^{(\mathbf{i})}, E_\alpha, E_{-\alpha}, E_\alpha^{(\mathbf{i})}, E_{-\alpha}^{(\mathbf{i})}\right\} \tag{2.2}$$

is not orthogonal. The **dual basis** with respect to the real trace form is

$$\left\{H'_\alpha, -H'^{(i)}_\alpha, E_{-\alpha}, E_\alpha, -E^{(i)}_\alpha, E^{(i)}_\alpha\right\} \text{ with } H'_\alpha = H_\alpha/2. \tag{2.3}$$

We now recall some facts about differential operators. Let $Z \in g$. We associate to Z the differential operator $\tilde{Z} = \mathscr{D}(Z)$ defined on a function f by

$$\left(\tilde{Z}f\right)(g) = \mathscr{D}(Z)\,f(g) = \frac{d}{dt}f\left(g\,\left|\exp\left(tZ\right)\right)\right|_{t=0}.$$

This operator is left-invariant, i.e., commutes with left translation $L_{g'}$ for any $g' \in G$, that is,

$$\mathscr{D}(Z)\,L_{g'} = L_{g'}\mathscr{D}(z)\,.$$

It will be useful to remember that $\mathscr{D}$ is a Lie homomorphism, that is,

$$\mathscr{D}([Z_1, Z_2]) = [\mathscr{D}(Z_1), \mathscr{D}(Z_2)] = \mathscr{D}(Z_1)\,\mathscr{D}(Z_2) - \mathscr{D}(Z_2)\,\mathscr{D}(Z_1)\,.$$

For example,

$$[E_\alpha, E_{-\alpha}] = H_\alpha \text{ implies } \left[\tilde{E}_\alpha, \tilde{E}_{-\alpha}\right] = \tilde{H}_\alpha.$$

We shall deal with the Casimir operator, whose definition is based on a universal property in multilinear algebra as follows. Let V be a vector space (over $\mathbf{R}$, say), and let $\check{V}$ be the dual space. Let B be a nonsingular symmetric bilinear form on V. Then B induces an isomorphism $V \to \check{V}$ of V with its dual. Let $\{v_1, \ldots, v_N\}$ be a basis of V, and let $\{v'_1, \ldots, v'_N\}$ be the B-dual basis, that is,

$$B\left(v_i, v'_j\right) = \delta_{ij}.$$

There is a unique linear isomorphism

$$V \otimes \check{V} \longrightarrow \mathrm{End}(V)$$

such that for $v \in V$ and $\lambda \in \check{V}$ the corresponding element $h_{v,\lambda} \in \mathrm{End}(V)$ is given by

$$h_{v,\lambda}\,(w) = \lambda\,(w)\,v.$$

In the tensor product $V \otimes \check{V} \cong \mathrm{End}(V)$, the element corresponding to the identity in $\mathrm{End}(V)$ is given in terms of the dual basis by

$$\omega_{V,B} = \sum v_i \otimes v'_i.$$

Thus all such elements are equal in the tensor product $V \otimes V$, independent of the choice of basis and B-dual basis. This element is called the **universal Casimir element** (depending on B).

Returning to the case $V = g = \mathrm{Lie}(G)$, $G = \mathrm{SL}_2(\mathbf{C})$, we have exhibited one basis $\{Z_1, \ldots, Z_6\}$ and its dual basis, with respect to the real trace form B. By the universality just described, the element

$$\omega = \omega_{G,B} = \sum_{i=1}^{6} \mathscr{D}(Z_i)\,\mathscr{D}\left(Z'_i\right)$$

is independent of the choice of basis, and will be called the (G,B)-**Casimir operator**, or simply the **Casimir operator** on G. By the G-conjugation invariance of the universal Casimir in the tensor product, it follows that ω is G-conjugation-invariant, aside from being left-translation-invariant. (For higher dimensions, cf. [JoL 01], Section 7.2, as well as Chapter 2.) Note that Casimir depends on the choice of B, but is independent of the choice of Haar measure.

Of course we also have Casimir on A, or $\mathbf{T}$ for that matter,

$$\omega_A = \tilde{H}_\alpha \tilde{H}'_\alpha \text{ and } \omega_{\mathbf{T}} = -\tilde{H}_\alpha^{(\mathrm{i})} \tilde{H}_\alpha'^{(\mathrm{i})} = -\frac{1}{2}\left(\tilde{H}_\alpha^{(\mathrm{i})}\right)^2.$$

In terms of the dual bases (2.2) and (2.3), Casimir on G has the decomposition

$$\begin{aligned}\omega &= \omega_A + \omega_T + \left(\tilde{E}_\alpha \tilde{E}_{-\alpha} + \tilde{E}_{-\alpha}\tilde{E}_\alpha\right) - \left(\tilde{E}_\alpha^{(\mathrm{i})}\tilde{E}_{-\alpha}^{(\mathrm{i})} + \tilde{E}_\alpha^{(\mathrm{i})}\tilde{E}_\alpha^{(\mathrm{i})}\right) \\ &= \omega_A + \omega_T + \omega_{unip}, \end{aligned} \tag{2.4}$$

where ω_{unip} will be called the unipotent part of ω. The next two lemmas will allow us to determine the effect of the two terms on the right, namely they will show that

$$\omega_{\mathrm{unip}} \equiv -2\tilde{H}_\alpha + 2\tilde{E}_\alpha^2 + 2\left(\tilde{E}_\alpha^{(\mathrm{i})}\right)^2 \bmod \tilde{E}_\alpha \tilde{k} + \tilde{E}_\alpha^{(\mathrm{i})}\tilde{k}. \tag{2.4a}$$

Putting (2.4) and (2.4a) together yields the expression for Casimir showing how it acts on right K-invariant functions, i.e., on G/K, namely since $\omega_{\mathbf{T}} \in \tilde{k}^2$,

$$\omega_G \equiv \omega_A - 2\tilde{H}_\alpha + 2\tilde{E}_\alpha^2 + 2(\tilde{E}_\alpha^{(\mathrm{i})})^2 \bmod \tilde{k}^2 + \tilde{E}_\alpha\tilde{k} + \tilde{E}_\alpha^{(\mathrm{i})}\tilde{k}. \tag{2.5}$$

The terms following mod annihilate a function on G/K.

Now we come to the lemmas proving (2.4a) and therefore (2.5).

Lemma 2.4.1. *We have*

$$\tilde{E}_\alpha \tilde{E}_{-\alpha} + \tilde{E}_{-\alpha}\tilde{E}_\alpha \equiv 2\tilde{E}_\alpha^2 - \tilde{E}_\alpha \bmod \tilde{E}_\alpha \tilde{k}$$

Proof. We reproduce here a standard computation. We have

$$\begin{aligned}\tilde{E}_{-\alpha}\tilde{E}_\alpha = \left(\tilde{E}_{-\alpha}\tilde{E}_\alpha - \tilde{E}_\alpha\tilde{E}_{-\alpha}\right) + \tilde{E}_\alpha\tilde{E}_{-\alpha} &= \left[\tilde{E}_{-\alpha}.\tilde{E}_{-\alpha}\right] + \tilde{E}_\alpha\tilde{E}_{-\alpha} \\ &= -\tilde{H}_\alpha + \tilde{E}_\alpha\tilde{E}_{-\alpha}.\end{aligned}$$

Furthermore,

$$E_{-\alpha} = {}^tE_\alpha = \left({}^tE_\alpha - E_\alpha\right) + E_\alpha,$$

and ${}^t\tilde{E}_\alpha - \tilde{E}_\alpha \in \tilde{k}$. Hence

$$\tilde{E}_\alpha\tilde{E}_{-\alpha} \equiv \tilde{E}_\alpha^2 \bmod \tilde{E}_\alpha\tilde{k}.$$

This proves the lemma.

Lemma 2.4.2. *We have*

$$\tilde{E}_{\alpha}^{(\mathbf{i})}\tilde{E}_{-\alpha}^{(\mathbf{i})}+\tilde{E}_{-\alpha}^{(\mathbf{i})}\tilde{E}_{\alpha}^{(\mathbf{i})}\equiv -2\left(\tilde{E}_{\alpha}^{(\mathbf{i})}\right)^{2}+\tilde{H}_{\alpha} \bmod \tilde{E}_{\alpha}^{(\mathbf{i})}\tilde{k}.$$

Proof. It will be convenient to use the operator θ (**Cartan involution**) defined on the Lie algebra by

$$\theta Z = -{}^{t}\bar{Z}.$$

Then k consists of those $Z \in g$ such that $\theta Z = Z$ (fixed point set of θ). We write

$$\tilde{E}_{-\alpha}^{(\mathbf{i})}\tilde{E}_{\alpha}^{(\mathbf{i})} = \left[\tilde{E}_{-\alpha}^{(\mathbf{i})},\tilde{E}_{\alpha}^{(\mathbf{i})}\right]+\tilde{E}_{\alpha}^{(\mathbf{i})}\tilde{E}_{-\alpha}^{(\mathbf{i})} = \tilde{H}_{\alpha}+\tilde{E}_{\alpha}^{(\mathbf{i})}\tilde{E}_{-\alpha}^{(\mathbf{i})}.$$

Furthermore,

$$\tilde{E}_{-\alpha}^{(\mathbf{i})} = \theta\left(\tilde{E}_{\alpha}^{(\mathbf{i})}\right) = \left(\theta\tilde{E}_{\alpha}^{(\mathbf{i})}+\tilde{E}_{\alpha}^{(\mathbf{i})}\right)-\tilde{E}_{\alpha}^{(\mathbf{i})},$$

so

$$\tilde{E}_{\alpha}^{(\mathbf{i})}\tilde{E}_{-\alpha}^{(\mathbf{i})} \equiv -\left(\tilde{E}_{\alpha}^{(\mathbf{i})}\right)^{2} \bmod \tilde{E}_{\alpha}^{(\mathbf{i})}\tilde{k}.$$

This proves the lemma.

We shall deal with the notion of direct image in the special case that concerns us here. Let $\pi = \pi_A$ be the Iwasawa projection from G to A. Given a function f on A, we can compose it with π. If D is a differential operator on G, or G/K, we define its **direct image** $\pi_* D$ by the formula

$$(\pi_* D) f = (D(f^{\circ}\pi))_A,$$

where the subscript A means restriction to A. Note that $\tilde{k}$ annihilates $f \circ \pi$.

Proposition 2.4.3. *Let $g = uak$ (Iwasawa decomposition of $g \in G$). Let f be a $\mathbf{C}^{\infty}$ function on A. Then $\pi_*\omega_G = \omega_A - 2\tilde{H}_{\alpha}$, but more generally,*

$$(\omega_G(f\circ\pi))(g) = \omega_G(f\circ\pi)(a) = ((\omega_A - 2\tilde{H}_{\alpha}) f)(a).$$

Proof. The expression on the left is $\omega_G(f\circ\pi)$ composed with left translation L_u and right translation R_k, evaluated at a. Since ω_G is G-invariant (left, conjugation, and right), it follows that the expression on the left is equal to

$$(\omega_G(f\circ\pi\circ L_u R_k))(a),$$

and $f\circ\pi\circ L_u\circ R_k = f\circ\pi$ concludes the proof of the first equality.

As to the second equality, let f be a function on A. Its pullback $f\circ\pi$ to G is U-invariant on the left and K-invariant on the right. Now we use the fact that A normalizes U, and for $Z \in n$ (for instance $Z = E_{\alpha}$ or $\mathbf{i}E_{\alpha}$), and $a \in A$ we have $aZa^{-1} \in n$, so

$$f\left(a\cdot\exp(tZ)a^{-1}a\right) = f\left(\exp\left(taZa^{-1}\right)a\right) = f(a).$$

Taking d/dt yields 0. Similarly, the square terms $\tilde{E}_\alpha^2$ and $(\tilde{E}_\alpha^{(\mathrm{i})})^2$ are also seen to give 0; they come from differentiating the expression

$$f\,(a.\exp(t_1 Z_1)\exp(t_2 Z_2)) \text{ with } Z_1,\, Z_2 \in n.$$

The normalization property shows that this expression is constant in $t_1,\ t_2$. Since $\tilde{k}$ annihilates $f \circ \pi$, all the terms in (2.5) which don't come from differentiation $\tilde{H}$ with $H \in a$ are in the kernel of the projection. The remaining terms are precisely

$$\omega_A - 2\tilde{H}_\alpha.$$

This concludes the proof of Proposition 2.4.3.

Remark. In higher dimension or rank, the expression $2H_\alpha$ becomes the trace of the Lie (a,n)-representation.

Eigenvalues and eigenfunctions. Let D be a differential operator and f an eigenfunction of D, meaning that there is a complex number denoted by $\mathrm{ev}(D,f)$ such that

$$Df = \mathrm{ev}\,(D,f)\,f.$$

Example. Characters $\chi = \chi_\zeta$ on A are eigenfunctions of ω_A. Indeed, on $\check{a}_{\mathbf{C}}$, let $\check{B}_{\mathbf{C}}$ be the symmetric $\mathbf{C}$-bilinear product extended by duality from the trace form on a. Thus for $\zeta = s\alpha,\ s \in \mathbf{C}$, we have

$$\check{B}_{\mathbf{C}}\,(s\alpha, s\alpha) = 2s^2.$$

Then for $H \in a$, directly from the definition, we get

$$\tilde{H}\chi_\zeta = \zeta\,(H)\,\chi_\zeta. \tag{2.6a}$$

Hence

$$\omega_A\chi_\zeta = \tilde{H}_\alpha\tilde{H}'_\alpha\chi_\zeta = \frac{1}{2}\tilde{H}_\alpha^2\chi_\zeta = \frac{1}{2}\zeta\,(H_\alpha)^2\,\chi_\zeta.$$

But $\zeta(H_\alpha)^2 = \check{B}_{\mathbf{C}}(s\,\alpha,\ \alpha)^2 = (2s)^2 = 4s^2$, so

$$\mathrm{ev}\,(\omega_A, \chi_\zeta) = \check{B}_{\mathbf{C}}\,(\zeta,\zeta) = 2s^2. \tag{2.6b}$$

Applying Proposition 4.2.3 and letting $\tau = 2\rho$ be the trace of the (a,n) representation yields

$$\mathrm{ev}\,(\omega_G, \chi_\zeta \circ \pi_A) = \check{B}_{\mathbf{C}}\,(\zeta,\zeta) - \check{B}_{\mathbf{C}}\,(\tau,\zeta) = 2s\,(s-2). \tag{2.6c}$$

Note that $\chi_{s\alpha}(a) = a^{s\alpha} = y^s$, with $y = a^\alpha$ as usual. So y^s is an eigenfunction of Casimir on G. This will be used in connection with Eisenstein series in Chapter 11. We may rewrite (2.6c) using the notation χ_s instead of $\chi_{s\alpha} \circ \pi_A$, in the following form:

Proposition 2.4.4.

$$\omega_G\chi_s = \mathrm{ev}\,(\omega_G, \chi_s)\,\chi_s = 2s\,(s-2)\,\chi_s.$$

We extend this relation to the spherical kernel as eigenfunctions of Casimir. We use DUTIS (Differentiation Under The Integral Sign) on the spherical kernel

$$\Phi_\zeta(g) = \int_K (kg)_A^{\zeta+\rho}\, dk, \text{ and } \rho = \alpha.$$

Proposition 2.4.5. *The spherical kernel is an eigenfunction of Casimir ω_G, with eigenvalue*

$$\begin{aligned} \mathrm{ev}\left(\omega_G, \Phi_\zeta\right) &= \mathrm{ev}\left(\omega_A - 2\tilde{H}_\alpha, \chi_{\zeta+\rho}\right) = \check{B}_{\mathrm{C}}(\zeta,\zeta) - \check{B}_{\mathrm{C}}(\rho,\rho) \\ &= 2s^2 - 2 \; if \; \zeta = s\alpha. \end{aligned}$$

Proof. The integral defining the spherical kernel being over a compact group, we can DUTIS with ω_G without problem. We are then in the situation of Proposition 2.4.3. We use the fact that ω_G is left invariant. The function f is $\chi_{\zeta+\rho} = \chi_{\zeta+\alpha}$. By the computation in the example preceding the proposition, we find that the eigenvalue is

$$\check{B}_{\mathrm{C}}(\zeta+\alpha, \zeta+\alpha) - 2\check{B}_{\mathrm{C}}(\alpha \cdot \zeta+\alpha) = \check{B}_{\mathrm{C}}(\zeta,\zeta) - \check{B}_{\mathrm{C}}(\alpha,\alpha),$$

as was to be shown.

In Section 3.4, we deal with the eigenfunction property under convolution with the heat kernel. In Chapter 11 we apply this to convolution between the heat kernel and Eisenstein series. These can be viewed as a continuation of the present considerations.

Remark 1. Note that the eigenvalue of Proposition 2.4.5 on the line $\mathrm{Re}(s) = 0$ is the same as the eigenvalue of Proposition 2.4.4 on the line $\mathrm{Re}(s) = 1$.

Remark 2. Proposition 2.4.5 is the first instance in which we use DUTIS. It will be systematically used all the way, in cases in which it becomes more elaborate to justify its legitimacy. In the next section we meet a slightly more involved case, using the inverse spherical transform rather than **S** itself. Then having defined the heat kernel, we take convolutions with the heat kernel, and DUTIS with the heat kernel forever after.

For the convenience of the reader, we reproduce the DUTIS lemma. For the (short) proof, see (if necessary) [Lan 93] Chapter 8, Lemma 8.2.2.

Lemma 2.4.6. DUTIS. *Let M be a measured space with positive measure μ. Let V be an open subset of $\mathbf{R}^n$. Let F be a function on $V \times M$. Assume:*

(i) For each $x \in V$ the function $w \mapsto F(x,w)$ is in $L^1(\mu)$.
(ii) The partial derivatives $\partial_j F$ with respect to the variables of $\mathbf{R}^n$ exist, uniformly in the strong sense that there exists a function $F_1 \in L^1(\mu)$ such that for all $x \in V$, $w \in M$,

$$\left|\partial_j F(x,w)\right| \leqq \left|F_1(w)\right|.$$

Let

$$\Phi(x) = \int_M F(x,w)\,d\mu(w).$$

Then $\partial_j\Phi$ *exists and we have*

$$\partial_j\Phi(x) = \int_M \partial_j F(x,w)\,d\mu(w).$$

2.4.1 Scaling

Casimir was defined by using the dual basis with respect to the trace form. Formula (2.5) gives an expression that determines its direct image to G/K, which has the Iwasawa coordinates

$$(x,y) = (x_1,x_2,y) \text{ from the matrices } \begin{pmatrix} 1 & x \\ 0 & 1 \end{pmatrix}\begin{pmatrix} y^{1/2} & 0 \\ 0 & y^{-1/2} \end{pmatrix} = u(x)\,a_y.$$

A function f on G/K can be lifted to G via the natural projection on the (x,y) space, and can be written as $f(x,y)$ in terms of these coordinates. Then

$$(\omega f_G)(u_x a_y) = 2\Delta f(x,y),$$

where

$$\Delta = y^2\left(\left(\frac{\partial}{\partial x_1}\right)^2 + \left(\frac{\partial}{\partial x_2}\right)^2 + \left(\frac{\partial}{\partial y}\right)^2\right) - y\frac{\partial}{\partial y},$$

which is the usual Laplacian. The factor 2 disappears in the formula for the eigenvalues if one uses the Laplacian instead of Casimir. This won't be needed until Section 12.2, where we give the proof.

Quite generally, the Casimir element depends on the original choice of the scalar product on the Lie algebra. As in Section 2.4; we take B to be the real trace form. Thus the question arises, how does Casimir change under scaling, i.e., if we multiply B by a positive constant c? Let $\{Z_1,\dots,Z_N\}$ be a basis of g, and let $\{Z_1',\dots,Z_N'\}$ be the B-dual basis, so

$$B(Z_i,Z_i') = 1.$$

Then the cB-dual basis is $\{c^{-1}Z_1',\dots,c^{-1}Z_N'\}$. Hence the Casimir operator scaling formula becomes the following:

Proposition 2.4.7. $\omega_{cB} = c^{-1}\omega_B$.

Thus if we use $2B$ instead of B for the basic bilinear form, then Casimir with respect to $2B$ is the above Laplacian. The factor 2 may be interpreted as the multiplicity $m(\alpha) = 2 = [\mathbf{C}:\mathbf{R}]$ in the complex case with which we are dealing. In addition, the factor 2 in the eigenvalue $2s(s-2)$ disappears if we replace B by $2B$. These are the normalizations that are used in a standard way by the differential geometers on $\mathbf{H}^3$.

On a Riemannian manifold, there is no "Haar" measure, and a volume form is determined by the Riemannian metric. In the present case of G/K, one may view the form B as defining a Riemannian metric, in which case the Riemannian metric defined by $2B$ at the origin is the one usually taken on $\mathbf{H}^3$.

2.5 The Heat Equation

We combine the last two sections to get the heat Gaussian. A **heat Gaussian** is a Weierstrass–Dirac family $\{\mathbf{g}_t\}$ of Gaussians depending on a parameter $t > 0$ $(t \to 0)$ satisfying the **heat equation**, namely putting $\omega = \omega_{G,B}$,

$$-\omega \mathbf{g}_t + \partial_t \mathbf{g}_t = 0.$$

The operator $-\omega + \partial_t$ is called the **heat operator**. We are not doing physics; this is just a code name for historical reasons. Mathematically, this heat equation and its solutions have a controlling effect in many, if not all, areas of mathematics. For the fundamental properties, especially the characterization of the heat kernel by initial conditions, see Dodziuk [Dod 83]. For our purposes, we need only adopt a direct definition, based on the preceding sections, as follows. We let E_t be the Euclidean gaussian defined by

$$E_t(\zeta) = e^{ev(\omega,\Phi_\zeta)t} = e^{(2s^2-2)t} \tag{2.1}$$

if $\zeta = s\alpha$, $s \in \mathbf{C}$. On the imaginary axis, putting $\zeta = \mathbf{i}r\alpha$, we get

$$E_t(\mathbf{i}r\alpha) = e^{-(2r^2+2)t}.$$

The significant aspect of this expression is that the coefficient of t in the exponent is the eigenvalue of Casimir on the spherical kernel (Proposition 2.4.5). In Section 3.4 we get the additional information that $E_t(\zeta)$ is the eigenvalue of convolution with the heat kernel on the spherical function.

In Theorem 2.1.2, we gave an explicit formula for the Gaussian $\mathbf{g}_t$ such that under the spherical transform,

$$\mathbf{S}(\mathbf{g}_t) = E_t,$$

and this family satisfied the WD property with respect to the corresponding measure; cf. Section 2.1. We now come to the second fundamental condition pertaining to the heat equation.

Theorem 2.5.1. *Let $\{\mathbf{g}_t\}$ be the unique WD Gaussian family such that*

$$\mathbf{S}(\mathbf{g}_t) = E_t.$$

Then $\mathbf{g}_t$ satisfies the heat equation.

Proof. Changing the Haar measure by a constant changes the heat Gaussian by a constant, so it suffices to prove the theorem for one measure, which we take as the polar measure. By Chapter 1, Theorem 7.4,

$$\mathbf{g}_{t,p}(g) = \left(\mathbf{S}_{\mathbf{p}}^{-1}E_t\right)(g) = \int_{-\infty}^{\infty} e^{-2(r^2+1)t}\Phi(\mathbf{i}r\alpha, g)\, d\mu_{\mathrm{HC}}(r).$$

Differentiating under the integral sign in the variable g presents no problem since g can be restricted to a compact set and the integral converges uniformly rapidly. By DUTIS and Proposition 2.4.5, $\Phi_{\mathbf{i}r\alpha}$ is an eigenfunction of Casimir with the eigenvalue $-(2r^2+2)$. Taking $\partial/\partial t$ under the integral sign and subtracting gives 0, thereby proving the theorem.

Corollary 2.5.2. *Let $\mathbf{K}_t$ be the point-pair invariant associated with $\mathbf{g}_t$. Then $\mathbf{K}_t$ satisfies the heat equation in each variable. In fact,*

$$\partial_t \mathbf{K}_t(z,w) = \omega_z \mathrm{K}_t(z,w) = \omega_w \mathrm{K}_t(z,w).$$

The last stage of the construction of the heat kernel in Theorem 2.5.1 follows a general pattern as in the next theorem.

Theorem 2.5.3. *Let $z \mapsto F(z)$ be a $\mathbf{C}^\infty$ eigenfunction of Casimir, with eigenvalue $-\lambda$. Define the* **heated function**

$$F^{\#}(t,z) = e^{-\lambda t}F(z).$$

Then $F^{\#}(t,z)$ satisfies the heat equation.

This is immediate. Note that the function $g \mapsto \Phi_\zeta(g)$ is an eigenfunction of Casimir, and the expression under the integral sign in the proof of Theorem 2.5.1 is precisely the corresponding heated function. We shall meet further examples of Theorem 2.5.3 in Section 3.4 and beyond.

Theorem 2.2.4 (giving the Weierstrass–Dirac property) and Theorem 2.5.1 (giving the heat equation) show that it is legitimate to call $\{\mathbf{K}_t\}$ the **heat kernel**, and it was legitimate to have called $\{\mathbf{g}_t\}$ the **heat Gaussian**, because these two properties define a heat kernel on a Riemannian manifold. Actually, on a Lie group, we have the possibility of using jointly or separately the normalization of the Haar measure, and the scalar product used to define the dual basis in the Lie algebra, whence the Casimir element, whence the heat equation. Scaling the Haar measure, i.e., changing it by a constant factor, does not affect the property that the Gaussian WD family satisfies the heat equation. On a Riemannian manifold, we choose the scalar product to be the one defining the Riemannian metric, and then a volume form (mod ± 1) is also determined by this metric. If the Riemannian manifold is G/K, then the metric also determines a Haar measure. On a general Riemannian manifold, the metric determines the Laplacian, which in the case of G/K is $\pm$Casimir (depending on whether one wants the eigenvalues to be positive or negative). In any case, the construction we have followed gives ipso facto the properties relating the heat kernel to the spherical transform.

It can be shown that on a Riemannian manifold satisfying some very broad properties (e.g., Ricci curvature bounded from below), applying to the spaces G/K of interest to us, the two conditions (WD and heat equation) determine the family uniquely. Grigoryan told us that a proof can be extracted from Dodziuk's paper [Dod 83]. The two conditions can therefore be taken ab ovo as the conditions defining *the* heat family. We arrived at its existence and other properties by using the specific context of Gaussians in addition to the spherical inversion map to get immediately an explicit determination of this heat family. The arguments given here are expected to work for complex reductive groups in general. To get the real case in addition, cf. [FlJ 78] and [JLS 03]. Originally Gangolli did the general real case (including the complex case) directly, using the Harish-Chandra spherical inversion theory [Gan 68], and we followed this main idea. Actually, Gangolli's paper constructs a "fundamental solution" of the heat equation in the sense of functional analysis. To show that such a solution satisfies the Dirac condition as we formulated it takes some additional arguments. Gangolli told us that one can use a lemma of Gording [Gar 60] for this purpose.

2.5.1 Scaling

We conclude this section with the scaling property for the heat equation, analogous to what we did for the dependence on the Haar measure, and the dependence of Casimir on the basic bilinear form B on the Lie algebra. Instead of using the same function for the Jacobian factor and for the metric, we keep these functions separate. *As before, let B be the trace form on a.*

Let us write $H_{B,\beta}$ for the vector in a representing a functional $\beta \in \check{a}$ with respect to B. Then

$$\textbf{Sc1}.\ H_{cB,\beta} = c^{-1}H_B,\beta.$$
$$\textbf{Sc2}.\ (cB)(H_{cB},\beta_1,H_{cB,\beta_2}) = c^{-1}\beta_1\left(H_{B,\beta_2}\right).$$

The form B on a induces the bilinear form on the dual $\check{a}$ that we have been using with the same notation, but we must now make the distinction in the notation, so we write $\check{B}$ for the form on $\check{a}$, so that

$$\check{B}(\xi,\lambda) = B\left(H_{B,\xi},H_{B,\lambda}\right).$$

For $c > 0$, we have the scaled form cB on a, so $(c\check{B})$ on $\check{a}$, and then

$$\textbf{Sc3}.\ (c\check{B}) = c^{-1}\check{B}.$$

The form B determines Casimir ω_B, and together with the Iwasawa measure (say) they determine the heat Gaussian

$$\mathbf{g}_{t,\mu,B} = \mathbf{g}_{t,B} \text{ for } \mu = \mu_{\mathrm{Iw}},$$

which we now write for $b = \exp H$ in the form

$$\mathbf{g}_{t,B}{}^{(b)} = (8\pi t)^{-3/2} e^{-\check{B}(\rho,\rho)t} e^{-B(H,H/4t)} \frac{\log y}{\sinh(\log y)}.$$

Of course, we have $B(H,H) = |\log b|_B^2$, the norm being with respect to the scalar product B. We now view the above formula as a function of B. We need another symbol to deal with changes in this new variable, so we let $\mathbf{g}_{t,\mu,cB}$ be the fundamental solution of the heat equation, i.e., the Gaussian family that is a WD family with respect to the measure μ, and satisfies the heat equation with respect to ω_{cB}.

On the other hand, let $\mathbf{f}_{t,\mu,cB}$ be the function obtained by replacing B with cB in the above formula, so $\mathbf{g}_{t,B} = \mathbf{f}_{t,\mu,B}$ with $\mu = \mu_{\mathrm{Iw}}$.

Proposition 2.5.4. *Let* $c > 0$. *Then*

$$\mathbf{f}_{t,\mu,cB} = c^{-3/2}\mathbf{f}_{t/c,\mu,B} = c^{-3/2}\mathbf{g}_{t/c,\mu,B}.$$

Proof. This is immediate from **Sc 3** and the formula.

Theorem 2.5.5. *The fundamental solution of the heat equation with the scaled metric is given by*

$$\mathbf{g}_{t,\mu,cB} = \mathbf{g}_{t/c,\mu,B}.$$

Proof. If a function satisfies the heat equation, then any scalar multiple of the function satisfies the heat equation. The Gaussian $\mathbf{g}_{t/c,\mu,B}$ has total integral 1, so $f_{t,\mu,cB}$ has total integral $c^{-3/2}$. Hence multiplying by $c^{3/2}$ gives us back the WD probabilistic condition. The others are trivially valid under scaling. This proves the theorem.

The most important scaling is with $c = 2$, because

$$\omega_{2B} = \frac{1}{2}\omega_B = \Delta.$$

which is the Laplacian on $G/K = \mathbf{H}^3$ = hyperbolic 3-space (cf. Sections 6.1 and 12.2) for the Riemannian metric determined by 2 times the trace form. In this case, we have the normalization of the geometers.

Corollary 2.5.6. *Recalling that* $2B(H,H) = (\log\, y)^2$ *for* $H = \log\, b$, $y = b^{\alpha}$, *and* $\sigma(b) = \log\, y$ *is the distance from b to the origin for* $y = y_b > 1$, *we have*

$$\mathbf{g}_{t,\mu,2B}(b) = (4\pi t)^{-3/2} e^{-t} e^{-(\log y)^2/4t} \frac{\log y}{\sinh(\log y)}.$$

We note explicitly that although $\check{B}(\alpha,\alpha) = 2$, we have

$$(2\check{B})(\alpha,\alpha) = 1.$$

Thus we see all the way through how taking the basic scalar product to be 2 times the trace form rather than the trace form itself gets rid of the factor 2. However, from the point of view of comparing formulas on groups embedded in each other, it is not clear for the long run whether there is a normalization optimal for all contexts.

Chapter 3
QED, LEG, Transpose, and Casimir

Gaussians have a specially useful order of decay, which allows us to convolve them with test functions that have a nontrivial order of growth and that will include characters, spherical functions, and later, Eisenstein functions. This chapter lays down the analytic foundations, starting with a description of the key spaces with quadratic exponential decay and linear exponential growth.

3.1 Growth and Decay, QED^∞ and LEG^∞

Let's go to a Lie group G. Polynomial differential operators (with constant coefficients) on Euclidean space correspond to invariant differential operators defined as follows. Let $g = \mathrm{Lie}(G)$ and let $Z \in g$. The **associated differential operator (derivative in the direction of** Z**)** $\mathscr{D}(Z)$ or $\tilde{Z}$ is defined by

$$(\mathscr{D}(Z)f)(x) = \left(\tilde{Z}f\right)(x) = \frac{d}{dt}f(x \cdot \exp(tZ))\bigg|_{t=0}.$$

We assume that the reader is acquainted with the basic properties of such operators as in [JoL 01], Chapter 2. We shall work on $G = \mathrm{SL}_2(\mathbf{C})$ as usual.

Let $S(g)$ be the symmetric algebra (universal commutative algebra of polynomials in elements of g). There is a natural extension of $\mathscr{D}$ to an isomorphism also denoted by $\mathscr{D}$,

$$\mathscr{D} : S(g) \to \mathrm{IDO}(G)$$

between this symmetric algebra and the algebra of invariant differential operators on G. Furthermore $\mathrm{IDO}(G)$ is generated by the elements $\tilde{Z}$, $Z \in g$. See [JoL 01], Chapter 2, Theorem 2.1.1, Corollary 2.1.3.

We shall introduce certain spaces defined in a manner that immediately applies to Gaussians. These spaces in several ways are more serviceable than either the C_c^∞ spaces or the ordinary Schwartz spaces that impose conditions on derivatives.

On Euclidean space $\mathbf{R}^m$ we say that a function f has (at most) **polynomial growth of order** N if

$$|f(x)| = O\left(\|x\|^N\right) \text{ for } \|x\| \to \infty.$$

Here, $\|x\|$ is the scalar product norm or the sup norm on the coefficients of $\mathbf{R}^m$. We say that f has (at least) **polynomial decay of order** N if

$$|f(x)| = O\left(\|x\|^{-N}\right) \text{ for } \|x\| \to \infty.$$

We say that f has **superpolynomial decay** (abbreviated SPD) if it has polynomial decay of order N for all positive integers N. So that the notation will be functorial with respect to the ideas, we may denote the vector space of functions with superpolynomial decay by SPD. If the function is $\mathbf{C}^\infty$ and for every monomial differential operator

$$D = D_1^{j_1} \cdots D_m^{j_m}$$

the function Df has superpolynomial decay, we say that the function is **Schwartzian**, and we denote the **Schwartz space** of such functions by Sch or $\mathbf{SPD}^\infty$.

Let $G = \mathrm{SL}_2(\mathbf{C})$ for simplicity. We let $y_\mathbf{I}$ be the Iwasawa coordinate, so if $g \in G$, $g = uak$. Then $y_\mathbf{I}(g) = y(a)$. Similarly, we have the polar coordinate. If $g = k_1 b k_2$ then $y_\mathbf{p}(g) = y(b) = b^\alpha$. The function $y_\mathbf{p}$ extends to a K-bi-invariant function on G, except for the ambiguity of the action of the Weyl group $(g \mapsto g^{-1})$. A C^∞ even function of $y_\mathbf{p}$ is then C^∞ on G. We let

$$v_\mathbf{p} = \log y_\mathbf{p},$$

defined up to sign on G, and also subject to the preceding ambiguity.

Lemma 3.1.1. *Every $g \in G$ is such that its polar coordinate $y_\mathbf{p}$ and $y_\mathbf{p}^{-1}$ can be written as a polynomial in its real Iwasawa coordinates. Furthermore,*

$$y_\mathbf{p}, y_\mathbf{p}^{-1} \in \mathrm{Poly}^\infty\left(y_\mathbf{I}, y_\mathbf{I}^{-1}, x_1, x_2, k\right),$$

meaning that all derivatives of $y_\mathbf{p}$ and $y_\mathbf{p}^{-1}$ (with respect to $y_\mathbf{I}$, $y_\mathbf{I}^{-1}$, x_1, x_2, and components of k) are polynomials in the real Iwasawa coordinates of g.

Proof. By writing g in both polar and Iwasawa coordinates, we obtain the expression

$$a_\mathbf{p}^2 = k^* u a_\mathbf{I}^2 u^* k.$$

From this equation, the first assertion follows immediately since

$$a_\mathbf{p}^2 = \begin{pmatrix} y_\mathbf{p} & 0 \\ 0 & y_\mathbf{p}^{-1} \end{pmatrix}.$$

The second assertion regarding the derivatives of $y_\mathbf{p}$ and $y_\mathbf{p}^{-1}$ then follows by induction.

We define

$$B(G) = \text{space of bounded function on } G;$$
$$B^\infty(G) = \text{space of } C^\infty\text{-functions such that } Df \in B(G) \text{ for all } D \in \mathrm{IDO}(G).$$

We have the polar height function σ. We can define **polynomial growth, polynomial decay, superpolynomial decay with respect to** σ, in a manner exactly the one used define these concepts on Euclidean space.

We denote the spaces of such functions by PG_σ, PD_σ, SPD_σ respectively, and the spaces requiring the corresponding behavior for all *Df* with invariant differential operators *D* by putting the superscript ∞, giving rise to $\mathrm{PG}_\sigma^\infty$, $\mathrm{PD}_\sigma^\infty$, and $\mathrm{SPD}_\sigma^\infty$.

In Section 2.1 we encountered the spaces QED_σ and LEG_σ, of functions with quadratic exponential decay and linearly exponential growth with respect to σ. We recall that LEG_σ is the space of functions f satisfying

$$|f(g)| = O\left(e^{c\sigma(g)}\right) \text{ for } \sigma(g) \to \infty.$$

with some constant $c > 0$. Note that bounded functions are in LEG_σ. For QED_σ we replace $c\sigma$ by $-c\sigma^2$.

For each positive integer p, we define

$$\mathrm{LEG}_\sigma^P(G) = \text{space of } C^P\text{-functions on } G \text{ such that } Df \in \mathrm{LEG}_\sigma(G) \text{ for all } D \in \mathrm{IDO}(G) \text{ of order at most } P.$$
$$\mathrm{QED}_\sigma^P(G) = \text{similar definition with } Df \in \mathrm{QED}_\sigma(G).$$

We define $\mathrm{LEG}_\sigma^\infty(G)$ and $\mathrm{QED}_\sigma^\infty(G)$ as the subspaces that are LEG^p and QED^p for all p, respectively.

We now show that the functions making up Gaussians are in $\mathrm{QED}_\sigma^\infty$ or $\mathrm{LEG}_\sigma^\infty$ in a way that shows that Gaussians are in $\mathrm{QED}_\sigma^\infty$.

Proposition 3.1.2. *The spaces* $\mathrm{QED}_\sigma^\infty$ *and* $\mathrm{LEG}_\sigma^\infty$ *are closed under multiplication. We also have*

$$\mathrm{QED}_\sigma^\infty \cdot \mathrm{LEG}_\sigma^\infty \subset \mathrm{QED}_\sigma^\infty.$$

Proof. Immediate from the rule for the derivative of a product, and induction.

For simplicity, we omit the subscript σ in what follows.

Lemma 3.1.3. *For any* $k \in K$, *let* $L_k(g) = kg$ *and* $R_k(g) = gk$, *viewed as functions on G.*

(a) If $f \in \mathrm{LEG}^\infty$, *then for each* $k \in K$, $f \circ L_k$, *and* $f \circ R_k$ *are in* LEG^∞.
(b) If $f \in \mathrm{LEG}^\infty$, *then the functions*

$$\int_K (f \circ L_k)\, dk \text{ and } \int_K (f \circ R_k)\, dk$$

are in LEG^∞.

(c) If $f \in \mathrm{QED}^\infty$, *then for each* $k \in K$, $f \circ L_k$ *and* $f \circ R_k$ *are in* QED^∞.
(d) If $f \in \mathrm{QED}^\infty$, *then the functions.*

$$\int_K (f \circ L_k)\,dk \text{ and } \int_K (f \circ R_k)\,dk$$

are in QED^∞.

Proof. We will give the proofs of (a) and (b), with the proofs of (c) and (d) following the same pattern.

Let $D \in \mathrm{IDO}$, which, in polar coordinates can be written in terms of derivatives with respect to $y_{\mathbf{p}}$, derivatives with respect to real K-coordinates, with coefficients that are polynomials in $y_{\mathbf{p}}$, $y_{\mathbf{p}}^{-1}$, and real K-coordinates. Clearly, polynomials in $y_{\mathbf{p}}$ and $y_{\mathbf{p}}^{-1}$ are in LEG^∞. Therefore, for any $f \in \mathrm{LEG}^\infty$, the chain rule from elementary calculus then implies that $f \circ L_k$ and $f \circ R_k$ are in LEG^∞. From this, part (a) follows.

For any $D \in \mathrm{IDO}$ and any $f \in \mathrm{LEG}^\infty$, the functions $D(f \circ L_k)$ and $D(f \circ R_k)$ are continuous functions of K. Furthermore, for each fixed $k \in K$, the functions $D(f \circ L_k)$ and $D(f \circ R_k)$ are in LEG^∞. Since K is compact, the implied constant in the LEG^∞ condition, for each D, can be taken to be uniform for all K. From this, (b) follows since the conditions for DUTIS are satisfied.

Proposition 3.1.4. *The following functions are elements of the indicated function spaces:*

(a) $\sigma^2 \in \mathrm{LEG}_\sigma^\infty$, *and so is the function* $(\log y_{\mathbf{p}})/(y_{\mathbf{p}} - y_{\mathbf{p}}^{-1})$.
(b) $e^{-\sigma^2} \in \mathrm{QED}_\sigma^\infty$;
(c) If φ *is a Gaussian, then* $\varphi \in \mathrm{QED}_\sigma^\infty$;
(d) $\chi_s \in \mathrm{LEG}_\sigma^\infty$;
(e) $\Phi_s \in \mathrm{LEG}_\sigma^\infty$;
(f) $\mathbf{K}_t \in \mathrm{QED}_\sigma^\infty$, *uniformly for* $t \geq t_0 > 0$ *for any* t_0, *and viewing* $\mathbf{K}_t$ *as a function of one variable, uniformly when the other ranges over a compact set.*

Proof. Each assertion follows by first considering derivatives involving the polar coordinate variables, and then, in general, using Lemma 3.1.1. For part (e), one uses (d) together with the uniformity considerations as in the proof of Lemma 3.1.3.

3.2 Casimir, Transpose, and Harmonicity

We formulate general lemmas concerning differential operators and convolution, especially with the heat kernel. These lemmas will be used especially in Chapters 8 and 11. The general idea was already expressed in [JoL 01], p. 145: Although using the space $C_c^\infty(X)$ is often formally efficient, it is mathematically insufficient because we actually want to apply certain results to functions that do not have compact support, such as characters and their K-projections to spherical functions. Hence, in practice, one has to rely on a point-pair invariant that has sufficiently rapid decay at infinity, compared to the growth of the function and its derivatives to which the operators are applied. Thus some standard formal properties that are proved without much ado for C^∞ functions with compact support can be extended when one of the functions may have some positive order of growth as long as the other functions have a corresponding sufficiently faster order of decay. Of course, these orders have to be made precise. The heat kernel has quadratic exponential decay. This will allow some formulas to be valid when we convolve the heat kernel with functions that have exponential linear growth.

The DUTIS in the proof of Chapter 2, Theorem 2.5.1, is already slightly more complicated than that of Chapter 2, Proposition 2.4.5. The general pattern is that if some function $\Psi(z,w)$ satisfies a differential equation in the variable z, then the convolution defined by

$$(\Psi * f)(z) = \int \Psi(z,w) f(w) dw$$

satisfies the same differential equation, provided DUTIS is legitimate. The preceding section determined some bounds that legitimate DUTIS with the heat kernel, for a suitable class of functions f. In addition, if the differential equation is in the variable w, then we meet the problem of determining its transpose on f. This is especially the case when f is an eigenfunction, in several contexts. The subsequent sections already give examples taking convolution with the heat kernel. In later chapters, we carry out a similar analysis on $\Gamma\backslash G/K$ rather than G or G/K, with our discrete group Γ, which is used to periodize functions on G.

From the basic properties of invariant differential operators, abbreviated IDO, we know that they have a transpose locally, i.e., on $\mathbf{C}_c^\infty$. This transpose ${}^tD$ is uniquely determined by satisfying the **transpose formula** for any Haar measure $d\mu$:

TR 1.
$$\int_G Df_1 \cdot f_2 \, d\mu = \int_G f_1 \cdot {}^tDf_2 \, d\mu \quad \text{for } f_1, f_2 \in C_c^\infty.$$

In fact, if $Z \in g$, then

TR 2.
$${}^t\tilde{Z} = -\tilde{Z}.$$

TR 3.
$$\text{For } D_1,\ D_2 \in \mathrm{IDO}(G), \text{we have } {}^t(D_1D_2) = {}^tD_2{}^tD_1.$$

For a detailed treatment, see for instance [JoL 01], Chapter 2, Propositions 6.1 and 6.3. **TR 3** is of course trivial. We reproduce a proof of **TR 1** and **TR 2**.

Proof of **TR 1**, **TR 2**. It suffices to prove the formula when $D = \tilde{Z}$ with $Z \in g$, since an arbitrary IDO is a (noncommutative) polynomial in operators of the form $\tilde{Z}$. Trivially,

$$\tilde{Z}(f_1 f_2) = \tilde{Z}f_1 \cdot f_2 + f_1 \cdot \tilde{Z}f_2.$$

It suffices to prove that

$$\int_G \tilde{Z}(f_1 f_2)\, d\mu = 0 \quad \text{or} \quad \int_G \tilde{Z}(f)\, d\mu = 0,$$

that is, for $f = f_1 f_2$, or just $f \in C_c^\infty$,

$$\int_G \frac{d}{dr} f(g \cdot \exp(rZ))\Big|_{r=0} d\mu(g) = 0.$$

Shah once showed one of us the following argument avoiding Stokes's theorem. We take the derivative outside the integral sign, and then use the right invariance of Haar measure $d\mu(g)$ to cancel the factor $\exp(rZ)$ in the integral expression, so that this expression is constant with respect to r. Hence one gets the value 0 after differentiating with respect to r, as desired.

We want to extend the transpose formula to a larger class of pairs of functions f_1, f_2.

Lemma 3.2.1. *The transpose formula holds if f_1, f_2 are both C^∞ and one of them has compact support.*

Proof. Immediate since the domain of the convolution integral is contained in this support.

Next we formulate a very general L^1 condition under which the transpose formula holds. We call conditions (i) and (ii) below the **L^1-transpose conditions**.

Theorem 3.2.2. *Let $D \in \mathrm{IDO}(G)$ have degree m. Let f_1, $f_2 \in C^m(G)$. Assume:*

(i) $(Df_1)f_2 \in L^1(G)$.
(ii) For every $D_2 \in \mathrm{IDO}(G)$ of degree $\leqq m$, the function $f_1 D_2 f_2$ is in $L^1(G)$. Then the transpose formula holds.

Proof. We need a general lemma concerning the existence of a cutoff function; cf. [JoL 01]. For $R > 0$, let $\bar{\mathbf{B}}_R$ be the closed ball of radius R in a, with respect to the distance function σ (Chapter 1 Theorem 1.8.2), and let

$$G_R = K\left(\exp \bar{\mathbf{B}}_R\right)K = \{z \in G \text{ such that } \sigma(x) \leqq R\}.$$

Then for R, $R' > 0$,

$$G_R = G_R^{-1},\ G_R G_{R'} \subset G_{R+R'} \text{ (triangle inequality for } \sigma).$$

Lemma 3.2.3. *Given $s > 0$, there exists a function $h_{R,s} \in C_c^\infty(K\backslash G/K)$ such that:*

(i) $0 \leqq h_{R,s} \leqq 1$;
(ii) $h_{R,s} = 1$ on G_R and $= 0$ outside G_{R+s};

(iii) Given a differential operator $D \in \mathrm{IDO}(G)$, there exists a constant $c_{D,s}$ such that for all $R > 0$,

$$||Dh_{R,s}||_\infty \leqq c_{D,s}.$$

Proof. Let φ_R be the characteristic function of $\mathbf{B}_{R+s/2}$ and let h be an element of $C_c^\infty(K\backslash G/K)$ such that

$$h(z) = h(z^{-1}) \geqq 0 \text{ for all } z \in G, \int_G h(z)\,dz = 1,$$

$$\text{and } \operatorname{supp}(h) \subset G_{s/4}.$$

Thus h is a standard bump function around the origin, of the type used to make up Dirac families. Let

$$h_R = h_R * \varphi_R.$$

Then we see at once that $h_R \in C_c^\infty(K\backslash G/K)$ and is bounded by 1, so (i) is satisfied. Property (ii) follows from the triangle inequality for σ and the formula

$$h_R(z) = \int_{G_{S/4}} \varphi_R(zw^{-1})\,h(w)\,dw.$$

As for (iii), it follows at once from the standard formula (DUTIS)

$$D(\varphi_R * h) = \varphi_R * Dh \quad \text{using} \quad h_R(z) = \int_G \varphi_R(w^{-1})\,h(wz)\,dw.$$

We can then take $c_{D,s} = ||Dh||_1$. This proves the lemma.

In the application of the lemma, we take $s = 1$, say. Then by the dominated convergence theorem and Lemma 3.2.1,

$$\begin{aligned}\int (Df_1)f_2 &= \lim_{N\to\infty} \int (Df_1)h_N f_2 = \lim_{N\to\infty} \int f_1{}^tD(h_N f_2)\\ &= \lim_{N\to\infty} \int f_1 h_N{}^tDf_2 + \sum_i \int f_1 (D_{1,i}h_N)(D_{2,i}f_2),\end{aligned}$$

where $D_{1,i}$, $D_{2,i} \in \mathrm{IDO(G)}$, the sum of their degrees is deg D, and $D_{1,i}h_N$ has support in the annulus $G_{N,N+1}$. The limit of the first integral is what we want. By Lemma 3.2.3, each term of the sum is estimated by integrals of the form

$$c_{D_{2,i}} \int_{G_{N,N+1}} |f_1||D_{2,i}f_2|,$$

which tend to 0 as $N \to \infty$ because the integrand is in L^1, and the annulus goes to infinity. This proves Theorem 3.2.2.

Corollary 3.2.4. *Suppose $f_1 \in C^m(G)$, $D \in \mathrm{IDO}(G)$ of degree $m \geqq 1$, and without constant term (as a noncommutative polynomial), and $Df_1 \in L^1(G)$. Then*

$$\int_G DF_1 \, d\mu = 0.$$

Proof. Take $f_2 = 1$ in Theorem 3.2.2.

Remark. Some people may see Corollary 3.2.4 as some form of Stokes's theorem without boundary.

In practice, one deals with two function spaces $\mathscr{S}_1$ and $\mathscr{S}_2$ such that every pair of functions $f_1 \in \mathscr{S}_1$ and $f_2 \in \mathscr{S}_2$ satisfy the transpose formula with every invariant differential operator. It may therefore sometimes be convenient to define a notion of a **transpose admissible** pair $(\mathscr{S}_1, \mathscr{S}_2)$ of vector spaces of functions if these spaces satisfy the following conditions:

Adm 1. The spaces are algebras (closed under multiplication) and $\mathscr{S}_1\mathscr{S}_2 \subset \mathscr{S}_1$.
Adm 2. The spaces $\mathscr{S}_1$ and $\mathscr{S}_2$ are modules over $\mathrm{IDO}(G)$, $\mathrm{IDO}(G)\mathscr{S}_i \subset \mathscr{S}_i$ for $i = 1, 2$.
Adm 3. $\mathscr{S}_1 \subset L^1(G)$.

In the examples with which we deal, much stronger conditions are satisfied, reflecting fast decay of functions in $\mathscr{S}_1$ with respect to slower growth for functions in $\mathscr{S}_2$. Instead of the admissibility condition **Adm 3** as stated, which we could call **L^1-admissibility**, one could give other conditions describing the two different rates of decay and growth respectively.

Examples. Lemma 3.2.1 says that $\mathscr{S}_1 = C^\infty$ and $\mathscr{S}_2 = C_c^\infty$ are L^1-admissible.

Let $\mathscr{S}_1 = L^{1,\infty}$ be the space of C^∞ functions all of whose invariant derivatives are in L^1. Let $\mathscr{S}_2 = B^\infty$ be the space of C^∞ functions all of whose derivatives are bounded. This pair is admissible.

One may define such pairs in various contexts of Schwartzification, notably with the Harish-Chandra Schwartz space as $\mathscr{S}_1$, and the space of linear exponential growth LEG^∞ as $\mathscr{S}_2$. For $\mathscr{S}_1$ one could also take the space of functions all of whose invariant derivatives decay linear exponentially with arbitrarily high linear factor in the exponent; cf. [GaV 88] and [JoL 01] Chapter 10, aside from Harish-Chandra's collected papers.

For us, the pair QED^∞ and LEG^∞ is the most natural and useful.

Note that the superscript ∞ is used functorially with respect to the ideas to mean that all derivatives satisfy the condition indicated by the accompanying superscripted letter.

After this more general discussion, we apply Theorem 3.2.2 to the Casimir operator.

Theorem 3.2.5. *Let* ω *be the Casimir operator. Then* ${}^t\omega = \omega$.

Proof. Let $\{Z_i\}$ be a basis of $\mathfrak{g}$, with dual basis $\{Z_i'\}$. Then

$$\omega = \sum \tilde{Z}_i \tilde{Z}_i', \text{ so } {}^t\omega = \sum {}^t\tilde{Z}_i'{}^t\tilde{Z}_i = \sum \tilde{Z}_i'\tilde{Z}_i$$

because $(-1)^2 = 1$. By the fact that the expression for Casimir does not depend on the choice of basis, we get $\omega = {}^t\omega$.

We recall the expression for Casimir from Section 2.4, (5):

$$\omega = \frac{1}{2}\tilde{H}_\alpha^2 - 2\tilde{H}_\alpha + 2\tilde{E}_\alpha^2 + 2(\tilde{E}_\alpha^{(\mathbf{i})})^2 \bmod \mathrm{IDO}\cdot\tilde{\mathfrak{k}}. \tag{3.1}$$

We state the analogue of Chapter 2, Proposition 2.4.3.

Proposition 3.2.6. *Let* $f \in C^2(G/K)$. *Let* $\pi = \pi_{\mathrm{UA}} : G \to \mathrm{UA}$ *be the Iwasawa projection. Let* $\mathscr{L}$ *be the sum expression on the right side of (3.1). Then for* $g \in G$, $g = uak$ *(Iwasawa decomposition), we have*

$$(\omega(f\circ\pi))(g) = (\mathscr{L}f)(ua).$$

Proof. As in Chapter 2, we note that ω is right K-invariant, so the assertion is clear.

Except for the term $-2\tilde{H}_\alpha$, the operator $\mathscr{L}$ is a sum of squares. We may write $\omega(f)$ instead of $\omega(f\circ\pi)$. We shall need a formula for $\omega(f^2)$. For any element Z in the Lie algebra of G, we have

$$\tilde{Z}(f^2) = 2f\tilde{Z}f \text{ and } \tilde{Z}^2 f^2 = 2\left(\tilde{Z}f\right)^2 + 2f\tilde{Z}^2 f.$$

Hence letting $Z_1 = H_\alpha/\sqrt{2}$, $Z_2 = E_\alpha$, $Z_3 = E_\alpha^{(\mathbf{i})}$, we get

$$\begin{aligned}\omega(f^2) &= \left(\tilde{H}_\alpha f\right)^2 + f\tilde{H}_\alpha^2 f - 4f\tilde{H}_\alpha f + 2\tilde{E}_\alpha^2 f^2 + 2(\tilde{E}_\alpha^{(\mathbf{i})})^2 f^2 \\ &= 2\left(\tilde{Z}_1 f\right)^2 + 4(\tilde{Z}_2 f)^2 + 4(\tilde{Z}_3 f)^2 + 2f\omega f.\end{aligned} \tag{3.2}$$

In the next theorem, the basic hypothesis is that $\omega f = 0$. The L^1-hypothesis is just one choice, to have some control on how large the function and its first two derivatives can be, so the method of proof works.

Proposition 3.2.7. *Let* f *be a* C^2*-function on G/K, such that* $\omega(f^2) \in L^1(G)$ *and* $\omega f = 0$. *Then f is constant.*

Proof. By Theorem 3.2.2,

$$\int_G \omega(f^2)\,d\mu = 0.$$

By formula (3.2), this implies that the integral of the sum of the three squares in 0. Without loss of generality, we may suppose that f is real, since the real and imaginary parts of f satisfy the two hypotheses. Then the squares are $\geqq 0$, and the integral being 0 implies that each derivative is 0. By assumption of K-invariance, the derivatives $\tilde{Z}f$ are also 0 for $\tilde{Z} \in \tilde{k}$. Hence there is a basis of g for which all the derivatives with respect to the basis elements are 0, so f is constant, as was to be shown.

Remark. The above type of argument is used in the literature to obtain a lower bound for the first nonzero eigenvalue [Roe 56], [DeI 82] (see their Theorem 3, for which they give a short proof due to Vigneras). See also Szmidt [Szm 82], who works on $SL_2(\mathbf{C})$ instead of $SL_2(\mathbf{R})$, and gets $\lambda_1 > \pi^2$.

The "Gammafication" of Proposition 3.2.7 will be given in Theorem 3.2.7 of Chapter 5. Proposition 3.2.7 gives a basic result about the zero-eigenvalue and eigenfunction of Casimir. For the nonzero case, see Section 3.4 for a start.

3.3 DUTIS

Lemma 3.3.1. *Let f be measurable and in* LEG_σ. *Let* $\mathbf{K} = \mathbf{K}(z,w)$ *be symmetric* C^∞ *on* $G \times G$, *and* $\mathrm{QED}_\sigma^\infty$ *in each variable, uniformly for the other variable in a compact set. Let* $D \in \mathrm{IDO}(G)$. *Then* $\mathrm{K} * f$ *is* C^∞, *and for all* $D \in \mathrm{IDO}(G)$,

$$D(\mathbf{K} * f)(z) = (D\mathbf{K}) * f(z),$$

where $D\mathbf{K}$ is the derivative with respect to the first variable.

Proof. Special case of the DUTIS lemma, taking into account Section 3.1.

The most important special case of the above lemma is that in which $\mathbf{K} = \mathbf{K}_t$ is the heat kernel, with built-in uniformities.

Theorem 3.3.2. *Let f be measurable, bounded. Then for all t,* $\mathbf{K}_t * f \in B^\infty$.

Proof. The DUTIS lemma applies, taking into account Proposition 3.1.3(f).

We then have three possible variables (t,z,w). The derivative ∂_t can also be taken in and out of the integral even more easily than the derivatives D_z and D_w arising from invariant differential operators, and notably the Casimir operator. We state the result specifically for Casimir, which is the most important case for us, and for the heat kernel $\mathbf{K}_t$.

Theorem 3.3.3. *Let f be measurable on G and in* LEG_σ. *Then* $\mathbf{K}_t * f$ *is* C^∞ *and satisfies the heat equation. We have*

$$\omega(\mathbf{K}_t * f) = (\omega K_t) * f = (\partial_t \mathbf{K}_t) * f = \partial_t (\mathbf{K}_t * f).$$

Proof. The differential operator ω can be taken inside the convolution to act on $\mathbf{K}_t$ by Lemma 3.3.1. We then use the heat equation directly on $\mathbf{K}_t$. Finally, we can take ∂_t out of the integral sign even more easily by the DUTIS lemma, to conclude the proof.

Remarks. Theorem 3.3.3 is a variation of Gårding's lemma 1 [Gar 60], Section 8. Instead of convolving only with $\mathbf{K}_t$, Gårding convolves with an extra test function in C_c^∞.

We find it convenient to use differential operators on G, and to develop the formalism of convolutions on G. Right K-invariance of the functions being considered plays no role in the above. However, by using systematically right K-invariant functions, and viewing them as functions on G/K, one can use another approach. The space G/K has a natural Riemannian metric (up to a constant factor) that gives it negative curvature. The direct image of the Casimir element is the Laplacian for this metric, and is elliptic. Casimir itself is not elliptic on G.

Alternatively, Zuckerman pointed out to us a technique used by Nelson [Nel 59]. One adds a suitable element from $S(k)^\sim$ to Casimir, to obtain an elliptic operator on G (with suitable Riemannian metric) having the same direct image to G/K as Casimir. Then one can work on G itself without going to G/K.

The two years 1959–1960 gave rise to several papers concerning the differential operators on Lie groups. Harish-Chandra [Har 58a] carried out the direct image from G to A using the Iwasawa decomposition. Helgason [Hel 59] saw how Harish-Chandra's work "leads" to formulating this direct image to the homogeneous space G/K; cf. [Hel 84], Chapter 2, Theorems 4.8–4.10 and the exposition in [JoL 01], Section 2.2, for the isomorphism

$$\mathrm{IDO}\,(G)^K \xrightarrow{\approx} \mathrm{DO}\,(G/K)^G.$$

Coifman pointed out to us that one can replace the analysis arising from Section 3.1 and the DUTIS lemma leading to Theorem 3.3.3 by the Schwartzian technique of distributions, using the elliptic Laplacian on G/K, and the regularity theorem. We note that Gårding used this distribution argument to prove the above-mentioned lemma [Gar 60]. Gangolli in 1999 wrote one of us about the possibility of using this lemma to prove properties of the heat kernel to complement his construction.

Readers will find a proof of the regularity theorem in books on distributions, and also Appendix A4 of [Lan 75/85], Theorems 4 and 5. We state the theorem for the convenience of the reader. For a function ψ, we denote by T_ψ the associated distribution, i.e., the functional on C_c^∞ given by the formula

$$T_\psi(h) = \int h\psi, \text{ denoted by } [h,\psi].$$

Regularity Theorem. *Let L be a C^∞ elliptic operator, and let T be a distribution on an open set V of Euclidean space. Assume that there is a function $\psi \in C^\infty(V)$ such that $T \circ L = T_\psi$. Then there exists $\Psi \in C^\infty(V)$ such that $T = T_\Psi$, and ${}^tL\Psi = \psi$.*

Then the distribution argument proving Theorem 3.3.3 runs as follows:

$$\begin{aligned}[\omega h, \mathbf{K}_t * f] &= \int (\omega h)(z) \int \mathbf{K}_t(z,w) f(w) dw \cdot dz \\ &= \iint (\omega h)(z) \mathbf{K}_t(z,w)\, dz \cdot f(w)\, dw \text{ [interchanging order of integration]} \\ &= \iint h(z)\, \omega_z \mathbf{K}_t(z,w)\, dz \cdot f(w)\, dw \text{ [by Lemma 3.2.1]} \\ &= [h, (\omega \mathbf{K}_t) * f] = [h(\partial_t \mathbf{K}_t) * f].\end{aligned}$$

The regularity theorem can be applied to conclude that $\mathbf{K}_t * f$ is C^∞, and also that $\omega(\mathbf{K}_t * f) = (\omega \mathbf{K}_t) * f$.

3.4 Heat and Casimir Eigenfunctions

We shall see that the heat and Casimir eigenfunctions are the same (under conditions that make integrals converge). The following results are related to the general pattern of [JoL 01], Chapter 4, Propositions 4.2.1, 4.2.2, etc., made explicit in the present case depending on the structure of the heat kernel.

A function f on G will be called a **heat eigenfunction** if the convolution integral $\mathbf{K}_t * f$ is absolutely convergent, and if there exists a scalar $\eta(t)$ for each t such that $\mathbf{K}_t * f = \eta(t) f$ for all t. If f is not the 0 function, then the scalar is uniquely determined and is called the **eigenvalue**. Three cases are of special interest: when the eigenvalue is 0, so f is in the kernel of convolution (Chapter 2, Theorem 2.3.2); when the eigenvalue is 1 (Proposition 3.2.7), so f is a fixed point of convolution; other cases, when we shall get further information about the eigenvalues.

Theorem 3.4.1. *Let $f \neq 0$ be in* $\mathrm{LEG}_\sigma^\infty$ *(G/K) and an eigenfunction of Casimir, with eigenvalue $-\lambda \in \mathbf{C}$, $\lambda \neq 0$, that is*

$$\omega f = -\lambda f.$$

Then for all $t > 0$, f is a heat eigenfunction, and more precisely,

$$\mathbf{K}_t * f = e^{-\lambda t} f.$$

Proof. Using only the elementary DUTIS lemma for the first step, we get for each z,

$$\begin{aligned}\partial_t (\mathbf{K}_t * f)(z) &= \partial_t \int_G \mathbf{K}_t(z,w) f(w)\, dw = \int_G \partial_t \mathbf{K}_t(z,w) f(w)\, dw \\ &= \int \omega \mathbf{K}_t(z,w) f(w)\, dw \qquad \text{[by the heat equation]} \\ &= \int \mathbf{K}_t(z,w)\, \omega f(w)\, dw \qquad \text{[by Theorem 3.2.2]} \\ &= \int \mathbf{K}_t(z,w)(-\lambda) f(w)\, dw \\ &= -\lambda (\mathbf{K}_t * f)(z).\end{aligned}$$

This implies that there is a constant $C(z)$ such that

$$(\mathbf{K}_t * f)(z) = e^{-\lambda t} C(z).$$

Letting $t \to 0$ by Chapter 2 Theorem 2.6, we conclude that $C(z) = f(z)$, thereby proving Theorem 3.4.1.

Remark 1. The theorem applies to $\mathbf{K}_{t,\mathrm{Iw}}$, $\mathbf{K}_{t,\mathbf{p}}$ with the corresponding Iwasawa and polar Haar measures respectively.

Remark 2. We use notation for the eigenvalue so that $-\omega$ will be seen to be positive. The notation reflects the property that $-\omega$ is a positive operator, which amounts to the positive Laplacian in the context of differential geometry.

Theorem 3.4.2. *Let f be measurable, in* LEG_σ*, and a heat eigenfunction* $\neq 0$*, with nonzero eigenvalues, that is, for all* $t > 0$,

$$\mathbf{K}_t * f = \eta(t) f, \quad \eta(t) \neq 0.$$

Then f is a C^∞ *eigenfunction of Casimir, and if its Casimir eigenvalue is* $-\lambda$*, then*

$$\eta(t) = e^{-\lambda t}.$$

Proof. By Lemma 3.3.1 and Theorem 3.3.3, we know that $\mathbf{K}_t * f$ is C^∞. Since f is assumed not to be the 0 function, from a given z such that $f(z) \neq 0$, we deduce from Chapter 2, Theorem 2.2.2, that $t \mapsto \eta(t)$ is a continuous homomorphism into the multiplicative group, so of the form $e^{-\lambda t}$. That $-\lambda$ is an eigenvalue of Casimir then follows directly from Theorem 3.3.3.

If f is bounded, we can add to the conclusions of Theorem 3.4.2.

Theorem 3.4.3. *Let f be a bounded measurable heat eigenfunction, nonconstant. Let* $\eta(t) = e^{-\lambda t}$ *be its heat eigenvalue. Then* λ *is real* $\geqq 0$*. If in addition* $\omega(f^2) \in L^1(G)$ *then* $\lambda > 0$.

Proof. Since both real and imaginary parts of f are bounded, we may assume that f real. We have $\omega f = -\lambda f$, and ωf is real, so λ is real. As Coifman pointed out to us, that $\lambda \geqq 0$ is then immediate because the convolution integral is bounded,

$$\left| \int_G \mathbf{K}_t(z, w) f(w)\, dw \right| \leqq \|f\|_\infty,$$

and the integral on the left is also equal to $e^{-\lambda t} f$. We let $t \to \infty$ to conclude the proof that $\lambda \geqq 0$. For $\lambda > 0$ we use Proposition 3.2.7.

Note that we are now at the next stage, continuing Section 2.5.

Let $f(z) = \chi_s(z) = y_{\mathrm{Iw}}^s$ be the Iwasawa character. We know from Chapter 2, Proposition 3.4.3, Example, that χ_s is an eigenfunction of ω, with eigenvalue

$$\mathrm{ev}(\omega, \chi_s) = 2s(s-2).$$

Thus we now know its heat eigenvalue:

$$\mathbf{K}_t * \chi_s = \eta_s(t)\chi_s \text{ with } \eta_s(t) = e^{2s(s-2)t}.$$

Ultimately, we shall consider the vertical line $s = 1 + \mathbf{i}r$ (r real), so that the eigenvalue has the negative expression $-2(1+r^2)$, corresponding to the fact that ω is a negative operator relative to the Hermitian integral scalar product.

Instead of the character itself, we also deal with the spherical function obtained as its K-invariantization on the left. In turn, characters and spherical functions may be viewed as examples of the more general K-invariantization operation; cf. [JoL 01], Chapter 2, Proposition 2.5.2, and Chapter 3, Proposition 3.3.4. Here we stick to the standard example as follows.

Theorem 3.4.4. *Let $\Phi(s,z) = \Phi_s(z)$ be the spherical kernel (the K-invariantization of the character χ_{s+1}). Then Φ_s is a heat eigenfunction, and precisely*

$$\mathbf{K}_t * \Phi_S = \eta_{\Phi,s}(t)\,\Phi_s \textit{ with eigenvalue } \eta_{\Phi,s}(t) = e^{ev(\omega,\chi_{s+1})t} = e^{ev(\omega,\Phi_s)}.$$

Note that $\chi_{s+1} = \chi_s\chi_1 = \chi_{s\alpha+\rho}$, which corresponds to the right notation for general groups.

Proof. By definitions,

$$(\mathbf{K}_t * \Phi_s)(z) = \int_G \int_K \mathbf{K}_t(z,w)\chi_{s+1}(kw)\, dk\, dw,$$

because $\alpha = \rho$ in the case of $SL_2(\mathbf{C})$. We may then repeat the proof of Theorem 3.4.1 in the present context, using the fact that ω commutes with translations, in the present case left multiplication by k. The integral over compact K adds no further convergence problem to the integral over G.

The Eisenstein functions of Chapters 11 and 12 will provide examples at the next level. We shall meet bounded eigenfunctions in Chapter 8, Theorem 8.5.4.

Onward

The next step is to periodize with respect to the discrete group Γ, which for us here, concretely, is $SL_2(\mathbf{Z}[\mathbf{i}])$. Chapter 7 mostly can be read immediately as a continuation of the present chapter, although it may be useful to read also Chapter 6 on the fundamental domain, which gives a chart for integration over $\Gamma\backslash G$. We put Chapters 4 and 5 earlier because they show immediately how Γ-periodization is used for its ultimate purpose of getting a theta inversion identity. Furthermore, from Chapter 7 onward, we meet convolutions with the periodized heat kernel, especially in Chapter 8, acting on the cuspidal part of $L^2(\Gamma\backslash G)$, and in Section 11.4 acting on the Eisenstein functions that represent the continuous part of the eigenfunction expansion on $\Gamma\backslash G$.

Part II
Enter Γ: The General Trace Formula

Chapter 4
Convergence and Divergence of the Selberg Trace

Let $o = \mathbf{Z}[\mathbf{i}]$ be the Gaussian integers. The discrete group

$$\Gamma = \mathrm{SL}_2(\mathbf{Z}[\mathbf{i}]) = \mathrm{SL}_2(o)$$

will now come in, never to leave us. We also use the subgroup $\Gamma_U = \Gamma \cap U$.

This brief chapter describes the essentially different situations in which certain series over certain subsets of Γ will be integrable or not over $\Gamma\backslash G$. The first section describes the standard ways to estimate the "height" or size of a matrix. In Sections 4.2 and 4.3, we give the fundamental divergence and convergence theorems. The conditions under which these theorems are proved are adjusted to fit the main application we have in mind, namely the heat kernel.

We shall meet several instances of the following pattern. Let φ be a K-bi-invariant function on G. We define its associated **point-pair invariant** (following Selberg) by letting

$$\mathbf{K}_\varphi(z,w) = \varphi\left(z^{-1}w\right).$$

Then $\mathbf{K}_\varphi$ is G-invariant, that is, $\mathbf{K}_\varphi(gz, gw) = \mathbf{K}_\varphi(z,w)$ for all $g \in G$, and is defined for z, $w \in G/K$. Let Γ' be a subset of Γ. Under appropriate conditions, we consider the trace series

$$\mathbf{K}_\varphi^{\Gamma'}(z,w) = \sum_{\gamma\in\Gamma'} \varphi\left(z^{-1}\gamma w\right).$$

In this chapter and the next, we consider mostly the case $z = w$, so the series is a conjugation Γ'-trace

$$\sum_{\gamma\in\Gamma'} \varphi\left(g^{-1}\gamma g\right).$$

Starting in Chapter 7, we shall meet the more general case $z \neq w$. The subsets Γ' will reflect the group structure of Γ and its relation to the geometry of G/K. In the present chapter, we consider the parabolic subgroup $\Gamma_U = \Gamma \cap U$ and the nontriangular subset (containing the noncuspidal subset) exhibiting divergence and convergence respectively for the above series integrated over $\Gamma\backslash G$.

We are striving toward the trace formula in Chapter 5. However, there is another type of estimate that would also fit in the present chapter, namely an estimate of the order of growth of the function $\mathbf{K}_\varphi^\Gamma$ when φ has quadratic exponential decay. We shall give such an estimate in Section 7.1 (polynomial growth). We postpone the estimate to Chapter 7 because it fits well at this later point, with other estimates involving the phenomenon of cuspidality.

For further comments, see the "What Next" preceding Chapter 5.

4.1 The Hermitian Norm

This section describes another basic estimating function besides the polar height. The group $G = \mathrm{SL}_2(\mathbf{C})$ has an Iwasawa decomposition $G = UAK,\ g = uak$, and a polar decomposition

$$G = KAK,$$

whereby an element $g \in G$ can be written as a product $g = k_1 b k_2$ with $b \in A$ and $k_1,\ k_2 \in K$. The A-component b is unique up to permutation of its diagonal components because $b_1^2,\ b_2^2$ are the eigenvalues of ${}^t\bar{g}g = gg^*$. We repeat the definition of the polar height σ on G, mentioned in Section 1.6, Remark, and Section 1.8, namely

$$\sigma(g)^2 = 2\,[\log b]^2 = 2\,(\log b_1)^2 + 2\,(\log b_2)^2 = \sigma(b)^2,$$

where $b_1,\ b_2$ are the diagonal components of b. Thus $b_2 = b_1^{-1}$, but we wrote the sum to fit its higher-dimensional generalization.

We recall the multiplicative **Iwasawa character** α: $A \to \mathbf{R}^+$, defined by

$$b^\alpha = b_1/b_2 = b_1^2.$$

We also write α additively on a, so if $H = \mathrm{diag}(h_1,\ h_2)$ with $h_1,\ h_2 \in \mathbf{R}$, then

$$\alpha(H) = h_1 - h_2 \quad \text{and} \quad b^\alpha = e^{\alpha(\log b)}.$$

We use the coordinate y on A by defining

$$y = y(b) = b^\alpha.$$

Thus

$$\sigma^2(b) = (\log y)^2.$$

If b is the A-polar component of g, we write $y(b) = y_{\mathbf{p}}(g)$ or $y_{\mathrm{Pol}}(g)$.

For the following basic definitions on semisimple Lie groups, see [Bor 69] and [Har 68].

For a matrix $Z \in \mathrm{Mat}_2(\mathbf{C})$ we let $Z^* = {}^t\bar{Z}$. We define the **positive definite Hermitian trace form** by

$$\langle Z, w\rangle = \mathrm{tr}(ZW^*).$$

In terms of coordinates, $\operatorname{tr}(ZW^*) = \sum z_{ij}\bar{w}_{ij}$. We note the adjointness properties

$$\langle ZW, W' \rangle = \langle W, Z^*W' \rangle \quad \text{and} \quad \langle WZ, W' \rangle = \langle W, W'Z^* \rangle, \tag{4.1a}$$

$$\langle Z, W \rangle = \langle W^*,\ Z^* \rangle = \langle W,\ Z \rangle. \tag{4.1b}$$

We define the corresponding **Hermitian trace norm** by

$$||Z|| = (\operatorname{tr}(ZZ^*))^{1/2} \text{ and } ||z||^2 = \sum |z_{ij}|^2.$$

Trivially, we have

$$||z|| = ||z^*||.$$

If b_1^2, b_2^2 are the eigenvalues of ZZ^*, with $b_i \geqq 0$, then

$$||z||^2 = b_1^2 + b_2^2.$$

Then trivially,

$$2\,(\det ZZ^*)^{1/2} \leqq \operatorname{tr}(ZZ^*). \tag{4.2}$$

For $g \in G = \mathrm{SL}_2(\mathbf{C})$, we have four basic properties of the Hermitian norm:

HN 1. $||g|| \geqq \sqrt{2}$ for all $g \in G$, and $||k|| = \sqrt{2}$ for $k \in K$.
HN 2. $||gg'|| \leqq ||g||\ ||g'||$.
HN 3. $||g|| = ||g^{-1}||$.
HN 4. The norm is K-bi-invariant, $||k_1 g k_2|| = ||g||$.

HN 1 is immediate from (2). **HN 2** is the Schwarz inequality. **HN 4** is immediate from (1a). **HN 3** is immediate from **HN 4** and the symmetry of the trace on diagonal elements.

Next we compare the Hermitian norm with the polar height. Recall Section 1.8.

Proposition 4.1.1. *As functions on G,*

$$\log ||g|| \gg\ll \sigma(g) + 1.$$

Hence a function has polynomial growth with respect to the Hermitian norm if and only if it has linear exponential growth with respect to σ.

Proof. This is immediate. Recall that the notation $\gg\ll$ means that each size is less than a positive constant times the other.

Let c, $Y > 0$. We define a **Siegel set** $S_{c,Y}$ to consist of those elements $g = uak$ (Iwasawa decomposition) with $u = u(x)$, $a = \operatorname{diag}(a_1, a_2)$, $k \in K$, such that $|x| \leqq c$ and $y = y(a) = a_1^2 \geqq Y$. If Ω_U is a compact set in U, we also define the **Siegel set** $S_{\Omega_U,Y}$ to be the set of those elements g such that $u \in \Omega_U$ and $y \geqq Y$. We note that a Siegel set has finite measure $(dx_1\, dx_2\, dy/y^3)$. The significance of a Siegel set will be described after we state and prove the next result.

Proposition 4.1.2. *In a Siegel set, with $g = k_1 b k_2$, we have $\|a\| \gg\ll \|b\|$, that is, the Iwasawa projection and the polar projection on A have the same order of magnitude.*

Proof. Taking the quadratic map gives

$$u a^{2t} \bar{u} = k_1 b^2 k_1^{-1}.$$

Using the fact that $\|u\|$ is bounded for g in a Siegel set, we get the stated inequality.

Now we come to the significance of a Siegel set. The next lemma shows that a Siegel set contains representatives of the orbits of Γ in G as soon as Y is sufficiently small and c sufficiently large. Thus some Siegel set contains a fundamental domain, to be considered more precisely in Chapter 6, Theorem 6.6.1 (beginning of the proof). For our purposes here, the following lemma suffices.

Lemma 4.1.3. *Let $g \in G$. There exists $\gamma \in \Gamma$ such that*

$$y(\gamma g) \geqq 1/\sqrt{2} \text{ and } |x(\gamma g)| \leqq 1/\sqrt{2}.$$

Proof. Let $\|.\|$ be the Euclidean norm on $\mathbf{C}^2$, so $\|(x_1, x_2)\|^2 = |x_1|^2 + |x_2|^2$. Let e_1, e_2 be the unit vectors (1,0) and (0,1). Let $g \in G$, and let $g = uak$ be its Iwasawa decomposition, with $a = \operatorname{diag}(a_1, a_2)$. Suppose g is chosen such that

$$a_2 = \|{}^t e_2 g\| = \min_{\gamma \in \Gamma} = \|{}^t e_2 \gamma g\|.$$

It will suffice to prove that $a_2 \leqq \sqrt{2} a_1$, that is, for some γ, $y(\gamma g) = a_1/a_2 \geqq 1/\sqrt{2}$. Let $u = u(x)$. Possibly after a Γ_U-translation, we may assume that $|x| \leqq 1/\sqrt{2}$. Let $\gamma \in \Gamma$ interchange ${}^t e_1$ and ${}^t e_2$ (acting on the right). Then

$$a_2^2 \leqq \|{}^t e_2 \gamma g\|^2 = \|{}^t e_1 g\|^2 = a_1^2 + a_2^2 |x|^2 \leqq a_1^2 + \frac{1}{2} a_2^2.$$

The desired inequality $a_2^2 \leqq 2a_1^2$ follows.

In one lower dimension, the picture of the Siegel set is that of a rectangle whose bottom is at $y = 1/\sqrt{2}$.

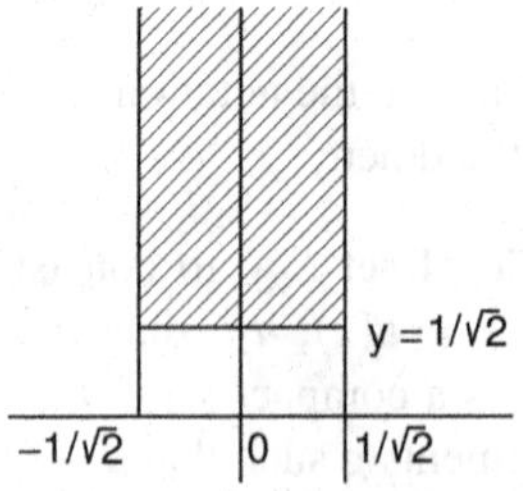

In general, it is a tube starting at the same height, going up to infinity, lying above the disk of radius c. By using the sup norm on $x = (x_1, x_2)$, one could replace the bottom disk by a square of side length 1, centered at 0.

4.2 Divergence for Standard Cuspidal Elements

We shall give a quantitative, more extensive analysis of some phrases in some papers of Jim Arthur, starting with [Art 78], p. 913, and essentially repeated in [Art 89], p. 16: "Experiments with examples suggest that the contribution to the integral of a conjugacy class in $G(\mathbf{Q})$ diverges when the conjugacy class intersects a proper parabolic subgroup defined over $\mathbf{Q}$. The more parabolics it meets, the worse will generally be the divergence of the integral." We are on SL_2. See Theorems 4.2.1 and 4.3.2. We elaborate on this statement, after making appropriate definitions of a group-theoretic nature.

4.2.1 Cuspidal and Parabolic Subgroups

We call a matrix **triangular** if it is upper triangular or lower triangular. We let

$$\mathrm{Tri}\,(G) \text{ or } G_{\mathrm{Tri}} = \text{subset of } G \text{ consisting of the triangular elements;}$$
$$\mathrm{Tri}\,(\Gamma) \text{ or } \Gamma_{\mathrm{Tri}} = \mathrm{Tri}\,(G) \cap \Gamma.$$

The upper triangular subgroup of G will be called the **standard parabolic subgroup**, denoted by G_∞. It contains UA as a normal subgroup, with a restriction on the diagonal elements (positive). We call UA the **reduced standard parabolic**. Letting D denote the full diagonal group, then $UD = G_\infty$ is the standard parabolic. A **parabolic subgroup of** G is defined to be a conjugate of the standard parabolic.

Remark. Our definition of parabolic is in line with the definition used by the Lie industry in higher dimension, namely conjugate to an algebraic subgroup containing the upper triangular group (containing what's called a Borel subgroup in invariant terminology). That definition is different from a definition originating earlier in the century, and still sometimes used today by some people, classifying elements into parabolic, elliptic, loxodromic, hyperbolic, which give us the hiccups.

The intersection

$$\Gamma_\infty = G_\infty \cap \Gamma = UD \cap \Gamma$$

will be called the **standard Γ-cuspidal subgroup**. Let R be the matrix

$$R = \begin{pmatrix} \mathbf{i} & 0 \\ 0 & -\mathbf{i} \end{pmatrix}.$$

Then it is immediately verified that Γ_U has index 4 in Γ_∞, and

$$\Gamma_\infty = (\pm I, \pm R)\,\Gamma_U.$$

Indeed, by the condition of determinant 1, the two diagonal components of an element $\gamma \in \Gamma_\infty$ have to be units in $\mathbf{Z}[\mathbf{i}]$ inverse to each other.

An element of Γ will be called **Γ-cuspidal** if it is Γ-conjugate to an element of Γ_∞. We want the smallest $\mathbf{c}(\Gamma)$-invariant subset of Γ containing Γ_∞. We let

$\mathrm{Cus}(\Gamma) = \Gamma$-**cuspidal subset** of $\Gamma = \mathbf{c}(\Gamma)\Gamma_\infty$;
$\mathrm{NCus}(\Gamma) =$ complement of $\mathrm{Cus}(\Gamma)$ in Γ, also called the noncuspidal subset of Γ.

Note that $\mathrm{Cus}(\Gamma)$ contains the lower triangular subgroup of Γ, so the full triangular subset (upper and lower triangular). They behave quite differently with respect to the fundamental domain; cf. Theorems 4.2.3 and 4.3.2 below.

For invariant formulations of certain results, such as Corollary 3.3, we note:

Both $\mathrm{Cus}(\Gamma)$ *and* $\mathrm{NCus}(\Gamma)$ are $\mathbf{c}(\Gamma)$*-invariant.*

An element of G is called **regular** if its eigenvalues are distinct. Note that the cuspidal element R above is regular. We study regular elements more systematically in the next chapter. It will be proved in Chapter 7, Corollary 7.2.3, that if $\gamma \in \Gamma$ and γ is not cuspidal, then γ is regular.

Remark. With regard to the above definitions involving conjugacy classes, note that two elements of Γ may be G-conjugate to each other but not Γ-conjugate to each other, for instance the two elements

$$R = \begin{pmatrix} \mathbf{i} & 0 \\ 0 & -\mathbf{i} \end{pmatrix} \text{ and } \begin{pmatrix} \mathbf{i} & b \\ 0 & -\mathbf{i} \end{pmatrix} = g_{\mathbf{i},b}$$

with odd $b \in \mathbf{Z}[\mathbf{i}]$. Note that $g_{\mathbf{i},b} \in \Gamma_\infty$ is cuspidal, regular, and has order 4 (its square is $-I$). We suggest that you carry out the computation of conjugating R and $g_{\mathbf{i},b}$ by an element $u(x)$ to see what's going on.

We now apply these group-theoretic distinctions to questions of divergence (and later convergence) for certain sums on G. Invariantly on G and $\Gamma\backslash G$ we have to deal with $\mathbf{c}(\Gamma)$-invariant subsets. Let Γ' be a $\mathbf{c}(\Gamma)$-invariant subset of Γ. Let φ be a function on G. For $g \in G$ define

$$f(g) = \sum_{\gamma \in \Gamma'} \varphi(g^{-1}\gamma g).$$

Then f is left Γ-invariant, that is, for all $\gamma_1 \in \Gamma$, we have

$$f(\gamma_1 g) f(g) \quad \text{for all} \quad g \in G.$$

This is immediate, since $\mathbf{c}(\gamma_1)$ induces a permutation of Γ'. In particular, this function f is invariant under translations by Γ_U. This will give the possibility of using a Fourier series expansion on $\Gamma_U\backslash U$ as in Chapters 10 and 11, to deal with the nonconvergence phenomenon.

To handle the invariant formulation, we first formulate concretely a divergence result on a Siegel set in terms of coordinates.

Theorem 4.2.1. *Let $S_{c,Y}$ be a Siegel set with $Y \geqq 1$. Let φ be a positive measurable function on G that is K-bi-invariant, so even, and has* **at most quadratic exponential decay**, *that is, there exists $c' > 0$ such that for $b \in A$,*

$$\varphi(b) \gg e^{-c'\sigma^2(b)}.$$

Then

$$\int_{S_{c,Y}} \sum_{\gamma \in \Gamma_U} \varphi(g^{-1}\gamma g)\, dg = \infty.$$

Proof. Let $E_{12} = E_\alpha$ be the matrix $\begin{pmatrix} 0 & 1 \\ 0 & 0 \end{pmatrix}$, which is a $\mathbf{C}$-basis of the α-eigenspace for the conjugation action of A. For $m \in \mathbf{Z}[\mathbf{i}]$, $m \neq 0$, abbreviate

$$u(m) = u_m = I + mE_\alpha. \tag{4.3}$$

We have

$$a^{-1}u_m a = I + (m/y)\, E_\alpha \text{ with } y = a^\alpha = a_1^2. \tag{4.4}$$

Therefore

$$||a^{-1}u_m a|| \ll 1 + |m|/y. \tag{4.5}$$

Thus we obtain the estimate

$$\begin{aligned} \sigma^2\left(a^{-1}u_m a\right) &= \sigma^2\left(I + (m/y)E_\alpha\right) \\ &= O(1) \text{ for all } m \text{ and } |m| \leqq y. \end{aligned} \tag{4.6}$$

Since $||.||$ is K-bi-invariant, for $g = uak$ in a Siegel set, we have

$$\log ||g^{-1}u_m g|| = \log ||a^{-1}u_m a|| = O(1) \text{ for } |m| \leqq y, \tag{4.7}$$

because u commutes with m. (In the higher-dimensional case, one has to argue that the elements $a^{-1}u^{-1}a$ and $a^{-1}ua$ lie in compact sets.)

Now for the proof of Theorem 4.2.1 proper, it suffices to prove that the A-integral is infinity, and it suffices to show that given $c' > 0$,

$$\int_Y^\infty \sum_{Y \leqq |m|} e^{-c'\left(\log ||a^{-1}u_m a||\right)^2} \frac{dv}{y^3} = \infty.$$

We replace the integrand by the smaller sum taken over $Y \leqq |m| \leqq y$. By what we saw in (3), each term of the sum is bounded from below. The number of terms is the number of lattice points from $\mathbf{Z[i]}$ in the annulus between radii Y (fixed) and y (tending to ∞), which amounts to the lattice points in the big disk. That is $\gg y^2$ for $y \to \infty$. Hence the whole integral diverges essentially like $\int_Y^\infty dy/y$, thus concluding the proof.

Corollary 4.2.2. *The integral*

$$\int_{\Gamma\backslash G} \sum_{\gamma \in \mathbf{c}(\Gamma)\Gamma_U} \varphi(g^{-1}\gamma_g)\, dg$$

diverges.

For the record, we mention the matrix representing the other coset of Γ_∞/Γ_U, namely

$$Ru_m = \begin{pmatrix} \mathbf{i} & \mathbf{i}m \\ 0 & -\mathbf{i} \end{pmatrix} \quad \text{with } m \in o$$

Note that a regular upper triangular element of Γ is of the form $\pm Ru_m$.

Essentially the same argument as for Theorem 4.2.1 proves the following result:

Theorem 4.2.3. *Assumptions being as in Theorem 4.2.1, the integral*

$$\int_{S_{C,y}} \sum_{m \in o} \varphi(g^{-1} Ru_m g)\, dg$$

diverges.

Thus to ensure convergence, we shall have to exclude upper triangular matrices and their conjugates. We carry this convergence out in the next section.

Using the fundamental domain defined later, one can get an asymptotic description of the logarithmic divergence of the integral in Theorems 4.2.1, 4.2.2, 4.2.3, which will be of fundamental importance later. See Chapter 13, which is a direct continuation of this section. It gives an asymptotic description for the divergence, by truncating the fundamental domain from above.

4.3 Convergence for the Other Elements of Γ

We turn to the convergence result, staying away in some fashion from triangular elements. First we have a universal estimate for all matrices, about the sum without the manifold integral on top of it.

Theorem 4.3.1. *The following series converges with any constant $c > 0$.*

$$\sum_{0 \neq \gamma \in \mathrm{Mat}_2(\mathbf{Z}[\mathbf{i}])} e^{-c(\log\|\gamma\|)^2} < \infty.$$

Proof. The sum on the left can be estimated by the double sum

$$\sum_{N=1}^{\infty} \sum_{N \leqq \|\gamma\| < N+1} \ll \sum_{N=1}^{\infty} N^8 e^{-c'} (\log N)^2$$

with some constant $c' > 0$, so the convergence is clear.

Note that the exponential quadratic decay is much stronger than what is needed for convergence. Superpolynomial decay would suffice.

We write a matrix $\gamma \in \Gamma$ as we did in Section 4.2, namely

$$\gamma = \begin{pmatrix} a_\gamma & b_\gamma \\ c_\gamma & d_\gamma \end{pmatrix}.$$

The next theorem gives the convergence result avoiding the standard cuspidal elements.

Theorem 4.3.2. *Let S be a Siegel set and $c' > 0$. There exists $C > 0$ such that for all $g \in S$,*

$$\sum_{\gamma \notin \Gamma_\infty} e^{-c'(\log \|g^{-1}\gamma g\|)^2} \leqq C.$$

Therefore

$$\int_S \sum_{\gamma \notin \Gamma_\infty} e^{-c'(\log \|g^{-1}\gamma g\|)^2} dg < \infty.$$

Proof. The second assertion follows from the first because the measure of a Siegel set is finite. As to the first, by Theorem 4.3.1 it suffices to show that

$$-(\log \|g^{-1}\gamma g\|)^2 \ll -(\log \|r\|)^2.$$

Let $g = uak$ be the Iwasawa decomposition, so $\|u\|$ and $\|u^{-1}\|$ are bounded. Write

$$\|g^{-1}\gamma g\| = \|a^{-1}u^{-1}\gamma_{ua}\| = \|a^{-1}u^{-1}a \cdot a^{-1}\gamma a \cdot a^{-1}ua\|.$$

We have $a^{-1}u(x)a = u(x/y_a)$. Since y_a is bounded from below, we conclude that $\|a^{-1}ua\|$ and $\|a^{-1}u^{-1}a\|$ are bounded. Hence it suffices to prove that

$$-(\log \|a^{-1}\gamma a\|)^2 \ll -(\log \|\gamma\|)^2 \text{ for } a \in A,$$

or equivalently,

$$\log \|\gamma\| \ll \log \|a^{-1}\gamma a\|. \tag{4.8}$$

But

$$a^{-1}\gamma a = \begin{pmatrix} a_\gamma & b_\gamma y^{-1} \\ c_\gamma y & d_\gamma \end{pmatrix}, \text{ where } y = y_\alpha = a^\alpha,$$

so

$$\|a^{-1}\gamma a\|^2 = |a_\gamma|^2 + (|c_\gamma|y)^2 + (|b_\gamma|y^{-1})^2 + |d_\gamma|^2.$$

However, by $C^2 + B^2 \geqq 2CB$, we have

$$|c_\gamma|^2 y^2 + |b_\gamma|^2 y^{-2} \geqq 2|c_\gamma||b_\gamma| \text{ and } \gg |c_\gamma|^2 \text{ if } b_\gamma = 0.$$

Since $|c_\gamma| \geqq 1$, we get a uniform lower bound (which is all that we need),

$$\begin{aligned}\log \|a^{-1}\gamma a\|^2 &\gg \log \max(|a_\gamma|, |b_\gamma|, |c_\gamma|, |d_\gamma|)\\ &\gg \log(|a_\gamma|^2 + |b_\gamma|^2 + |c_\gamma|^2 + |d_\gamma|^2),\end{aligned}$$

which is (4.8), and concludes the proof.

Remark. The complement of Γ_∞ does contain cuspidal elements. Even with such elements we have convergence in Theorem 4.3.2. For the analogous situation with $z^{-1}\gamma w$, $z \neq w$, see Chapter 8, Lemma 2.1 and Proposition 2.3.

To formulate Theorem 4.3.2 on $\Gamma\backslash G$, we need to work with a $\mathbf{c}(\Gamma)$-invariant subset. The next theorem deals with the subset of noncuspidal elements, defined at the beginning of Section 4.2.

Corollary 4.3.3. *Let φ be a semipositive measurable function on G that is K-bi-invariant. Suppose that there exists $c > 0$ such that*

$$\varphi(b) \gg e^{-c\sigma^2(b)} \text{ for } b \in A, \sigma(b) \to \infty.$$

Then

$$\int_{\Gamma\backslash G} \sum_{\gamma \in \mathrm{NCus}(\Gamma)} \varphi(g^{-1}\gamma g)\, dg$$

converges.

The main point of this corollary is its application to Chapter 5, Theorem 4.2.1.

We define a function φ on G to be **admissible** if it is measurable, K-bi-invariant, so even, and having quadratic exponential decay, that is, there exists $c > 0$ such that for all $b \in A$,

$$|\varphi(b)| \ll e^{-c\sigma^2(b)}.$$

Of course the main case for us is that of the Gaussians. *We see that Corollary 3.3 applies to the absolute value of an admissible function.* In particular, we have given conditions under which Theorems 4.2.1 and 2.4 of Chapter 5 below are valid.

In Section 5.3 we shall give an explicit determination of the integral of Corollary 3.3 for the noncuspidal trace of the heat kernel (Gaussian).

What Next?

The previous chapter has set up the basic distinction between two classes of elements of Γ that have to be handled separately in the rest of the book. Here is the plan:

1. Take care of a trace formula for the noncuspidal elements, with convergence.
2. Use the fundamental domain to get some estimates for divergence phenomena in connection with the cuspidal trace.
3. Apply these estimates to settle the case in which convergence can be recovered using a special class of test functions, namely "cuspidal" functions.
4. Handle the continuous part of the eigenfunction expansion by introducing Eisenstein functions, playing a role similar to the functions e^{ixy} for the Fourier transform, but in the more sophisticated context of a noncompact noncommutative situation.
5. Combine the Eisenstein series and the cuspidal series in an affair that produces just the right cancellations to land us back in a convergent situation giving rise to a computable term in the trace formula.

Thus the Eisenstein series play at least two roles. The whole thing then results in an explicit formula that is the version on the noncompact space $\Gamma\backslash G/K$ of a theta inversion formula whose origins on the reals date back to Poisson. The formula that we give in Section 12.1 and the last page of this book is the starting point for the development of new zeta functions, as mentioned in the introduction.

Chapter 5
The Cuspidal and Noncuspidal Traces

Let φ be a function on $G = \mathrm{SL}_2(\mathbf{C})$, and $\Gamma = \mathrm{SL}_2(\mathbf{Z}[\mathbf{i}])$. We have seen that the series

$$\sum_{\gamma \in \Upsilon} \varphi\left(g^{-1}\gamma g\right)$$

may converge absolutely, but its integral over $\Gamma\backslash G$ may not because of certain elements that we want to isolate and leave out of the trace in the manner of [JoLu 95] and [DodJ 98]. Three structures are going to intermingle:

- Some pure group theory concerning conjugacy classes, centralizers, and the dichotomy between cuspidal and noncuspidal elements. See Section 4.2.
- Integration on $\Gamma\backslash G$.
- The divergence–convergence theorems from Chapter 4.

The first one will lead us to separate the cuspidal elements from the others in the trace series written down above. Cuspidal elements will be combined with Eisenstein series in a later chapter.

The relations to be established in a context when analysis is done in a group-theoretic setting will be done in the setting of conjugacy classes in G. There exist already special cases in which such results are given in terms of objects associated with the base number field, i.e., more in terms of number-theoretic invariants. These two alternatives are reminiscent of theories that establish a bijection between conjugacy classes in a Galois group and objects associated with the base (class field theory, covering space theory, etc). This direction deserves a separate treatment elsewhere; see Zagier [Zag 82], Szmidt [Szm 87], [Wal 84]. In Zagier's method, the objects associated to the number field are quadratic forms. A major direction is open to describe precisely the correspondence with conjugacy classes, describe precisely the way the terms of the theta inversion are equal to each other, and to find the analogous structure in the more general groups, starting with SL_n. From this point of view, the Zagier method and the conjugacy classes method are not alternative ways of doing the same thing. They are a part of a common geometric-arithmetic nonabelian class-field-theoretic structure.

5.1 Some Group Theory

In the previous chapter we have already seen a dichotomy between cuspidal and noncuspidal elements. There are other dichotomies for elements of G, according to their Jordan normal form, or whether they have distinct eigenvalues or not. We go into a systematic tabulation of these possibilities.

Let $g \in G$. Suppose the characteristic roots are equal. Then g is conjugate to an upper triangular matrix whose diagonal elements are the eigenvalues. There are two cases:

Case 1. $$g = \pm \begin{pmatrix} 1 & 0 \\ 0 & 1 \end{pmatrix}$$

Case 2. $$g \sim \pm \begin{pmatrix} 1 & b \\ 0 & 1 \end{pmatrix} \text{ with } b \in \mathbf{c}, b \neq 0.$$

Note that an element in Case 2 is of infinite order.

Recall that a regular element of G is one with distinct eigenvalues. Quite generally, a regular element is conjugate in G to a diagonal element

$$g \sim \begin{pmatrix} w_1 & 0 \\ 0 & w_1^{-1} \end{pmatrix} \text{ with } w_1 \neq \pm 1,\ |w_1| \geqq 1.$$

One uses conjugation by the Weyl element

$$\begin{pmatrix} 0 & 1 \\ -1 & 0 \end{pmatrix}$$

if necessary to satisfy the condition $|w_1| \geqq 1$.

It may be convenient to write w_1 in the form $w_1 = \varepsilon\lambda$, with $|\varepsilon| = 1$ and $\lambda \geqq 1$. If $\lambda > 1$ then we distinguish two cases:

Reg 1. $$g \sim \pm \begin{pmatrix} \lambda & 0 \\ 0 & \lambda^{-1} \end{pmatrix}$$

or with ε not real,

Reg 2. $$g \sim \pm \begin{pmatrix} \varepsilon\lambda & 0 \\ 0 & \bar{\varepsilon}\lambda^{-1} \end{pmatrix}.$$

If $\lambda = 1$, then

Reg 3. $$g \sim \pm \begin{pmatrix} \varepsilon & 0 \\ 0 & \bar{\varepsilon} \end{pmatrix} \text{ and } \varepsilon \neq \pm 1, \text{ so } \varepsilon \text{ is not real.}$$

Thus the elements of **Reg 3** are parametrized by the nonreal points on the circle.

We are interested in the intersection of Γ with some of the above subsets. Following the same notation used in Section 5.2 for $\mathrm{Cus}(\Gamma)$ and $\mathrm{NCus}(\Gamma)$, we denote the set of regular elements in Γ by

$$\mathrm{Reg}(\Gamma) = \Gamma_{\mathrm{Reg}} = \Gamma \cap \mathrm{Reg}.$$

Proposition 5.1.1. *Elements of type* **Reg 1** *and* **Reg 2** *are of infinite order. On the other hand, elements of* Γ_{Reg3} *are of finite order, and this order can be only* 3, 4, *or* 6.

Proof. Let g be as in **Reg 3**, and (g) the group generated by g. Since Γ is discrete, (g) can be conjugated to a discrete subgroup of the circle, which is necessarily finite.

As to the limitation on the order, let

$$g = \begin{pmatrix} a & b \\ c & d \end{pmatrix}, \text{ so } a,b,c,d \in \mathbf{Z}[\mathbf{i}].$$

The characteristic polynomial has degree 2, so a root ε has degree $\leqq 4$ over the rationals. An nth root of unity with $\phi(n) \leqq 4$ has necessarily $n \leqq 12$. The cases with $n = 12$ and $n = 8$ cannot occur, because if ε is a 12th or 8th root of unity, then $\varepsilon + \bar{\varepsilon}$ involves $\sqrt{3}$ or $\sqrt{2}$, but $\mathrm{tr}(g) = a + d \in \mathbf{Z}[\mathbf{i}]$, contradiction. The case $n = 7$ cannot occur because $\phi(7) = 6$. Finally $n \neq 5$, because the fifth roots of unity are not quadratic over **Q(i)**. This concludes the proof.

We let $\mathbf{c}\colon G \to \mathrm{Aut}(G)$ be the **conjugation representation** of G, so by definition

$$\mathbf{c}(g)g_1 = gg_1g^{-1} \text{ and } \mathbf{c}\left(g^{-1}\right)g_1 = g^{-1}g_1g = g_1^g.$$

Proposition 5.1.2. *The regular subset* $\mathrm{Reg}(\Gamma)$ *is* $\mathbf{c}(\Gamma)$*-invariant. So are the subsets of regular elements of infinite order, and also of finite order. There is only a finite number of* $\mathbf{c}(G)$*-conjugacy classes of finite order.*

Proof. This is immediate from the definitions, and the finiteness comes from Proposition 5.1.1, together with the fact that the group of roots of unity of given order is finite.

Remark. For the finiteness of Γ-conjugacy classes and related questions, i.e., a fixed-point theorem for finite subgroups, see Chapter 6, Theorem 6.3.3.

Next we look at centralizers. Given $g_0 \in G$, we denote its **centralizer** in a subgroup H by $\mathrm{Cen}_H(g_0)$, or also H_{g_0}. Trivially

$$\mathrm{Cen}_G\left(gg_0g^{-1}\right) = g\mathrm{Cen}G\left(g_0\right)g^{-1}.$$

In other words, $\mathbf{c}(g)$ commutes with taking the centralizer. Similarly with the normalizer. Thus to see the structure of a centralizer, we may suppose the element is in Jordan normal form, i.e., diagonalized for a regular element.

Proposition 5.1.3. *Let* $g_0 \in G$ *be regular diagonal. Then*

$$\mathrm{Cen}_G\left(g_0\right) = \textit{Diagonal group}.$$

In particular, the centralizer is abelian.

Proof. Let E_{ij} $(i \neq j)$ be the matrix with ij-component $= 1$ and 0 elsewhere. Then E_{ij} is an eigenvector of $\mathbf{c}(g_0)$, with eigenvalues equal to the quotients of the diagonal elements of g_0. So if g_0 commutes with a matrix, then the nondiagonal components of the matrix have to be 0, and the matrix is diagonal, as stated.

Note that the centralizer of an element g_0 in G is the isotropy subgroup of g_0 (the subgroup leaving g_0 fixed) under the conjugation representation. Let $c_G(g_0)$ be the G-conjugacy class of g_0. Then we have a bijection

$$\mathrm{Cen}_G(g_0)\backslash G \longrightarrow c_G(g_0) \text{ by } g \mapsto g^{-1}g_0 g.$$

Similarly for Γ, and an element $\gamma_1 \in \Gamma$, we have a bijection

$$\mathrm{Cen}_\gamma(\Gamma_1)\backslash\Gamma \to c_\Gamma(\gamma_1) \text{ by } \gamma \mapsto \mathbf{c}(\gamma^{-1})\gamma_1 = \gamma^{-1}\gamma_1\gamma.$$

Proposition 5.1.4. *Let $\gamma \in \Gamma$ be noncuspidal. Then γ is regular, and there is a finite group $Tor(\Gamma_\gamma)$ such that*

$$\mathrm{Cen}_\Gamma(\gamma) \cong \mathbf{Z} \times \mathrm{Tor}(\Gamma_\gamma).$$

Proof. Suppose first γ is regular. Under some conjugation, Γ goes to a discrete subgroup of G and γ is diagonalized. As we just saw in Proposition 5.1.3, the centralizer is contained in the diagonal group, and is a subgroup of a discrete group. The full diagonal group in G is isomorphic to $\mathbf{C}^* \cong \mathbf{R}^+ \times \mathbf{C}_1 \cong \mathbf{R} \times \mathbf{R}/\mathbf{Z}$. Let S be a discrete subgroup. Then the $\mathbf{R}$-coordinates of elements (r,t) (t mod $\mathbf{Z}$) have to form a discrete subgroup of $\mathbf{R}$, so infinite cyclic or trivial. Then the $\mathbf{R}/\mathbf{Z}$-coordinates are contained in a finite set because an infinite discrete subgroup of $\mathbf{R}/\mathbf{Z}$ is not compact. So far we have proved that

$$\mathrm{Cen}_\Gamma(\gamma) \approx \mathbf{Z} \times \mathrm{Tor}(\Gamma_\gamma) \text{ or } \mathrm{Tor}(\Gamma_\gamma) \text{ with } \mathrm{Tor}(\Gamma_\gamma) \text{ finite}.$$

We shall prove that γ is regular, and we shall eliminate the possibility that $\mathrm{Cen}_\Gamma(\gamma)$ is a finite group later, in Corollaries 2.2 and 2.3, via analysis. Note that the proposition is false for the cuspidal regular element $\gamma = R$, as well as for the nonregular $\pm I$.

Lemma 5.1.5. *Let $g_1, g_2 \in G$ be regular elements that commute. Then*

$$\mathrm{Cen}_G(g_1) = \mathrm{Cen}_G(g_2) \text{ and } \mathrm{Cen}_\Gamma(g_1) = \mathrm{Cen}_\Gamma(g_2).$$

If g_1, g_2 are regular but do not commute, then

$$\mathrm{Cen}_G(g_1) \cap \mathrm{Cen}_G(g_2) = \{\pm 1\}.$$

Proof. We diagonalize g_1 by a conjugation, so for the first equality, without loss of generality, we may assume g_1 diagonal. Then $\mathrm{Cen}_G(g_1)$ is the diagonal group by Proposition 5.1.3, so g_2 is diagonal. But then $\mathrm{Cen}_G(g_2)$ is the diagonal group also, proving the first equality. The second is immediate since

$$\mathrm{Cen}_\Gamma(g_1) = \mathrm{Cen}_G(g_1) \cap \Gamma,$$

and similarly for g_2. For the final assertion, note that every diagonal element is regular except ± 1, so the first equality implies the final conclusion.

A regular element $\gamma_0 \in \Gamma$ of infinite order will be called **primitive** if γ_0 generates $\mathrm{Cen}_\Gamma(\gamma_0)$ mod torsion. By Lemma 1.4, to an infinite order regular $\gamma \in \Gamma$, there is a primitive γ_0 that generates $\mathrm{Cen}_\Gamma(\gamma)$ mod torsion. So by Proposition 5.1.4 and Lemma 5.1.5,

$$\mathrm{Cen}_\Gamma(\gamma_0) = \mathrm{Cen}_\Gamma(\gamma) \cong (\gamma_0) \times \mathrm{Tor}\left(\Gamma_{\gamma_0}\right). \tag{5.1}$$

Here and elsewhere we denote by (γ_0) the group generated by γ_0. By Lemma 5.1.5, we have

$$\mathrm{Cen}_G(\gamma) = \mathrm{Cen}_G(\gamma_0) \tag{5.2}$$

and the similar equality replacing G by Γ.

Furthermore, the element γ can be expressed uniquely as a product

$$\gamma = \gamma_0^n \varepsilon \text{ with } n \in \mathbf{Z},\ n \neq 0, \text{ and } \varepsilon \in \mathrm{Tor}(\Gamma_{\gamma_0}).$$

In more invariant form,

$$\mathrm{Tor}\left(\Gamma_{\gamma_0}\right) \backslash \mathrm{Cen}_\Gamma(\gamma_0) = (\gamma_0). \tag{5.3}$$

The factor group on the left side can be taken right or left, since $\mathrm{Cen}_\Gamma(\gamma_0)$ is abelian. We use the notation $\Gamma_{\mathrm{inf}} = \{\text{of infinite order}\}$ in Γ.

5.1.1 Conjugacy Classes

Let G be any group with a subgroup Γ. Let Γ' be a subset of Γ that is $\mathbf{c}(\Gamma)$-invariant. Such a subset is the disjoint union of the Γ-conjugacy classes contained in it. Let

$\mathrm{CC}_\Gamma(\Gamma')$ = set of Γ-conjugacy classes contained in Γ'. *For each such conjugacy class, select a representative γ_c in the class.*

For $\gamma \in \Gamma$, we deal with the centralizers

$$\Gamma_\gamma = \mathrm{Cen}_\Gamma(\gamma) \text{ and } G_\gamma = \mathrm{Cen}_G(\gamma).$$

For our purposes here, we take the conjugation action by an element g to be

$$[g]\gamma = g^{-1}\gamma g.$$

The isotropy group of an element is then the centralizer of this element. We have a bijection

$$\Gamma_{\gamma_c} \backslash \Gamma \to c \text{ under the map } \eta \mapsto \eta^{-1}\gamma_c\eta.$$

In other words,

$$c = [\Gamma]\gamma_c = \bigcup_{\eta \in \Gamma_{\gamma_c} \backslash \Gamma} \eta^{-1}\gamma_c\eta, \tag{5.4}$$

$$\Gamma' = \bigcup_{c \in CC_\Gamma(\Gamma')} \bigcup_{\eta \in \Gamma_{\gamma_c} \backslash \Gamma} \eta^{-1} \gamma_c \eta. \tag{5.5}$$

We specialize this decomposition to the case that G and Γ are the usual groups $SL_2(\mathbf{C})$ and $SL_2(\mathbf{Z}[\mathbf{i}])$, and $\Gamma' = \mathrm{Reg}(\Gamma_{\mathrm{inf}})$. For each conjugacy class $c \in \mathrm{Reg}(\Gamma_{\mathrm{inf}})$ we select a primitive element $\gamma_0(c)$, generating $\mathrm{Cen}_\Gamma(\gamma_0(c))$ mod torsion. We let

$$\mathrm{Reg}_0(\Gamma_{\mathrm{inf}}) = \text{set of such elements } \gamma_0(c) \text{ with } c \in \mathrm{Reg}(\Gamma_{\mathrm{inf}}).$$

Lemma 5.1.6. *The elements $\gamma_0^n \varepsilon$ with $\gamma_0 \in \mathrm{Reg}_0(\Gamma_{\mathrm{inf}})$, $\varepsilon \in \mathrm{Tor}(\Gamma_{\gamma_0})$, $n \in \mathbf{Z} \neq 0$, are distinct representatives γ_c for the Γ-conjugacy classes in* $\mathrm{Reg}(\Gamma_{\mathrm{inf}})$.

Proof. From the definition of $\mathrm{Reg}_0(\Gamma_{\mathrm{inf}})$, it suffices to show that for each $n \in \mathbf{Z}_{\neq 0}$ and $\varepsilon \in \mathrm{Tor}(\Gamma_{\gamma_0})$, $\varepsilon \neq 1$, the elements γ_0^n and $\gamma_0^n \varepsilon$ are not Γ-conjugate. Suppose there is $\eta \in \Gamma$ such that

$$\eta^{-1} \gamma_0^n \eta = \gamma_0^n \varepsilon.$$

Raising to the power 12 gives

$$\eta^{-1} \gamma_0^{12n} \eta = \gamma_0^{12n}.$$

By Lemma 5.1.5, $\eta \in \mathrm{Cen}_\Gamma(\gamma_0)$, so $\varepsilon = 1$, qed.

Specializing (5.5) to $\Gamma' = \mathrm{Reg}(\Gamma_{\mathrm{inf}})$ yields

$$\begin{aligned} \mathrm{Reg}\left(\Gamma_{\mathrm{inf}}\right) &= \text{set of distinct elements } \eta^{-1} \gamma_0^n \varepsilon \eta \\ &\quad \text{with } \gamma_0 \in \mathrm{Reg}_0(\Gamma_{\mathrm{inf}}),\ \eta \in \Gamma_{\gamma_0} \backslash \Gamma,\ \varepsilon \in \mathrm{Tor}\left(\Gamma_{\gamma_0}\right), n \in \mathbf{Z}_{\neq 0}, \end{aligned} \tag{5.6a}$$

which we may also write as the disjoint union

$$\mathrm{Reg}\left(\Gamma_{\mathrm{inf}}\right) = \bigcup_{\gamma_0 \in \mathrm{Reg}_0(\Gamma_{\mathrm{inf}})} \bigcup_{\eta \in \Gamma_{\gamma_0} \backslash \Gamma} \bigcup_{\varepsilon \in \mathrm{Tor}\left(\Gamma_{\gamma_0}\right)} \bigcup_{n \in \mathbf{Z}_{\neq 0}} \eta^{-1} \gamma_0^n \varepsilon \eta. \tag{5.6b}$$

5.2 The Double Trace and its Decomposition

The group-theoretic decomposition of Γ will now be interpreted in terms of the decomposition of the trace of a function. We just want to exhibit the formal structure. Convergence conditions have been determined in Chapter 4.

So let G be any locally compact group and Γ a discrete subgroup. We assume G unimodular. Let φ be a measurable function on G. Let Γ' be a subset of Γ that is $\mathbf{c}(\Gamma)$-invariant. Let $z \in G$. We define formally the Γ'-**trace**

$$\mathrm{Tr}_{\Gamma'}(\mathbf{c}(z)\varphi) = \sum_{\gamma' \in \Gamma'} (\mathbf{c}(z)\varphi)(\gamma') = \sum_{\gamma' \in \Gamma'} \varphi\left(z^{-1} \gamma' z\right).$$

On the other hand, using the notation of Section 5.1, we may write formally

$$\mathrm{Tr}_{[\Gamma']} = \sum_{c \in CC_\Gamma(\Gamma')} \sum_{\eta \in \Gamma_{\gamma_c} \backslash \Gamma} [\eta]. \tag{5.7a}$$

Applied to a function φ, this gives

$$\mathrm{Tr}_{[\Gamma']}(\varphi) = \sum_{c \in CC_\Gamma(\Gamma')} \sum_{\eta \in \Gamma_{\gamma_c} \backslash \Gamma} \varphi\left(\eta^{-1}\gamma_c \eta\right). \tag{5.7b}$$

Although for this chapter we deal with a one-variable function φ, we indicate how a two-variable formalism also occurs naturally. Let K be a compact subgroup of G. Suppose φ is K-bi-invariant, even ($\varphi(g) = \varphi(g^{-1})$). For $z,\ w \in G/K$, define

$$\mathbf{K}_\varphi(z,w) = \varphi\left(z^{-1}w\right).$$

Then $\mathbf{K}_\varphi$ is symmetric, and for $g \in G$,

$$\mathbf{K}_\varphi(gz,gw) = \mathbf{K}_\varphi(z,w) \text{ or equivalently } \mathbf{K}_\varphi(gz,w) = \mathbf{K}_\varphi\left(z,g^{-1}w\right).$$

This is what Selberg calls a **point-pair invariant**. For any subset Γ' of Γ that is $\mathbf{c}(\Gamma)$-invariant, we define formally the Γ'-**trace**

$$\mathbf{K}_\varphi^{\Gamma'}(z,z) = \mathrm{Tr}_{\Gamma'}(\mathbf{K}_\varphi)(z,z) = \sum_{\gamma' \in \Gamma'} \mathbf{K}_\varphi\left(\gamma' z, z\right) = \sum_{\gamma' \in \Gamma'} \varphi\left(z^{-1}\gamma' z\right) = \mathrm{Tr}_{\Gamma'}\left(\mathbf{c}(z)\varphi\right). \tag{5.8}$$

Furthermore, the Γ'-trace is left Γ-invariant, that is, for all $\gamma \in \Gamma$,

$$\mathrm{Tr}_{\Gamma'}(\mathbf{K}_\varphi)(\gamma_z,\gamma_z) = \mathrm{Tr}_{\Gamma'}(\mathbf{K}_\varphi)(z,z). \tag{5.9}$$

The subsets Γ' that are $\mathbf{c}(\Gamma)$-invariant of immediate concern to us are the cuspidal and the noncuspidal elements, and the further decomposition of noncuspidal elements into those of infinite order and those of finite order, that is,

$$\mathrm{NCus}(\Gamma) = \mathrm{NCus}(\Gamma_{\mathrm{inf}}) \cup \mathrm{NCus}(\Gamma_{\mathrm{fin}}). \tag{5.10}$$

This decomposition gives rise to an expression of the trace as a sum of two pieces

$$\mathrm{Tr}_\Gamma = \mathrm{Tr}_{\mathrm{Cus}\Gamma} + \mathrm{Tr}_{\mathrm{NCus}\Gamma}. \tag{5.11}$$

Of course, we also have the further decomposition into conjugacy classes. Let c denote Γ-conjugacy classes of noncuspidal elements in Γ, so take $\Gamma' = \mathrm{NCus}(\Gamma)$. Then the **noncuspidal trace** is

$$\mathrm{Tr}_{\mathrm{NCus}\Gamma} = \sum_{c \in CC_\Gamma(\Gamma')} \mathrm{Tr}_c = \sum_c \sum_{\gamma \in c} \gamma.$$

One might also call this trace the **regularized trace**, but note that the regular elements that are cuspidal are missing from this trace. The notion of regularization occurs in analysis, and applies to integrals from which certain terms are omitted or modified to make the integrals converge. Here, the terminology of noncuspidal is more precise in describing what terms are omitted. The **noncuspidal trace** is therefore

$$\sum_{\gamma \notin \mathrm{Cus}(\Gamma)} \varphi\left(g^{-1}\gamma g\right) = \mathrm{Tr}_{\mathrm{NCus}\Gamma}\left(\mathbf{K}_\varphi\right)(g,g) = \mathrm{NCuspTr}(\mathbf{K}_\varphi)(g,g).$$

We omit more than the nonregular elements in this sum, to make another kind of trace converge, as follows. Let f be a function on $\Gamma\backslash G \times \Gamma\backslash G$. By the **manifold trace** of f we mean the integral over the diagonal

$$\mathrm{Tr}_{\Gamma\backslash G}(f) = \int_{\Gamma\backslash G} f(g,g)\, dg_{\Gamma\backslash G}.$$

We encounter such functions f that are the Γ-trace of a function F on $G \times G$, or rather $G/K \times G/K$, i.e., right K-invariant functions, so

$$f = \mathrm{Tr}_\Gamma\left(\mathbf{K}_\varphi\right) \text{ on the diagonal.}$$

The main (by far) function of interest is the Γ-trace of the heat kernel on G/K, but for various considerations, it is not necessary to know the heat kernel in full. One needs to know only certain simply stated properties,

Remark. For some applications one may view f as the kernel function of an integral operator acting by convolution on a suitable space of functions. Then the manifold trace is the trace of this operator in the sense of functional analysis.

Quite generally, we are interested in the integral called the **Selberg trace**:

$$\int_{\Gamma\backslash G} \sum_{\gamma\in\Gamma} \varphi(g^{-1}\gamma g)dg = \int_{\Gamma\backslash G} f(g,g)dg = \int_{\Gamma\backslash G} \mathbf{K}_\varphi^\Gamma(g,g)dg. \tag{5.12}$$

This integral is the manifold trace of the group-theoretic Γ-trace. Even for some quite rapidly decaying functions φ, this integral does not converge. Thus we are led to take only part of the Γ-trace, namely the part restricted to certain $\mathbf{c}(\Gamma)$-invariant subsets Γ'. We call

$$\mathrm{Tr}_{[\Gamma\backslash G],\Gamma'}(\varphi) = \int_{\Gamma\backslash G} \sum_{\gamma\in\Gamma'} \varphi(g^{-1}\gamma g)dg = \int_{\Gamma\backslash G} \mathbf{K}_\varphi^{\Gamma'}(g,g)\, dg \tag{5.13}$$

the $([\Gamma\backslash G], \Gamma')$-**trace** of φ. When the subsets Γ' are such that the expression in (5.13) converges absolutely, we call it a **regularized trace**. In Chapter 4, Corollary 4.3.3, we gave a condition under which absolute convergence holds for the function φ, especially when $\Gamma' = \mathrm{NCus}\,(\Gamma)$ is the subset of noncuspidal elements in Γ.

A regularized trace can now be reworked as follows, assuming only that the series and integral are absolutely convergent. We use the twisted Fubini theorem on the following towers of subgroups:

$$G \supset \Gamma \supset \Gamma_{\gamma_c},$$
$$G \supset G_{\gamma_c} \supset \Gamma_{\gamma_c}.$$

Theorem 5.2.1. *Let G be locally compact unimodular, Γ a discrete subgroup. Let Γ' be a* $\mathbf{c}(\Gamma)$*-invariant subset of* Γ. *Suppose G_γ is unimodular for all $\gamma \in \Gamma'$. Let φ be measurable on G, and such that the expression* (5.13) *for the $(\Gamma\backslash G, \Gamma')$-trace is absolutely convergent. Then*

$$\mathrm{Tr}_{[\Gamma\backslash G]\Gamma'}(\varphi) = \sum_{c \in CC_\Gamma(\Gamma')} \int_{\Gamma_{\gamma_c}\backslash G} \varphi(g^{-1}\gamma_c g)dg$$
$$= \sum_{c \in CC_\Gamma(\Gamma')} \mathrm{Vol}\left(\Gamma_{\gamma_c}\backslash G_{\gamma_c}\right) \int_{G_{\gamma_c}\backslash G} \varphi(g^{-1}\gamma_c g)dg.$$

Proof. Straightforwardly, going around one side of the diagram,

$$\int_{\Gamma\backslash G} \sum_{\gamma \in \Gamma'} \varphi\left(g^{-1}\gamma g\right) dg = \sum_{c \in CC_\Gamma(\Gamma')} \int_{\Gamma\backslash G} \sum_{\eta \in \Gamma_{\gamma_c}\backslash\Gamma} \phi\left(g^{-1}\eta^{-1}\gamma_c \eta g\right) dg$$

gives the first formula. Going around the other side gives the equality

$$= \sum_{c \in CC_\Gamma(\Gamma')} \int_{G_{\gamma_c}\backslash G} \left(\int_{\Gamma_{\gamma_c}\backslash G_{\gamma_c}} \varphi\left(g_1^{-1} g_2^{-1} \gamma_c g_2 g_1\right) dg_2 \right) dg_1,$$

from which the second formula drops out because g_2 commutes with γ_c.

This theorem is a general version of the **Selberg trace formula**, taking into account the restriction to Γ'. The integral

$$\mathrm{Orb}\left(\varphi, \gamma\right) = \int_{G_\gamma\backslash G} \varphi\left(g^{-1}\gamma g\right) dg$$

occurring in Theorem 5.2.1 is the **orbital integral** of Chapter 1. In our concrete situation, let $z \in G_\gamma$ be regular. By Lemma 5.1.5, we know that $G_z = G_\gamma$, but it will be convenient to write the orbital integral in the form

$$\mathrm{Orb}_{G_\gamma\backslash G}\left(\varphi, z\right) = \int_{G_\gamma\backslash G} \varphi(g^{-1}zg)dg.$$

In practice, the function φ is continuous and the orbital integrals are not 0. Thus the volumes in Theorem 5.2.1 are finite:

$$\mathrm{Vol}(\Gamma_{\gamma_c}\backslash G_{\gamma_c}) \neq \infty.$$

Example. Let G, Γ be the usual groups, $G = \mathrm{SL}_2(\mathbf{C})$, $\Gamma = \mathrm{SL}_2(\mathbf{Z}[\mathbf{i}])$, and let $\gamma \in \Gamma$ be regular. By Proposition 5.1.3 there is a conjugation of G_γ with the diagonal group $\mathbf{D}$. Therefore we can finish the proof of Proposition 5.1.4 as follows.

Corollary 5.2.2. *Let $G = \mathrm{SL}_2(\mathbf{C})$ as usual, and let γ be a noncuspidal element in Γ. Then $\mathrm{Cen}_\Gamma(\gamma) = \Gamma_\gamma$ contains an element of infinite order and $\Gamma_\gamma \backslash G_\gamma$ is compact (a torus).*

Proof. We use a positive Gaussian-type function as in Section 1.6, Chapter 4, Corollary 4.3.3, and apply Theorem 5.2.1. Note that the corollary is false for the regular cuspidal element $\gamma = R$.

As to the orbital integral, it cannot be infinity, so directly from its definition we get the following:

Corollary 5.2.3. *If $\gamma \in \Gamma$ is noncuspidal, then γ is also regular.*

In the case of regular elements of infinite order, we know the structure of the isotropy group G_γ (conjugate to $\mathbf{D}$), and we have explicit representatives for the conjugacy classes in Section 5.1 (5.6a), (5.6b). Thus Theorem 5.2.1 can be expressed in terms of these representatives as follows.

Theorem 5.2.4. *Let φ be a measurable function on $G = SL_2(\mathbf{C})$ such that the following integral on the left is absolutely convergent. Then with respect to the representatives of Lemma 5.1.6, the following identity holds:*

$$\int_{\Gamma\backslash G} \sum_{\gamma \in \mathrm{Reg}(\Gamma_{\mathrm{inf}})} \varphi(g^{-1}\gamma g)dg = \sum_{\gamma_0 \in \mathrm{Reg}_0(\Gamma_{\mathrm{inf}})} \sum_{n \in \mathbf{Z}_{\neq 0}} \sum_{\varepsilon \in \mathrm{Tor}(\Gamma_{\gamma_0})} \int_{G_{\gamma_0}\backslash G} \varphi\left(g^{-1}\gamma_0^n \varepsilon g\right) dg.$$

We note that with the notation of Section 5.1, (5.9):

$$\int_{G_{\gamma_0}\backslash G} \varphi(g^{-1}\gamma_0^n \varepsilon g)dg = |\mathrm{Tor}(\Gamma_{\gamma_0})|^{-1} \int_{(\gamma_0)\backslash G} \varphi(g^{-1}\gamma_0^n \varepsilon g)dg.$$

5.3 Explicit Determination of the Noncuspidal Terms

This section describes the computation of each term of the noncuspidal trace, both in general and for the heat Gaussian. Let φ be an admissible function (see the end of Section 4.3). We saw in Theorem 5.2.1 how the noncuspidal trace computation is reduced to determining explicitly:

- the volume $\mathrm{Vol}(\Gamma_\gamma \backslash G_\gamma)$;
- the orbital integral $\displaystyle\int_{G_\gamma\backslash G} \varphi(g^{-1}\gamma g)dg$, partly determined in Sections 1.5 and 1.6.

5.3.1 The Volume Computation

By Proposition 5.1.4 (completed in Corollary 2.2), the centralizer Γ_γ is the centralizer of a regular element of infinite order, and we know its structure. The volume is unchanged by a conjugation, and there is a conjugation that transforms G_γ to the diagonal group $\mathbf{D}$, and Γ_γ to a group $(\tilde{\zeta}_0) \times \Phi(\tilde{\zeta}_0)$ with $\Phi(\tilde{\zeta}_0)$ being finite cyclic, and $\tilde{\zeta}_0$ being diagonal of the following form:

$$\tilde{\zeta}_0 = \begin{pmatrix} \zeta_0 & 0 \\ 0 & \zeta_0^{-1} \end{pmatrix} \text{ with } \zeta_0 = \lambda_0 = \lambda_0 \varepsilon_0, \lambda_0 > 1, |\varepsilon_0| = 1.$$

Such an element $\tilde{\zeta}_0$ will be called a **diagonalized primitive element** for Γ_γ. It is a regular element of infinite order.

We use the Iwasawa coordinates $G = AUK$, and the **α-Iwasawa measure** corresponding to a decomposition $g = auk$, defined as follows. Writing a first, we let $dg = da\, du\, dk$, so no further Jacobian factor is needed. If $a = \text{diag}(a_1, a_1^{-1})$, then $y = a_1^2 = a^\alpha$, and $da = dy/y$. We give K total measure 1. For the U-factor, $x \in \mathbf{C}$, $x = x_1 + \mathbf{i}x_2$,

$$u = u(x) = \begin{pmatrix} 1 & x \\ 0 & 1 \end{pmatrix} \text{ and } du = dx = dx_1\, dx_2.$$

The existence of a primitive element is essential for the next theorem.

Theorem 5.3.1. *Let $\tilde{\zeta}_0$ be a diagonalized primitive element for Γ_γ with noncuspidal $\gamma \in \Gamma$. Then with respect to the Iwasawa measure having $da = dy/y$,*

$$\text{Vol}((\tilde{\zeta}_0)\backslash\mathbf{D}) = 2 \cdot \log|\zeta_0|,$$

and

$$\text{Vol}(\Gamma_\gamma\backslash G_\gamma) = |\text{Tor}(\Gamma_\gamma)|^{-1} 2 \log|\zeta_0|.$$

Proof. Suppose first that $\varepsilon_0 = 1$. We put $y = e^v$, $v = \log y$, $dy/y = dv$, so

$$(\tilde{\lambda}_0)\backslash G = \tilde{\lambda}_0\backslash AUK \text{ and } \int_{(\tilde{\lambda}_0)\backslash A} da = \int_0^{2\log\lambda_0} dv = 2\log\lambda_0.$$

If ε_0 is not real, we have to adjust the Iwasawa decomposition as follows. We let

$$\mathbf{D} = \mathbf{D}_2(\mathbf{C}) = \text{complex diagonal group},\ \mathbf{D} = A\mathbf{T},$$

where $\mathbf{T}$ is the torus, $\mathbf{T} \cong \mathbf{C}_1$, consisting of diagonal matrices

$$\tilde{\varepsilon} = \begin{pmatrix} \varepsilon & 0 \\ 0 & \varepsilon^{-1} \end{pmatrix} \text{ with } |\varepsilon| = 1, \varepsilon \in \mathbf{C}.$$

Then $\mathbf{T} \subset K$. *We normalize Haar measure so that* $\mathbf{T}$ *as well as* K *has measure* 1. On $\mathbf{D}$ *we put the product measure of* $da = dy/y$ *and this normalized measure on* $\mathbf{T}$, *which we write* dw (w variable in $\mathbf{D}$). We call it the (T,K)-**normalized measure (Iwasawa)**.

Then for $f \in L^1(G/K)$, we have

$$\int_G f(g)\,dg = \int_U \int_{\mathbf{D}} f(wu)\,dw\,du.$$

The cyclic group $(\tilde{\zeta}_0)$ is discrete in $\mathbf{D}$ and the factor group is a torus,

$$\left(\tilde{\zeta}_0\right) \backslash \mathbf{D} = \mathbf{D} \backslash (\tilde{\zeta}_0).$$

Using the modified Iwasawa decomposition with $\mathbf{D}$ instead of A, we get the same value as before, with the measure dy/y on A, and total $\mathbf{T}$-measure 1, namely

$$\mathrm{Vol}((\tilde{\zeta}_0)\backslash \mathbf{D}) = 2\log \lambda_0 = 2\log |\zeta_0|,$$

giving the first formula in the theorem. The second is immediate because of the inclusion of subgroups

$$\Gamma_\gamma \supset (\gamma_0),$$

where γ_0 is primitive in Γ_γ conjugating to a diagonalized primitive element in a diagonalized conjugate of Γ, with the direct product decomposition

$$\Gamma_\gamma \cong (\gamma_0) \times \mathrm{Tor}(\Gamma_{\gamma_0}).$$

This concludes the proof.

5.3.2 The Orbital Integral

We recall the computation of the orbital integral from Sections 1.5 and 1.6 in the present context. Let $z \in G$ be a regular element commuting with γ. There is a conjugation (by an element of G) such that the conjugation of G_γ is $\mathbf{D}$, and the conjugation of z is a regular diagonal element

$$z \sim \tilde{\eta} = \begin{pmatrix} \eta & 0 \\ 0 & \eta^{-1} \end{pmatrix} \text{ with } |\eta| \geqq 1, \eta \neq \eta^{-1}, \eta = \eta(z).$$

Haar measure being invariant under conjugation, we get the **orbital integral**:

Theorem 5.3.2. *Let* $z \in G$ *and* $\gamma \in \Gamma$ *be commuting regular elements,* $\eta = \eta(z)$. *Then with the* (T,K)*-normalized measure on* $\mathbf{D}$,

$$\int_{G_\gamma\backslash G} \varphi(g^{-1}zg)dg = \int_{\mathbf{D}\backslash G} \varphi\left(g^{-1}\tilde{\eta}g\right)dg := \mathrm{Orb}_{\mathbf{D}\backslash G}(\varphi,\tilde{\eta})$$

$$= \frac{2\pi}{|\eta-\eta^{-1}|} \int_{\log|\eta|^2}^{\infty} \varphi_{\alpha,+}(v)\sinh(v)dv.$$

In particular, for the Gaussian φ_c, we get

$$\mathrm{Orb}_{G_\gamma\backslash G}(\varphi_c, z) = \frac{2\pi c}{|\eta-\eta^{-1}|^2} e^{-\left(\log|\eta|^2\right)^2/2c}.$$

Remark. This relation holds for all regular elements.

We now take $\varphi = \psi_c$, normalized as in Chapter 2, Proposition 5.1.1, and further normalized to be the Iwasawa heat Gaussian $\mathbf{g}_t$, whose formula is given in Section 2.1 formula (2.4_{Iw}). Plugging into the trace formula yields an inverted theta series as follows.

Theorem 5.3.3. *Let Γ' be the subset of noncuspidal elements in Γ. Then for $\mathbf{g}_t = \mathbf{g}_{t,\mathrm{Iw}'}$*

$$\mathrm{Tr}_{[\Gamma\backslash G],\Gamma'}(\mathbf{g}_t) = e^{-2t}(4t)^{-1/2} \sum_{c\in CC_\Gamma(\Gamma')} a_c e^{-|\log b_c|^2/4t}.$$

The constants a_c, b_c are as follows:

$$a_c = \frac{1}{|\mathrm{Tor}(c)|} \log|\eta_{c,0}|^2 \frac{(2\pi)^{-3/2}}{|\eta_c - \eta_c^{-1}|^2},$$

where

$\mathrm{Tor}(c)$ = *order of the torsion group of* Γ_{γ_c};
η_c = *eigenvalue of elements in c*;
$\eta_{c,0}$ = *eigenvalue of absolute value* > 1 *for a primitive element of* Γ_{γ_c};
b_c^2 = *polar component of* $\gamma_c\gamma_c^*$, where γ_c *is an element of c*.

Proof. This is just putting together Theorems 5.3.1 and 5.3.2, which evaluate the volume and orbital integral of Theorem 5.2.1.

We note that the series in Theorem 5.3.3 is an inverted theta series, so we call it the **noncuspidal inverted theta series**, and use the notation

$$\theta^{\mathrm{NC}}(1/t) = \sum_{c\in CC\Gamma(\Gamma')} a_c e^{-|\log b_c|^2/4\mathrm{t}}.$$

Let $\mathbf{K}_t^{\mathrm{NC}}$ be the noncuspidal trace of the heat kernel $\mathbf{K}_t$, that is,

$$\mathbf{K}_t^{\mathrm{NC}}(z,z) = \sum_{\gamma\in\Gamma'} \mathbf{K}_t\left(\gamma_{z,z}\right).$$

Then the formula of Theorem 5.3.3 may be rewritten in the following form:

Theorem 5.3.3. ′

$$\int_{\Gamma\backslash G} \mathbf{K}_t^{\mathrm{NC}}(z,z)d\mu(z) = e^{-2t}(4t)^{-1/2}\theta^{\mathrm{NC}}(1/t).$$

To give a taste of things to come, we point out right away that to get zeta functions out of the series of Theorem 5.3.3, one applies the **Gauss transform**, defined by

$$\mathrm{Gauss}(f)(s) = 2s\int_0^\infty e^{-s^2 t} f(t)dt,$$

with a suitable regularization; cf. [JoL 94], Section 5.2. The integral of each term involving t is then a K-Bessel integral precisely with the power $t^{-1/2}$, which causes the integral to degenerate to an exponential function (see Section 10.3, **K5**). This leads to the considerations of [JoL 96]. Alternatively, one could multiply the heat Gaussian $\mathbf{g}_t$ by e^{2t}, in which case one gets an inverted theta series on the right side. One can then apply [JoL 93] and [JoL 94] to go further. But all this belongs elsewhere, to go on and on.

5.4 Cuspidal Conjugacy Classes

Let $\mathrm{Cus}(\Gamma)$ be the cuspidal subset of Γ, so $\mathbf{c}(\Gamma)\Gamma_\infty$. This subset decomposes into the disjoint union of the **cuspidal nonregular** and **regular subsets**, namely

$$\mathrm{Cus}(\Gamma) = f\mathbf{c}(\Gamma)\Gamma_{\mathrm{U}} \cup \pm\mathrm{c}(\Gamma)(\mathrm{R}\Gamma_{\mathrm{U}}) = \mathrm{Cus}_{\mathrm{NR}}(\Gamma) \cup \mathrm{Cus}_{\mathrm{R}}(\Gamma),$$

with the usual $R = \mathrm{diag}(\mathbf{i}, -\mathbf{i})$. (But see Proposition 5.4.5.) We first determine the isotropy groups to have a clear view of the conjugacy classes, both in G and in Γ.

Proposition 5.4.1. *Let $u(x) \in U$ with $x \neq 0$, $x \in \mathbf{C}$. Then its centralizer in G is $\pm U$:*

$$\mathrm{Cen}_G(u(x)) = G_{u(x)} = \pm U.$$

Thus $\Gamma_{u(x)} = \pm\Gamma_U$.

Proof. Let

$$\gamma = \begin{pmatrix} a & b \\ c & d \end{pmatrix} \in G = \mathrm{SL}_2(\mathbf{C}).$$

One multiplies out $\gamma u(x)\gamma^{-1}$ by straight matrix multiplication, and one sees at once that this is equal to $u(x)$ if and only if $c = 0$, and $a^2 = 1$. For the record,

$$\gamma u(x)\gamma^{-1} = \begin{pmatrix} 1-cax & a^2x \\ -c^2x & 1+cax \end{pmatrix} \text{ and } \gamma^{-1} = \begin{pmatrix} d & -b \\ -c & a \end{pmatrix}. \tag{5.14}$$

Next we give a description of the nonregular cuspidal Γ-conjugacy classes. We use the following notation:

$c(\gamma) = \Gamma$-conjugacy class of γ.
$\gamma \sim \gamma'$ if and only if $c(\gamma) = c(\gamma')$.
$o = \mathbf{Z}[\mathbf{i}]$ and $o^\bullet = o_{\neq 0}$ is the subset of nonzero elements in o.
$U^\bullet =$ set of elements $u(x)$ with $x \in o^\bullet$.

Thus $U = U^\bullet \cup \{\mathrm{I}\}$ and $-U = -U^\bullet \cup \{-I\}$.

Note that trivially, the classes $c(u(m'))$ and $-c(u(m)) = c(-u(m))$ are disjoint for all $m, m' \in o$, because they have distinct eigenvalues (1 and -1 respectively).

Proposition 5.4.2. *For all $m \in o$, we have $u(m) \sim u(-m)$. For $m' \in o^\bullet$ and $m' \neq \pm m$, the Γ-conjugacy classes $c(u(m))$ and $c(u(m'))$ are disjoint. Thus letting $\{m\}$ be a family of representatives of $o^\bullet / \pm 1$, the map*

$$\gamma \mapsto \gamma u(m) \gamma^{-1}$$

is a bijection of $\Gamma / \pm \Gamma_U \to c(u(m))$. We have the disjoint union

$$\mathbf{c}(\Gamma)\Gamma_{U^\bullet} = \bigcup_{m \in o^\bullet \setminus \pm 1} c(u(m)).$$

Proof. First, observe that directly from (5.14), one has

$$Ru(m)R^{-1} = u(-m).$$

Next, suppose $\gamma u(m)\gamma^{-1} = u(m')$ with $m' \neq 0$. Then $c = 0$ by (5.14), so $a = \pm 1$ or $\pm \mathbf{i}$. If $a = \pm 1$, we are in the case of $m' \neq \pm m$. Otherwise,

$$\gamma \in \pm R \Gamma_U,$$

so conjugation by γ sends $u(m)$ to $u(-m)$. This proves the proposition.

Next we carry out the analogous description of conjugacy classes of RU. *As to R itself*, it is regular, so by Proposition 5.1.3, we know that

$$\mathrm{Cen}_G(R) = G_R = \text{diagonal group and in particular, } \Gamma_R = \{\pm I, \pm R\} = H.$$

Consider $Ru(x)$ with $x \neq 0$. We have

$$Ru(x) = \begin{pmatrix} \mathbf{i} & \mathbf{i}x \\ 0 & -\mathbf{i} \end{pmatrix}.$$

Note that any such element has order 4, and in fact for all $x \in \mathbf{C}$,

$$(Ru(x))^2 = \begin{pmatrix} \mathbf{i} & \mathbf{i}x \\ 0 & -\mathbf{i} \end{pmatrix} = -I. \tag{5.15}$$

With $\gamma \in \mathrm{SL}_2(o)$, we shall use the formula

$$\gamma Ru(x)\gamma^{-1} = \begin{pmatrix} \mathbf{i}(1+2bc-axc) & \mathbf{i}(a^2x-2ab) \\ \mathbf{i}(2cd-c^2x) & -\mathbf{i}(1+2bc-acs) \end{pmatrix}. \tag{5.16}$$

Note that $Ru(x)$ is regular, so Proposition 5.1.3 applies.

Proposition 5.4.3. (i) *Let $x \in \mathbf{C}$, $x \neq 0$. Then the centralizer of Ru(x) in G is contained in the upper triangular group G_∞, besides being conjugate to the full diagonal group. For $x = 0$, the centralizer of R is the diagonal group.* (ii) *Let $m \in o$. Then the centralizer of Ru(m) in Γ is the cyclic subgroup of order 4 generated by Ru(m), that is,*

$$\mathrm{Cen}_\Gamma(Ru(m)) = \Gamma_{Ru(m)} = \{\pm \mathrm{I}, \pm Ru(m)\}.$$

Proof. The case $x = 0$ was treated in Proposition 5.1.3, from which it follows at once that $\mathrm{Cen}_\Gamma(R)$ is the cyclic group H generated by R.

Suppose now that $x,\ m \neq 0$. Let $\gamma \in G$ be in the centralizer of $Ru(x)$, and look at (5.16). We must have $c = 0$; otherwise, the upper left component being $\mathbf{i}$ implies that $2b - ax = 0$, so the upper right component is 0, contradicting $x \neq 0$. Then γ is upper triangular. Suppose we are in the case $x = m \in o$, and $\gamma \in \Gamma$. Then $ad = 1$, a is a unit ± 1, $\pm \mathbf{i}$, and

$$a^2 m - 2ab = m.$$

If $a = \pm 1$, then $b = 0$ and $\gamma = \pm I$. Suppose $a = \mathbf{i}$. Then $-2ab = 2m$, so $b = \mathbf{i}m$, and similarly if $a = -\mathbf{i}$. This concludes the proof.

We are striving toward Theorem 5.4.6, which will describe exactly four conjugacy classes for elements of type $\pm Ru(m)$, $m \in o$, accounting for all regular elements in Γ_∞.

Proposition 5.4.4. *Let m', $m \in o$. Then the Γ-conjugacy classes $c(Ru(m))$ and $c(Ru(m'))$ are equal if and only if $m \equiv m'$ mod $2o$.*

Proof. Suppose first that m, $m' \neq 0$. Let $\gamma \in \Gamma$ be such that

$$\gamma Ru(m)\gamma^{-1} = RU(m').$$

As in the previous proposition, from the lower left corner, we must have $c = 0$, $ad = 1$, $a = \pm 1, \pm \mathbf{i}$, and

$$a^2 m - 2ab = m',$$

whence $m \equiv m'$ mod 2. Conversely, if this congruence is satisfied, we can solve for $b \in o$, for instance

$$b = \frac{m' - a^2 m}{2a}.$$

This concludes the proof in the present case.

Next, the conjugacy class of R is the same as the conjugacy class of $Ru(2)$, say by using formula (5.16), taking γ with $c = 0$, and solving for $b = a = am/2$ with $m = 2$. This proves the proposition.

By the proposition, there are four conjugacy classes of elements $Ru(m)$, because $(o:2o) = 4$. Representatives of $o/2o$ are given by the four elements

$$1,\ \mathbf{i},\ 1+\mathbf{i},\ 0\ (\text{or } 2).$$

The eigenvalues of each class are $\mathbf{i},\ -\mathbf{i}$.

Finally, we deal with the minus sign.

Proposition 5.4.5. *For all $m \in o$, we have*

$$-Ru(m) \sim Ru(m).$$

Proof. By Proposition 5.4.4, it suffices to prove the desired relation for $m = 0$, $1,\ \mathbf{i},\ 1+\mathbf{i}$. As to the case $m = 0$, so with -R, the conjugating element is the reversal element

$$S = \begin{pmatrix} 0 & -1 \\ 1 & 0 \end{pmatrix} \text{ such that } SRS^{-1} = -R.$$

Next suppose $m \neq 0$. We use formula (5.16) with $x = m$ to solve for

$$\gamma = \begin{pmatrix} a & b \\ c & d \end{pmatrix} \text{ having } a,b,c,d \in o,\ ad - bc = 1,$$

such that $-\gamma Ru(m)\gamma^{-1} = Ru(m')$ (with $m' \in o$), or equivalently

$$\gamma Ru(m)\gamma^{-1} = \begin{pmatrix} -\mathbf{i} & -\mathbf{i}m' \\ 0 & \mathbf{i} \end{pmatrix}.$$

From (5.16), we get necessary and sufficient conditions concerning the components $-\mathbf{i}$, 0 of the first column, namely

$$1 + 2bc - amc = -1 \text{ and } c(2d - cm) = 0.$$

We take $c = 2d/m$, and among other things, c must be in o. We determine solutions as follows, remembering that we are in the case $m \neq 0$.

If m is a unit 1 or $\mathbf{i}$, we take $a = 1/m,\ b = 0,\ c = 2,\ d = m$,

$$\gamma = \begin{pmatrix} 1/m & 0 \\ 2 & m \end{pmatrix}$$

If $m = 1+\mathbf{i}$, we take $a = 2-\mathbf{i},\ b = 1,\ c = 1-\mathbf{i},\ d = 1$, so

$$\gamma = \begin{pmatrix} 2-\mathbf{i} & 1 \\ 1-\mathbf{i} & 1 \end{pmatrix}$$

One gets $m' \equiv m \bmod 2$, because $m' \equiv 2ab - a^2m$. This concludes the proof.

We thus obtain four cuspidal regular conjugacy classes, the eigenvalues being **i**, −**i**. For definiteness, we let

$$\gamma_1 = Ru(1),\ \gamma_2 = Ru(\mathbf{i}),\ \gamma_3 = RU(1+\mathbf{i}),\ \gamma_4 = R.$$

We let the corresponding class be

$$c_j = \mathbf{c}(\Gamma)\gamma_j\ (j = 1,\ldots,4).$$

Let H_j be the cyclic group of order 4 generated by γ_j. By Proposition 5.4.3, we have a bijection

$$H_j\backslash\Gamma \leftrightarrow c_j \text{ by } \gamma \mapsto \gamma^{-1}\gamma_j\gamma = \mathbf{c}(\gamma^{-1})\gamma_j.$$

Two conjugacy classes are either equal or they are disjoint.

Summarizing Propositions 5.4.2, 5.4.4, 5.4.5, we get the following theorem:

Theorem 5.4.6. *The cuspidal* Γ*-conjugacy classes are*

$$c_0 = \{I\},\ \text{and } -c_0 = \{-I\};$$
$$c(u(m)) \text{ with } m \in o^{\bullet}/\pm 1, \text{ and } -c(u(m));$$
$$c_j(j = 1,\ldots,4).$$

Having a tabulation of the cuspidal conjacy classes, the situation is set up algebraically for the trace as in Section 5.2 (5.7a). We have tabulated natural representatives γ_c for all the cuspidal Γ-conjugacy classes. The next step is to carry out the integral over $\Gamma\backslash G$, but in the cuspidal case this integral diverges, as we know from Section 4.2. However, the integral can be viewed as an improper integral as in freshman calculus, and we shall develop an asymptotic expression for it in terms of a fundamental domain constructed in the next chapter. This fundamental domain will serve many purposes. The continuation of this section for the derivation of the theta inversion formula will be carried out in Chapter 13. Logically, it could have been carried out earlier, but several things will be derived logically independently of each other. In a fundamental sense, Chapters 7 and 13 form a natural pair, because two divergent integrals associated with Eisenstein series and the cuspidal trace turn out to have the same divergence asymptotic term.

Part III
The Heat Kernel on $\Gamma\backslash G/K$

Chapter 6
The Fundamental Domain

This chapter deals systematically with the construction of a good fundamental domain and some implications that are used later in connection with cuspidality and Eisenstein series. Fundamental domains give more precise information on the distribution of elements in Γ. Such distribution is used to analyze and estimate the periodization of certain important functions under the action of Γ, beginning in Chapter 7 and never ending. The more complicated aspects of the fundamental domain and the resulting tiling by its Γ-translations provide a way for number theory to come in through the back door. After all, we have to use the arithmetic definition of Γ, in this case the integral matrices of SL_2.

Quite generally, let Γ be a discrete group acting topologically on a space X. We define a **fundamental domain** for Γ to be a subset FD of X having the following properties.

FD 1. Given $z \in X$, there exists $\gamma \in \Gamma$ such that $\gamma z \in$ FD.

FD 2. If $\gamma \in \Gamma$, $\gamma \neq$ id on X, and both z, $\gamma z \in$ FD, then these points lie on the boundary of FD.

Remark. In practice, a fundamental domain is contained in the closure of its interior. *Under this condition,* **FD 2** *may be replaced by the weaker requirement that one of the points z, γz lie on the boundary, because this implies that both z, γz lie on the boundary.* Indeed, suppose z is in the interior of FD, and γz is on the boundary. There is a point w' in the interior close to γz, so $\gamma^{-1}w'$ is close to z, in the interior, contradiction. In this book, the fundamental domains will be equal to the closures of their interiors.

We shall describe a fundamental domain in our situation, with $\Gamma = SL_2(\mathbf{Z}[\mathbf{i}])$. Picard described such a domain, and invoked the reduction theory of quadratic forms to show that it is a fundamental domain [Pic 1884]. We give a different proof. We also prove several additional properties of the fundamental domain, amounting to finiteness statements of various sorts. Although our proof for the existence of the Picard fundamental domain is algebraic staying inside the group, some finiteness statements require going into estimates provided by differential geometry and a good fundamental domain satisfying conditions beyond those of the above definition.

In Section 6.4 we give implications for integration.
Further results about the fundamental domain will be given in Chapter 9.

6.1 $\mathrm{SL}_2(\mathbf{C})$ and the Upper Half-Space $\mathbf{H}^3$

In our case $G = \mathrm{SL}_2(\mathbf{C})$, one can represent the homogeneous space G/K in terms of the quaternions, cf. Kubota [Kub 68] for the present context, in analogy with the real case. It gives a convenient model that allows us to connect the geometric theory with classical Dedekind zeta functions.

We first recall the upper half-plane and $\mathrm{SL}_2(\mathbf{R})$. Let $\mathbf{H}^2$ be the upper half-plane of complex numbers $z = x + \mathbf{i}y$, with $x \in \mathbf{R}$, $y > 0$, so that

$$\mathbf{H}^2 = \mathbf{R} \times \mathbf{R}_{>0}.$$

This is the most special case of the construction that to any manifold X associates the upper half-space $X \times \mathbf{R}_{>0}$.

Let $G_{\mathbf{R}} = \mathrm{SL}_2(\mathbf{R})$. We suppose that the reader knows from complex analysis that $\mathbf{H}^2$ is a homogeneous space for $G_{\mathbf{R}}$. The action of an element

$$g = \begin{pmatrix} a & b \\ c & d \end{pmatrix} \text{ with } a,b,c,d \in \mathbf{R}$$

on $\mathbf{H}^2$ is defined by

$$g(z) = \frac{az+b}{ca+d}.$$

We have

$$\operatorname{Im} g(z) = \frac{\operatorname{Im}(z)}{|cz+d|^2} = \frac{y}{|cx+d|^2+|c|^2y^2}. \tag{6.1}$$

The isotropy group of $\mathbf{i} = \sqrt{-1}$, namely the subgroup of elements $k \in \mathrm{SL}_2(\mathbf{R})$ such that $k(\mathbf{i}) = \mathbf{i}$, is the real unitary group $K_{\mathbf{R}}$ consisting of those elements g whose transpose is equal to the inverse, that is,

$${}^t g = g^{-1}.$$

Then $\mathbf{H}^2$ can be viewed as the coset space

$$\mathbf{H}^2 = G_{\mathbf{R}}/K_{\mathbf{R}}.$$

The group $G_{\mathbf{R}}$ has the **Iwasawa decomposition**, as a manifold product,

$$G_{\mathbf{R}} = U_{\mathbf{R}} A_{\mathbf{R}} K_{\mathbf{R}},$$

where

$$U_{\mathbf{R}} = \text{group of unipotent matrices } u(b) = \begin{pmatrix} 1 & b \\ 0 & 1 \end{pmatrix} \text{ with } b \in \mathbf{R};$$

$$A_{\mathbf{R}} = \text{diagonal group of elements } \begin{pmatrix} a_1 & 0 \\ 0 & a_1^{-1} \end{pmatrix} = a_y \text{ with } a_1 > 0 \text{ and } y = a_1^2.$$

An element $g \in G_{\mathbf{R}}$ can be expressed uniquely as a product

$$g = uak = \begin{pmatrix} 1 & x \\ 0 & 1 \end{pmatrix} \begin{pmatrix} a_1 & 0 \\ 0 & a_1^{-1} \end{pmatrix} k = u(x)\, a_y k$$

with $k \in K$, $x \in \mathbf{R}$, $a_1 > 0$. We can prove this decomposition by relating it to the representation of G on $\mathbf{H}^2$. We note that if g is the above product, then

$$g(\mathbf{i}) = x + y\mathbf{i} \text{ with } y = a_1^2 = \text{Im}(z).$$

Conversely, if $g \in G_{\mathbf{R}}$ is arbitrary, write $g(\mathbf{i}) = x + \mathbf{i}y$ with $x \in \mathbf{R}$ and $y > 0$. Then g and the product $u(x)a_y k$ have the same image in $\mathbf{H}^2$, whence we get bijections

$$U_{\mathbf{R}} A_{\mathbf{R}} \leftrightarrow \mathbf{H}^2 \leftrightarrow G_{\mathbf{R}}/K_{\mathbf{R}},$$

and thereby also prove the Iwasawa decomposition and its uniqueness.

Let $G = \mathrm{SL}_2(\mathbf{C})$. Instead of $\mathbf{H}^2$ we consider the upper half-space $\mathbf{H}^3 = \mathbf{C} \times \mathbf{R} > 0$. Kubota [Kub 68] started using the quaternions with a matrix representation in connection with Eisenstein series, and [EGM 85] used the straight quaternions, which involve fewer symbols. Thus we let $\mathbf{H}^3$ be the set of quaternions with only the first three coordinates being possibly nonzero, so the quaternions of the form

$$z = x + y\mathbf{j} \text{ with } x \in \mathbf{C} \text{ and } y > 0.$$

We write $x = x_1 + x_2\mathbf{i}$ with $x_1, x_2 \in \mathbf{R}$. To denote the dependence on z, we also use the notation

$$y = y(z) \text{ or } y_z.$$

For a quaternion $q = x_1 + \mathbf{i}x_2 + \mathbf{j}x_3 + \mathbf{k}x_4$, we have $\bar{q} = x_1 - \mathbf{i}x_2 - \mathbf{j}x_3 - \mathbf{k}x_4$, and the norm is determined by

$$|q|^2 = q\bar{q}.$$

With this norm, one has at once for $z \in \mathbf{H}^3$ that

$$|z|^2 = |x|^2 + y^2, \text{ and } q^{-1} = \bar{q}/|q|^2.$$

The group $G = \mathrm{SL}_2(\mathbf{C})$ acts on $\mathbf{H}^3$ in a natural way as follows. Let

$$g = \begin{pmatrix} a & b \\ c & d \end{pmatrix} \in \mathrm{SL}_2(\mathbf{C}).$$

We define

$$g(z) = (az+b)(cz+d)^{-1},$$

As for the "imaginary" **j**-part, we note that for $z \in \mathbf{H}^3$ and $c, d \in \mathbf{C}$, we have

$$|cz+d|^2 = |cx+d|^2 + |c|^2 y^2.$$

For c, d not both 0, we **define**

$$\mathrm{y}(c,d;z) = \frac{y(z)}{|cz+d|^2}.$$

Then for $g \in \mathrm{SL}_2(\mathbf{C})$ as above, a straightforward calculation shows that

$$y(g(z)) = y(c,d;z) = \frac{y(z)}{|cz+d|^2}. \tag{6.2}$$

This is the analogue of (1) on $\mathbf{H}^3$. For the record, the x-coordinate is given by

$$x(g(z)) = \frac{(ax+b)\overline{(cx+d)} + a\bar{c}y^2}{|cz+d|^2}.$$

This formula will not be needed until Chapter 13.

In any case, we see that if $z \in \mathbf{H}^3$ then $g(z) \in \mathbf{H}^3$ also. Then the map

$$z \mapsto g(z)$$

defines an action of G on $\mathbf{H}^3$; in other words, the unit matrix acts as the identity, and for all g, $g' \in G$, we have

$$g(g'(z)) = (gg')(z).$$

The verification is by brute force, just as for $\mathbf{H}^2$.

The group G has the product **Iwasawa decomposition**

$$G = UAK,$$

where

K is the complex unitary group of elements g such that ${}^t\bar{g} = g^{-1} = g^*$;
$A = A_{\mathbf{R}}$ is the same group as for $G_{\mathbf{R}}$ (diagonal with positive components);
U is the upper triangular complex unipotent group of elements $u(b)$ with $b \in \mathbf{C}$.

Cf. Section 1.1. The map $b \mapsto u(b)$ gives an isomorphism of $\mathbf{C}$ with U.

One verifies directly that K is the isotropy group of $\mathbf{j}$.

A diagonal matrix $a = \mathrm{diag}(a_1, a_1^{-1}) = \mathrm{diag}(a_1, a_2)$ with $a_1 > 0$ acts on $\mathbf{j}$ just as in the real case, namely

$$\begin{pmatrix} a_1 & 0 \\ 0 & a_1^{-1} \end{pmatrix}(\mathbf{j}) = a_1^2\mathbf{j},$$

directly from the definition. On the group A we have the **Iwasawa character** α defined by $a^\alpha = a_1/a_2 = a_1^2$.

An element $u(b)$ $(b \in \mathbf{C})$ acts by translation,

$$u(b)(z) = z + b.$$

This shows that $G(\mathbf{j}) = \mathbf{H}^3$, and allows us to see the action of G in a manner completely similar to the real case. In particular, we have an isomorphism of G-homogeneous spaces

$$G/K \xrightarrow{\cong} \mathbf{H}^3 \text{ by } g \mapsto g(\mathbf{j}).$$

The Iwasawa decomposition follows as in the real case. For $g \in G$, we write

$$g(\mathbf{j}) = z = x + \mathbf{j}y \text{ with } x \in \mathbf{C},\ y > 0.$$

Then

$$g = u(x)a_y k \text{ with } a_y = \begin{pmatrix} y^{1/2} & 0 \\ 0 & y^{-1/2} \end{pmatrix}.$$

6.2 Fundamental Domain and Γ_∞

We start by giving a precise description of a fundamental domain for the action of $\Gamma = \mathrm{SL}_2(\mathbf{Z}[\mathbf{i}])$ on $\mathbf{H}^3$. We just saw that $\mathbf{H}^3$ is the set of quaternions

$$z = x + \mathbf{j}y \text{ with } x \in \mathbf{C},\ x = x_1 + \mathbf{i}x_2, \text{ and } y > 0,$$

so the $\mathbf{k}$-component is 0. We use the following elements of Γ:

$$T_1 = \begin{pmatrix} 1 & 1 \\ 0 & 1 \end{pmatrix},\ T_\mathbf{i} = \begin{pmatrix} 1 & \mathbf{i} \\ 0 & 1 \end{pmatrix},\ R = \begin{pmatrix} \mathbf{i} & 0 \\ 0 & -\mathbf{i} \end{pmatrix},\ S = \begin{pmatrix} 0 & -1 \\ 1 & 0 \end{pmatrix}.$$

As we saw in Section 1.1, the matrices T_1, $T_\mathbf{i}$ represent translations by 1 and $\mathbf{i}$. For $z \in \mathbf{H}^3$, *the matrix S gives the additive and multiplicative inversion*

$$S(z) = -z^{-1} = \frac{-\bar{z}}{|x|^2 + y^2};$$

and *the matrix R gives the reflection*

$$R(z) = -x + \mathbf{j}y.$$

We let Γ_0 be the subgroup of Γ generated by T_1, $T_\mathbf{i}$, R, and S. We shall prove $\Gamma_0 = \Gamma$.

Let H be the subgroup $\{\pm I, \pm R\}$ of Γ, so H has order 4. We define the **standard cuspidal subgroup**

$$\Gamma_\infty = \Gamma_U H = \Gamma_U(\pm I, \pm R),$$

with the index $(\Gamma_\infty : \Gamma_U) = 4$. This notation follows standard notation for the group of fractional linear transformations on the Riemann sphere leaving ∞ fixed, but we won't need this context for it. The elements of Γ_∞ are the matrices

$$\gamma = \begin{pmatrix} \varepsilon & b \\ 0 & \varepsilon^{-1} \end{pmatrix} \text{ with } \varepsilon = \pm 1,\ \pm\mathbf{i} \text{ and } b \in \mathbf{Z}[\mathbf{i}].$$

The action on $\mathbf{H}^3$ when $\varepsilon = \pm\mathbf{i}$ is given by

$$\gamma(z) = \gamma(x + \mathbf{j}y) = -x + \mathbf{j}y + \varepsilon b,$$

so γ is composed of R and the translation $T_{\varepsilon b}$.

The projection of a good fundamental domain on the plane $\mathbf{C}$ will turn out to be half of the standard square with sides 1, centered at the origin. There are four natural possibilities. We pick one of them.

We let $\mathscr{F}$ be the **Picard domain** of elements z defined by the three inequalities

$$-1/2 \leqq x_1 \leqq 1/2,\ 0 \leqq x_2 \leqq 1/2,\ |z|^2 = x_1^2 + x_2^2 + y^2 \geqq 1.$$

Note that the first two inequalities define a rectangle in the plane shown in the figure. We denote this rectangle by $\mathscr{F}_U$, and call it the **fundamental upper rectangle**.

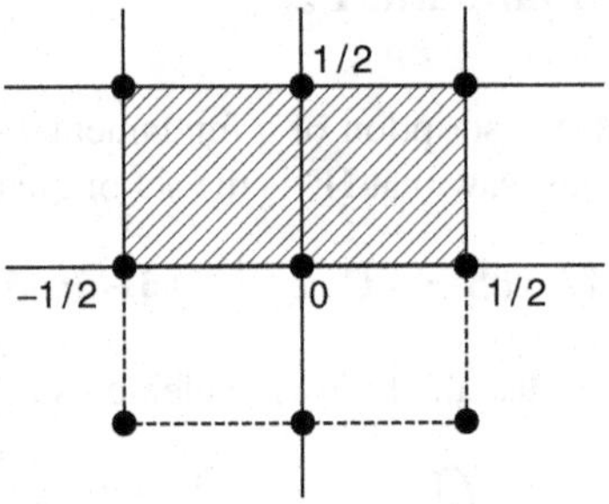

The corresponding square is a fundamental domain for $\Gamma_U \backslash \mathbf{C}$. The third inequality defines the exterior of the half-sphere of radius 1 in $\mathbf{H}^3$ lying above the rectangle. So for each x in $\mathscr{F}_U$, the points $(x,y) \in \mathscr{F}$ are those for which $y^2 \geqq 1 - |x|^2$.

Remark. Classically, instead of the upper rectangle, the upper triangle is taken, starting with Picard.

Note for future use some further inequalities on the coordinates of an element $z = x + \mathbf{j}y \in \mathscr{F}$. First

$$|x|^2 = x_1^2 + x_2^2 \leqq 1/2 \text{ so } y^2 \geqq 1/2. \tag{6.3}$$

Furthermore, either $y^2 > 1/2$, or $y^2 = 1/2$ in which case x is among the three boundary points at distance 1/2 from 0 and z is on the boundary of $\mathscr{F}$.

Since the action of R on $\mathbf{C}$ is to change the sign, R interchanges the lower and upper rectangles in the square. It then follows at once that the upper rectangle is a fundamental domain for the action of the upper triangular group Γ_∞ on $\mathbf{C}$.

Theorem 6.2.1. *The Picard domain $\mathscr{F}$ is a fundamental domain for the action of $\Gamma/\pm I$ on $\mathbf{H}^3$.*

Proof. First we remark that the proof will yield more than just the theorem. It will give explicit information useful afterward for Theorems 6.2.2, 6.3.2, and 6.3.4. Now to the proof: The formula giving the transformation of the $\mathbf{j}$-components (the y-function) under Γ shows that the $\mathbf{j}$-components of the elements in an orbit of Γ are bounded from above, and tend to 0 as $\max(|c|, |d|)$ goes to infinity. In an orbit $\Gamma_0(z)$, we can therefore select an element w whose $\mathbf{j}$-component is maximal. If $|w| < 1$, then $S(w) \in \Gamma_0(z)$ and $S(w)$ has greater $\mathbf{j}$-component, so $|w| \geqq 1$. So w satisfies the third inequality $|w|^2 \geqq 1$. Then applying positive or negative powers of T_1, $T_{\mathbf{i}}$, which do not change the $\mathbf{j}$-component, we can satisfy the inequalities

$$|x_1|,\ |x_2| \leqq 1/2,$$

so x lies in the square centered at 0, with sides of length 1. If x lies in the lower rectangle, we apply R to get an element w' whose x-component lies in the upper rectangle $\mathscr{F}_U$. This element w' has the same $\mathbf{j}$-component as w, maximal in a Γ_0-orbit, so as for w, we must have $|w'| \geqq 1$. This proves property **FD 1** and even more, namely every Γ_0-orbit contains an element of $\mathscr{F}$.

On the other hand, define the set

$$\mathscr{F}_{\max} = \{z = x + \mathbf{j}y \text{ such that } y(z) \text{ is maximal among } y(\gamma z) \text{ for } \gamma \in \Gamma, \\ \text{and } x \text{ lies in the upper rectangle } \mathscr{F}_U\}.$$

Then, as we concluded for Γ_0, we have $\mathscr{F}_{\max} \subset \mathscr{F}$, and every Γ-orbit contains an element of $\mathscr{F}_{\max}$. Therefore as a corollary of Theorem 6.2.1, we get:

Theorem 6.2.2. *We have $\mathscr{F} = \mathscr{F}_{\max}$.*

We return to the proof of Theorem 6.2.1.

Lemma 6.2.3. *Suppose z, $z' \in \mathscr{F}$ and z, z' are in the same Γ-orbit, $z' = \gamma(z)$,*

$$\gamma = \begin{pmatrix} a & b \\ c & d \end{pmatrix} \in \Gamma.$$

Say $y(z) \leqq y(z')$. Then we have one of the following three cases.

(i) $c = 0$, i.e., $\gamma \in \Gamma_\infty$.
(ii) $|c| = 1$, $y^2 \leqq 1$.
(iii) $|c|^2 = 2$, $y^2 = 1/2$, so x is an upper corner of $\mathscr{F}_U$, z is on the boundary of $\mathscr{F}$, $cx + d = 0$, $d = \pm 1$, $\pm\mathbf{i}$.

Proof. We have
$$y = y(z) \leqq y(\gamma z) = \frac{y}{|cx+d|^2 + |c|^2 y^2}, \tag{6.4}$$
or equivalently,
$$|cx+d|^2 + |c|^2 y^2 \leqq 1, \tag{6.5}$$
so $|c|^2 y^2 \leqq 1$. Then $c = 0$ or $y^2 \leqq 1/|c|^2$. But $c \in \mathbf{Z}[\mathbf{i}]$, so if $c \neq 0$, either $|c|^2 = 1$, or $|c|^2 \geqq 2$. We shall see immediately that in this last case, $|c|^2 = 2$.

If $|c|^2 \geqq 2$, then $y^2 \leqq 1/2$, so by (1), $y^2 = 1/2$, and the other properties stated in the lemma follow at once, thus proving the lemma.

Remark. In the analogous but simpler case of $\mathrm{SL}_2(\mathbf{R})$ and $\mathbf{H}^2$, one has only the cases $c = 0$ and $|c| = 1$. Since $|x| \leqq 1/2$ (and x is real), $y \geqq \sqrt{3}/2$, if $|c| = 1$, and $d \neq 0$, it follows from (3) that $x = \pm 1/2$.

If $|c|^2 = 1$, then c is a unit. By (3), $|cx+d|^2 \leqq 1/2$. Since $d = d_1 + \mathbf{i}d_2$ has ordinary integral coordinates d_1, $d_2 \in \mathbf{Z}$, it follows at once that x is a corner of the rectangle $\mathscr{F}_U$, unless $d = 0$. If $d = 0$, then $\gamma S \in \Gamma_\infty$ so $\gamma \in \Gamma_0$.

If $c = 0$, $\gamma \in \Gamma_\infty$, and γ is a translation $T_{\varepsilon b}$ possibly composed with the x-reflection. Let $\varepsilon = \pm 1$. Since both z, $\gamma(z)$ lie in $\mathscr{F}$, if $b \neq 0$, it follows that x is on the boundary of $\mathscr{F}_U$, so z is on the boundary of $\mathscr{F}$. If $\varepsilon = \pm\mathbf{i}$, then $\gamma(z) = -x + \mathbf{j}y + \varepsilon b$. Since $b = (b_1, b_2)$ has coordinates in $\mathbf{Z}$, and $|x_1|$, $|x_2| \leqq 1/2$, the case $b = 0$ cannot occur unless x is on the boundary of $\mathscr{F}_U$. If $b \neq 0$, we use the fact that $-x$ lies in the lower rectangle, and similarly the only way $-x + \varepsilon b$ is in the upper rectangle is that $-x$ is on the lower side of the square, so x is on the upper side of the fundamental rectangle, and in particular, $x \in \mathscr{F}_U$, so z is on the boundary. This proves Theorem 6.2.1.

As a first application of Theorem 6.2.1 and its proof, we have the following result:

Theorem 6.2.4. *We have* $\Gamma_0 = \Gamma$, *that is,* T_1, $T_{\mathbf{i}}$, R, S *generate* Γ.

Proof. Let z be an interior point and let $\gamma \in \Gamma$. At the beginning of the proof of Theorem 6.2.1, we showed that there exists $\gamma_0 \in \Gamma_0$ such that $\gamma_0\gamma(z) \in \mathscr{F}$. Since z is not on the boundary of $\mathscr{F}$, it follows from Theorem 6.2.1 that $\gamma_0\gamma(z) = z$, so $\gamma_0\gamma = \pm I$, whence $\gamma_0 = \pm\gamma^{-1}$, thus proving the theorem.

Note that the argument gives a general method for determining generators for a group in connection with the properties of a fundamental domain.

For more on the geometry of the fundamental domain under Γ-action see Chapter 9.

6.3 Finiteness Properties

We consider systematically certain finiteness properties for which the fundamental domain is used. First we consider the possibilities for which $\gamma\mathscr{F} \cap \mathscr{F}$ is not empty.

Lemma 6.3.1. *Let*

$$\gamma = \begin{pmatrix} a & b \\ c & d \end{pmatrix} \in \Gamma.$$

If $c \neq 0$ then $\gamma(\mathcal{F})$ is bounded (for the Euclidean metric).

Proof. For $z \in \mathcal{F}$, $|cz+d| \neq 0$ because $|c|^2 y^2 \neq 0$, and furthermore

$$|(az+b)(cz+d)^{-1}| = \frac{|a+b/z|}{|c+d/z|} \to |\frac{a}{c}| \text{ as } |z| \to \infty.$$

Thus the usual argument with $SL_2(\mathbf{R})$ acting on $\mathbf{h}_2$ works as well here.

Theorem 6.3.2. *There is only a finite number of elements $\gamma \in \Gamma$ such that $\gamma\mathcal{F} \cap \mathcal{F}$ is not empty. These elements can be determined explicitly from the proof.*

Proof. We first check the finiteness of (c,d)'s in each of the cases considered in the proof of Theorem 6.2.1. In the case $|c|^2 = 2$, there is only a finite number of (c,d) satisfying (6.3). The other two cases automatically restrict (c,d) to a finite number. Next fix γ. Let

$$\gamma = \begin{pmatrix} a & b \\ c & d \end{pmatrix} \text{ and } \gamma' = \begin{pmatrix} a' & b' \\ c & d \end{pmatrix}$$

have the same bottom row. Then by computation,

$$\gamma\gamma'^{-1} = \begin{pmatrix} 1 & m \\ 0 & 1 \end{pmatrix} \text{ with } m \in \mathbf{Z}[\mathbf{i}].$$

Suppose that $\gamma'\mathcal{F} \cap \mathcal{F}$ is not empty. Applying $\gamma\gamma'^{-1}$ we see that

$$\gamma\mathcal{F} \cap \gamma\gamma'^{-1}\mathcal{F} = \gamma\mathcal{F} \cap (\mathcal{F} + m) \text{ is not empty.}$$

We want to bound m. If $c \neq 0$, then γ maps $\mathcal{F}$ to a bounded set by Lemma 6.3.1, and we are done. If $c = 0$, then a, d are units ± 1, $\pm\mathbf{i}$, so γ is a pure translation up to a possible reflection $x \mapsto -x$. So the finiteness of possibilities for m is again clear, and explicit, as desired.

For more information on the explicit determination, see Chapter 9.

We can now settle a question we have skirted around previously. We rely on a special case of a general theorem due to Cartan–Bruhat–Tits–Serre.

Fixed-Point Theorem. *Let H be a finite group acting on $\mathbf{H}^3$. Then H has a fixed point.*

The result holds for a much wider class of groups acting on much more general spaces; cf. [Lan 99], Chapter 11, Theorem 3.2, for a detailed treatment and historical remarks. Roughly speaking, the space can be a complete manifold simply connected with seminegative curvature, or the corresponding notion of complete metric space satisfying the semiparallelogram law. The acting group needs only to have a bounded orbit. The circumcenter of such an orbit is then a fixed point. Spaces of type G/K, with G semisimple and K the unitary-type subgroup, fit this pattern. For a version and proof of the fixed-point theorem on $\mathbf{H}^3$, see Beardon [Bea 83], Theorem 4.3.7.

We follow Borel [Bor 62], Section 5, (3) in using Theorem 6.3.2, to get the following:

Theorem 6.3.3. *There is only a finite number of Γ-conjugacy classes of finite subgroups of Γ.*

Proof. Let H be a finite subgroup of Γ, so H acts on $\mathbf{H}^3$. Let z be a fixed point. Let $\gamma \in \Gamma$ be such that $\gamma z \in \mathcal{F}$. Then γz is a fixed point of $\gamma H \gamma^{-1}$, so by Theorem 6.3.2, $\gamma H \gamma^{-1}$ belongs to a finite set, thus proving the theorem.

We recall that a Siegel set S is a set of the form

$$S = U_c A_t K, \text{ with } c,\ t > 0,$$

where

$\mathrm{U_c}$ = set of U-elements $u(x)$ with $|x| \leqq c$;
A_t = set of A-elements a with $a^{\alpha} \geqq t$, that is, $y \geqq t$ in terms of the coordinate y.

Theorem 6.3.4. *A compact set is covered by a finite number of Γ-translations of the fundamental domain $\mathcal{F}$. So is a Siegel set.*

Proof. Actually, the finiteness property for a compact set implies the property for a Siegel set. Indeed, the Siegel set is the union of two sets defined by the inequalities

$$\{|x| \leqq c,\ t \leqq y \leqq 1\} \text{ and } \{|x| \leqq c,\ 1 \leqq y\}.$$

The first set is compact, and the second set is contained in a finite number of horizontal translations of the fundamental domain, i.e., translations by elements in Γ_U. The first set is just a cylinder of finite height.

The proof will need another property of the Picard fundamental domain, not necessarily satisfied by an arbitrary fundamental domain, and we owe the following considerations and proofs of Theorems 6.3.5, 6.3.6 below to Peter Jones.

We use slightly more differential geometry. We use the hyperbolic distance invariant under the action of G. It can be described most naively in the coordinates (x,y) that we have been using, where in elementary notation,

$$ds^2 = (dx_1^2 + dx_2^2 + dy^2)/y^2.$$

In group-theoretic terms, up to a constant factor, it is defined by the trace form on the space of Hermitian matrices viewed as a subspace of the tangent space of G at the origin. For our purposes here, the coordinate expression is easier to use. We let $B(z,r)$ be the open ball of radius r centered at a point z, with respect to this distance which we call the **Iwasawa distance**, $d_{\mathrm{Iw}}(z,z') = d(z,z')$. Then the Iwasawa distance and the polar distance $d_{\mathbf{p}}$ defined in Section 2.1 differ by a constant factor. One can see this a priori because they are both G-invariant. The above notation for the metric checks with the Iwasawa volume

$$dx_1\, dx_2\, dy/y^3.$$

We define a fundamental domain FD to be **good with respect to** $z_0 \in \mathrm{FD}$ if there exist $C_1,\ C_2 > 0$ such that for all $z \in \mathrm{FD}$, we have

$$d(z,z_0) \leqq C_1 d(z,\Gamma(z_0)) + C_2.$$

Note that if this condition is satisfied for every point $z_0 \in \mathrm{Int(FD)}$ (interior of FD), then it is also satisfied for z_0 on the boundary, by continuity.

Theorem 6.3.5. *The Picard domain is good with respect to every point* $z_0 \in \mathscr{F}$.

Proof. First observe the inequality

$$ds^2 \geqq \frac{dy^2}{y^2}.$$

Let z_0 be any point in the interior of $\mathscr{F}$. By Theorem 6.2.2, we know that

$$y(\gamma z_0) \leqq y(z_0). \tag{6.6}$$

For $z = x + \mathbf{i}y$ in $\mathscr{F}$, the above inequality yields

$$d(z,z_0) = \log y + O(1). \tag{6.7}$$

On the other hand, let C be a curve between z and γz_0. Then its length $L(C)$ satisfies the inequalities

$$\begin{aligned} L(C) = \int_C ds \geqq \int_C \frac{dy}{y} &= \int_{y(z)}^{y(\gamma z_0)} \frac{dy}{y} \\ &= |\log\, y(\gamma z_0) - \log\, y| \\ \text{[by (6.6), (6.7), and Theorem 6.2.2]} &\geqq |\log\, y - \log\, y(z_0)| \\ &= |\log\, y| - O(1) \\ \text{[by (2)]} &\geqq d(z,z_0) - O(1). \end{aligned}$$

This proves the goodness of Theorem 6.3.5.

We apply Theorem 6.3.5 to get the next finiteness theorem. Since every compact set is contained in an open ball $B(z_0,r)$, Theorem 6.3.4 is implied by the following Theorem:

Theorem 6.3.6. *Let z_0 be an interior point of $\mathscr{F}$. Given $r > 0$, the set of $\gamma \in \Gamma$ such that $\gamma(\mathscr{F})$ intersects $B(z_0, r)$ is finite.*

Proof. Let $\delta > 0$ be such that $B(z_0, \delta) \subset \mathscr{F}$. Then for all $\gamma \in \Gamma$,

$$B(\gamma z_0, \delta) = \gamma B(z_0, \delta) \subset \gamma \mathscr{F}.$$

If $\gamma' \neq \gamma$ on $\mathbf{H}_3$, then $B(\gamma' z_0, \delta) \cap B(\gamma z_0, \delta)$ is empty by condition **FD 2** defining a fundamental domain. Next we shall use the good condition. For each $\gamma \in \Gamma$ such that $\gamma\mathscr{F} \cap B(z_0, r)$ is not empty, there is an element $z_\gamma \in \mathscr{F}$ such that

$$d\left(\gamma z_\gamma, z_0\right) \leqq r.$$

Since $\Gamma(\gamma z_0) = \Gamma(z_0)$, by the good condition we have

$$\begin{aligned} d(\gamma z_\gamma, \gamma z_0) &\leqq C_1 d(\gamma z_\gamma, \Gamma(z_0)) + C_2 \\ &\leqq C_1 d(\gamma z_\gamma, z_0) + C_2 \\ &\leqq C_1 r + C_2. \end{aligned}$$

Therefore $B(\gamma z_0, \delta) \subset B(z_0, C_1 r + C_2 + \delta)$, by the triangle inequality

$$d(z, z_0) \leqq d(z, \gamma z_0) + d(\gamma z_0, \gamma z_\gamma).$$

The number of disjoint balls of equal volume that are contained in another ball is finite. More precisely,

$$\#\{\gamma \text{ in } \Gamma/\pm I \text{ such that } \gamma\mathscr{F} \cap B(z_0, \delta) \text{ is not empty}\} \leqq \frac{\operatorname{Vol} B(z_0, C_1 r + C_2 + \delta)}{\operatorname{Vol} B(z_0, \delta)},$$

which proves Theorem 6.3.6 and also gives an explicit bound.

Peter Jones also made the remark that the above inequality shows that the number of elements γ on the left is $\ll e^{cr}$ with some constant $c > 0$. Furthermore, this exponential growth in r is attained for the Picard domain. Finally, Peter Jones gave us an example of a fundamental domain that is not good, and for which the finite intersection property of Theorem 6.3.6 is not valid, so the appearance of some condition(s) beyond those defining a fundamental domain is not surprising in this light. The shape of the no-good fundamental domain is roughly as follows:

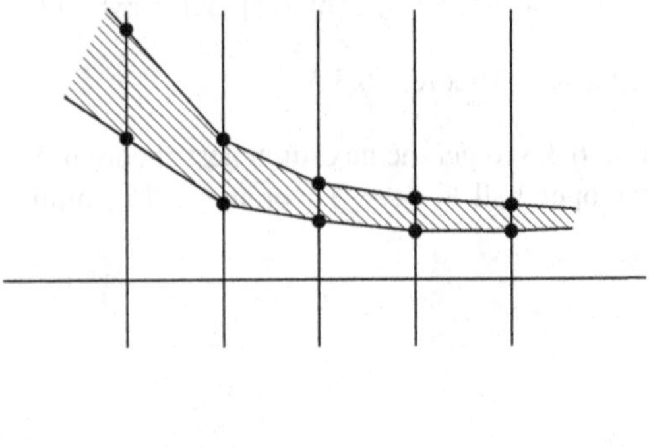

We are indebted to Eliot Brenner for bringing to our attention the book by S. Katok [Kat 92]. There she treats the more special notion of Dirichlet domain as follows, on the upper half-plane with $G = \mathrm{SL}_2(\mathbf{R})$, $\Gamma = \mathrm{SL}_2(\mathbf{Z})$. Let $p \in \mathbf{H}^2$ be a point not fixed by any element of Γ other than $\pm I$. The **Dirichlet domain** for Γ centered at p, denoted by $D(\Gamma, p)$, is the set of points $z \in \mathbf{H}^2$ such that

DD. $$d(z,p) \leqq d(z,\gamma p) \quad \text{for all } \gamma \in \Gamma.$$

Theorem 3.2.2 of [Kat 92] states that if p is not fixed by any element $\gamma \neq \pm I$, then the Dirichlet domain is a connected fundamental domain for Γ, and that the usual fundamental domain is equal to the Dirichlet domain for any point $p = y\mathbf{i}$, with $y > 1$ (Example A p. 55).

Brenner also pointed out the following more general context for Theorem 6.3.6. Let $\mathbf{X}$ be a complete metric space, with distance function d, such that every closed ball of finite radius is compact. Let Γ be a discrete group of isometries, **acting properly on X**, meaning that for every compact set Ω, the sets $\gamma\Omega$ and Ω are disjoint except for a finite number of elements $\gamma \in \Gamma$. For $p \in \mathbf{X}$ we define the **Dirichlet domain** $D(\Gamma, p)$ as the set of points $z \in \mathbf{X}$ satisfying **DD**.

Theorem 6.3.7. *Let $\bar{B}(p,r)$ be a closed ball in $\mathbf{X}$. Then only a finite number of translates $\gamma D(\Gamma,p)$ $(\gamma \in \Gamma)$ intersect $\bar{B}(p,r)$. The same holds for any compact subset of $\mathbf{X}$ instead of a closed ball. In particular, if $D(\Gamma,p)$ is also a fundamental domain, then a finite number of Γ-translates of $D(\Gamma,p)$ cover $\bar{B}(p,r)$.*

Proof. Replacing γ by γ^{-1}, suppose the first assertion false. Then there exists a sequence of distinct elements $\{\gamma_n\}$ in Γ and $\{z_n\}$ in $\bar{B}(p,r)$ such that $\gamma_n z_n \in D(p,r)$, all n. Passing to a subsequence if necessary, we may assume without loss of generality that the sequence $\{z_n\}$ converges to a point $w \in \bar{B}(p,r)$. Then by **DD** and isometry,

$$d(\gamma_n z_n, p) \leqq d(\gamma_n z_n, \gamma_n p) \leqq d(z_n, p). \tag{6.8a}$$

By the triangle inequality,

$$d(\gamma_n w, p) \leqq d(\gamma_n w, \gamma_n z_n) + d(\gamma_n z_n, p) \leqq d(w, z_n) + d(\gamma_n z_n, p). \tag{6.8b}$$

Combining these inequalities, we have

$$\begin{aligned} \limsup d(\gamma_n w, p) &\leqq \limsup d(w, z_n) + \limsup d(\gamma_n z_n, p) \\ &\leqq \limsup d(\gamma_n z_n, p) \ [\text{because } w = \lim z_n] \\ &\leqq \limsup d(z_n, p) \quad [\text{by (a)}] \\ &\leqq \limsup d(z_n, w) + \limsup d(w, p) \\ &= \limsup d(w, p) \leqq r \ [\text{because } w = \lim z_n]. \end{aligned}$$

Hence for infinitely many distinct γ_n, we have $d(\gamma_n w, p) \leqq 2r$. This contradicts the proper action of Γ, with respect to the compact set $\bar{B}(p, 2r)$, and concludes the proof of the first assertion. The others are then immediate.

Remark. The above finiteness statements provide an important condition allowing further results in Chapter 9. In the general context of a discrete group Γ acting topologically on a locally compact set $\mathbf{X}$, with fundamental domain $\mathscr{F}$, the action is called **locally finite** if given a compact set Ω, and letting Γ_{id} be the subgroup acting trivially on $\mathbf{X}$, there is only a finite number of cosets $\gamma\Gamma_{\text{id}}$ for which $\gamma\mathscr{F}$ intersects Ω. In our special case, $\Gamma_{\text{id}} = \{\pm I\}$. In defining proper action, we limited ourselves to conditions applying immediately to our special case, with a metric.

6.4 Uniformities in Lemma 6.2.3

The following lemmas give applications of Lemma 6.2.3 with uniformity properties.

Lemma 6.4.1. *Let $z \in \mathscr{F}$, $\gamma \in \Gamma$. If $y(z)$ and $y(\gamma(z)) > 1$, then $\gamma \in \Gamma_\infty$.*

Proof. Immediate from Lemma 6.2.3, since $y \leqq 1$ in the cases when $c_\gamma \neq 0$. We apply the lemma to $\gamma_\infty\gamma(z)$, with $\gamma_\infty \in \Gamma_\infty$ such that $\gamma_\infty\gamma(z) \in \mathscr{F}$, and one uses the fact that elements of Γ_∞ preserve the y-coordinate.

The next lemma concerns a Siegel subset of the fundamental domain. For $Y > 1$, we denote the part of the fundamental domain above Y by $\mathscr{F}_{\geqq Y}$, that is,

$$\mathscr{F}_{\geqq Y} = \{z \in \mathscr{F} \text{ such that } y_z \geqq Y\}.$$

Essentially as a corollary of Lemma 6.4.1 we get the following result:

Lemma 6.4.2. *Let $Y > 1$. There exists $\delta > 0$ such that if $z \in \mathscr{F}_{\geqq Y}$, $\gamma \in \Gamma$, and $d(z, \gamma z) < \delta$, then $y(\gamma z) > 1$ and $\gamma \in \Gamma_\infty$.*

Proof. For $z, z' \in \mathbf{H}^3$ we have $d(z,z') \geqq |\log y_z - \log y_{z'}|$. Hence for small δ, letting $z' = \gamma(z)$, we conclude that $y(\gamma z) > 1$, so we can apply Lemma 6.4.1 to conclude the proof.

The next lemma gives another type of uniformity for Γ-translates to lie in a ball of finite radius. We consider sets of pairs of points (z, w) and $\gamma \in \Gamma$ such that $d(z, \gamma w) \leqq 1$ (say). The smaller the set, the more possibilities there are for such γ. Thus in the next lemma, we can assume the set to be small.

Note that when y is large, then the distance between two opposite points of a horizontal segment across the fundamental domain is $1/y$, so there are $\gg y$ translations of this segment by elements of Γ_U that are at distance $\leqq 1$ (say) from a point on that segment. Thus Γ_∞ plays a special role. The next lemma asserts that there is no other such phenomenon.

Lemma 6.4.3. *For all Y sufficiently large, there exists $C_Y > 0$ such that for all z, $w \in \mathscr{F}_{\geqq Y}$ we have*

$$\#\{\gamma \in \Gamma - \Gamma_\infty \text{ such that } d(z, \gamma w) < 1\} \leqq C_Y.$$

Proof. We can pick Y so large that the condition $d(z,\gamma w) < 1$ implies $y(\gamma w) > 1$. Then Lemma 6.4.1 concludes the proof. Of course, instead of the ball of radius 1 around z, we could take a ball of any radius r, and take Y large depending on r.

We now formulate and prove the analogue of Theorem 6.3.5, getting goodness uniformly for pairs of points $z,\ w \in \mathcal{F}_{\geqq Y}$ with $\Gamma - \Gamma_\infty$ instead of Γ.

Lemma 6.4.4. *There exist constants $C_1 = C_1(Y)$ and $C_2 = C_2(Y)$ such that for all $z,\ w \in \mathcal{F}_{\geqq Y}$ we have*

$$d(z,w) \leqq C_1 d(z,(\Gamma - \Gamma_\infty)(w)) + C_2.$$

Proof. By Lemma 6.2.3, since $\gamma \notin \Gamma_\infty$, we must have $y(\gamma w) \leqq 1$.

Lemma 6.4.5. *For $z,\ w \in \mathcal{F}_{\geqq Y}$ and $\gamma \in \Gamma - \Gamma_\infty$ we have*

$$d(z,w) \leqq d(z,\gamma w) + |\log y(\gamma w) - \log y(w)| + O_Y(1).$$

Proof. By the differential-geometric formula for the metric, we know that

$$\begin{aligned} d(z,\gamma w) &\geqq |\log y(z) - \log y(\gamma w)| \\ &\geqq |\log y(z) - \log y(w)| - |\log y(\gamma(w)) - \log y(w)| \\ &= d(z,w) + O(1) - |\log y(\gamma w) - \log y(w)| \end{aligned}$$

as was to be shown.

6.5 Integration on $\Gamma\backslash G/K$

Let μ_{Iw} be the Iwasawa measure on G, which also fits the quaternion model of $\mathbf{H}^3$. Locally, G and $\Gamma\backslash G$ are C^∞ isomorphic and measure isomorphic. With this measure, which we call also the **Iwasawa measure** $\mu_{\Gamma\backslash G}$ on $\Gamma\backslash G$, we have the formula

$$\int_{\Gamma\backslash G} \sum_{\gamma\in\Gamma} \mathrm{F}(\gamma z) d\mu_{\Gamma/G}(z) = \int_G F(z) d\mu_{\mathrm{Iw}}(z), \tag{6.9}$$

for $F \in L^1(G)$. This is just what we call the **group Fubini theorem**, relating the Haar integral on a group, closed subgroup, and the coset space, when G is unimodular. In this case the subgroup is discrete, so the inner integral is replaced by the sum. We use the notation

$$F^\Gamma(z) = \sum_{\gamma\in\Gamma} F(\gamma z).$$

Here we are concerned with functions given on $\Gamma\backslash G$, K-right invariant, and their integral over $\Gamma\backslash G$ under the above measure. For $f \in L^1(\Gamma\backslash G/K)$ we have

$$\int_{\Gamma\backslash G} f(z) d\mu_{\Gamma\backslash G}(z) = \frac{1}{2} \int_{\mathcal{F}} f(z) d\mu_{\mathrm{Iw}}(z), \tag{6.10}$$

where in terms of the Iwasawa (x,y) coordinates, $d\mu_{\mathrm{Iw}}(z) = dx_1\, dx_2\, dy/y^3$. In particular,

$$\mathrm{Vol}(\Gamma\backslash G) = \frac{1}{2}\mathrm{Vol}(\mathscr{F}). \tag{6.11}$$

Proof. We test the formula on a function of type $f = F^{\Gamma}$. Then

$$\int_{\Gamma\backslash G} F^{\Gamma}(g)d\mu_{\Gamma\backslash G}(g) = \int_{G} F(g)d\mu_{\mathrm{Iw}}(g) = \int_{\mathbf{H}^3} F(z)d\mu_{\mathrm{Iw}}(z).$$

But $\mathbf{H}^3$ is the disjoint union of translates of $\mathscr{F}$ by the projective group $\mathbf{P}\Gamma = \Gamma/\pm I$ up to intersections of boundaries, so the last integral is

$$= \int_{\mathscr{F}} F^{\Gamma/\pm I} d\mu_{\mathrm{Iw}}(z),$$

which proves (6.2). The factor 1/2 may at first be surprising. Another way of understanding its presence is that Γ acts faithfully as a group of left translations on G, but only via the projective group $\mathbf{P}\Gamma = \Gamma/\pm I$ on $\mathbf{H}^3$. For our normalizations, we give priority to the group setting; cf. for instance Chapter 11, Theorem 4.3.

Letting $Y(x) = (1 - x_1^2 - x_2^2)^{1/2}$, we can rewrite the integral in the form

$$\begin{aligned}\frac{1}{2}\int_{\mathscr{F}} f(z)d\mu_{\mathrm{Iw}}(z) &= \frac{1}{2}\int_{\mathscr{F}_U}\int_{Y(x)}^{\infty} f(u(x)a(y))\,\frac{dy}{y^3}\,dx_1\,dx_2\\ &= \frac{1}{4}\int_{o\backslash\mathbf{C}}\int_{Y(x)}^{\infty} f(u(x)a(y))\,\frac{dy}{y^3}\,dx_1\,dx_2,\end{aligned}$$

whence

$$\int_{\Gamma\backslash G} f(g)\,d\mu_{\Gamma\backslash G}(g) = \frac{1}{4}\int_{o\backslash\mathbf{C}}\int_{Y(x)}^{\infty} f(u(x)a(y))\,\frac{dy}{y^3}\,dx_1\,dx_2. \tag{6.12}$$

We recall a basic lemma valid for an arbitrary closed subgroup, in the present context of a discrete group.

Lemma 6.5.1. *Let G be a unimodular Lie group and Γ a discrete subgroup. Let $f \in C_c^{\infty}(\Gamma\backslash G)$. Then there exists $F \in C_c^{\infty}(G)$ such that $F^{\Gamma} = f$. In particular,*

$$\int_{\Gamma\backslash G} f\, d\mu_{\Gamma\backslash G} = \int_{G} F\, d\mu_{G}.$$

Proof. Let $\pi:\ G \to \Gamma\backslash G$ be the projection. Let S_f = support of f. There exists a compact subset $S \subset G$ such that $\pi(S) = S_f$. Let $F_1 \in C_c^{\infty}(G)$ be such that $F_1 > 0$ on S. Then $F_1^{\Gamma} > 0$ on S_f. Let F_2 be defined by

$$F_2(z) = \begin{cases} f(\pi(z))/F_1(\pi(z)) & \text{if } F_1(z) \neq 0, \\ 0 & \text{if } F_1(z) = 0. \end{cases}$$

Let $F = F_1 F_2$. Then $F^\Gamma = f$, thus proving the lemma.

6.6 Other Fundamental Domains

The most classical fundamental domain is that for $\mathbf{Z}\backslash\mathbf{R}$ or $\mathbf{Z}[\mathbf{i}]\backslash\mathbf{C}$. For this latter case, the standard fundamental domain is the square centered at the origin, with sides of length 1. In terms of the group structure with which we are now dealing, this amounts to a fundamental domain for Γ_U/U. It is especially useful in dealing with Γ_U-periodic functions on U, i.e., functions on the torus $o\backslash\mathbf{C}$. Hence it may be convenient to deal with the **double fundamental domain**

$$\mathscr{F}^{(2)} = \mathscr{F} \cup R(\mathscr{F})$$

consisting of those elements $z = x + \mathbf{j}y$ with

$x \in \mathscr{F}_U^{(2)} =$ square of side 1 centered at 0 in $\mathbf{C} =$ fundamental square;
$y^2 \geqq Y(x)^2 = Y(-x)^2 = 1 - |x|^2$ as before.

Thus the region $\mathscr{F}^{(2)}$ is the region lying above the fundamental square and above the sphere.

We shall also deal with the **tubes**

$$\mathscr{T} = \mathscr{F}_U \times \mathbf{R}^+ \text{ and } \mathscr{T}^{(2)} = \mathscr{F}_U^{(2)} \times \mathbf{R}^+,$$

lying above $\mathscr{F}_U$ and $\mathscr{F}_U^{(2)}$ respectively. Given an element $z \in \mathbf{H}^3$ there exists $\eta \in \Gamma_\infty$ such that $\eta(z) \in \mathscr{T}$, and we can take $\eta \in \Gamma_U$ if we allow $\eta(z)$ to land in $\mathscr{T}^{(2)}$.

Even more, we get at once:

$\mathscr{T}$ is a fundamental domain for Γ_∞ and $\mathscr{T}^{(2)}$ is a fundamental domain for Γ_U.

Note that $\mathscr{T}$ and $\mathscr{T}^{(2)}$ are unions:

$$\mathscr{T} = \mathscr{F} \cup \mathscr{S} \text{ and } \mathscr{T}^{(2)} = \mathscr{F}^{(2)} \cup \mathscr{S}^{(2)}.$$

We shall also consider the corresponding regions lying *below the sphere*, that is we let

$$\mathscr{S} = \{z = x + \mathbf{j}y \text{ such that } x \in \mathscr{F}_U \text{ and } 0 < y \leqq Y(x)\},$$
$$\mathscr{S}^{(2)} = \mathscr{S} \cup R(\mathscr{S}) = \{z = x + \mathbf{j}y \text{ such that } x \in \mathscr{F}_U^{(2)} \text{ and } 0 < y \leqq Y(x)\}.$$

The present chapter will be complemented by Chapter 9, giving a broader understanding how the fundamental domain $\mathscr{F}$ is related to these other domains, culminating in the theorem that $\mathscr{T}$ is a union of Γ-translates of $\mathscr{F}$. We shall start with the case of $\mathbf{H}^2$, because an inductive procedure takes hold.

Chapter 7
Γ-Periodization of the Heat Kernel

The main point of this chapter is to periodize the heat kernel on G/K with respect to Γ, and obtain what we shall prove is a heat kernel on $\Gamma\backslash G/K$, namely satisfying the Dirac property and the heat equation, for K-right-invariant functions on $\Gamma\backslash G$. Thus we shall prove properties corresponding to those in Chapters 2, 3, where there was no Γ. In [Tam 60], Tamagawa gives some properties concerning convolution on $\Gamma\backslash G$ in a quite general context, except for the serious restriction that $\Gamma\backslash G$ is compact.

7.1 The Basic Estimate

In Chapter 5 we considered G-conjugations and functions of type $\varphi(g^{-1}g_1g)$ or traces $\sum_{\gamma}\varphi(g^{-1}\gamma g)$. We shall now consider functions of the form $\varphi(z^{-1}\gamma w)$ with $z,\ w \in G$ and take the sum over $\gamma \in \Gamma$ or subsets. We suppose φ is K-bi-invariant. As in Chapter 4, we deal with the point-pair invariant defined by

$$\mathbf{K}_{\varphi}(z,w) = \varphi(z^{-1}w) = \mathbf{K}_{\varphi}(w,z).$$

Ultimately, φ will be the heat Gaussian. We shall give estimates for various objects associated with φ. The main function of interest to us is $\varphi = \mathbf{g}_t$, i.e., φ is the heat Gaussian, so the corresponding point-pair invariant is the heat kernel.

We let the Γ-periodization $\mathbf{K}_{\varphi}^{\Gamma}$ be defined by

$$\mathbf{K}_{\varphi}^{\Gamma}(z,w) = \sum_{\gamma\in\Gamma}\mathbf{K}_{\varphi}(\gamma z,w) = \sum_{\gamma\in\Gamma}\mathbf{K}_{\varphi}(z,\gamma w).$$

The main object of this chapter is to show that when $\varphi = \mathbf{g}_t$, and we denote $\mathbf{K}_{\mathbf{g}_t}$ by $\mathbf{K}_t$, then the periodization $\mathbf{K}_t^{\Gamma}$ satisfies properties proved for $\mathbf{K}_t$ in Chapters 2 and 3, but on $\Gamma\backslash G$.

7.1.1 Convolution

The **convolution operator** with $\mathbf{K}_\varphi$ is defined as usual by

$$(\mathbf{K}_\varphi *_G f)(z) = \int_G \mathbf{K}_\varphi(z,w)f(w)dw = \int_G \varphi(z^{-1}w)f(w)dw. \tag{7.1}$$

The integral is assumed absolutely convergent. We have the analogous convolution on $\Gamma\backslash G$, with the function $\mathbf{K}_\varphi^\Gamma$ and a function f measurable on $\Gamma\backslash G$, namely

$$(\mathbf{K}_\varphi^\Gamma *_{\Gamma\backslash G} f)(z) = \int_{\Gamma\backslash G} \mathbf{K}_\varphi^\Gamma(z,w)f(w)dw. \tag{7.2}$$

The following relation is fundamental, and applies to a unimodular locally compact group G, a discrete subgroup Γ, just under the condition of absolute convergence.

Theorem 7.1.1. *Let f be measurable on $\Gamma\backslash G$. The integral of* (1) *is absolutely convergent if and only if the integral of* (2) *is absolutely convergent, in which case they are equal, that is, for $z \in G$, and any Haar measure dw,*

$$\int_G \varphi(z^{-1}w)f(w)dw = \int_{\Gamma\backslash G} \sum_{\gamma\in\Gamma} \varphi(z^{-1}\gamma w)f(w)dw;$$

or in convolution notation,

$$\mathbf{K}_\varphi^\Gamma *_G f = \mathbf{K}_\varphi^\Gamma *_{\Gamma\backslash G} f.$$

Proof. This is merely a special case of what we call the **group Fubini theorem**, valid for a closed subgroup H, namely

$$\int_G F(\varphi)dw = \int_{H\backslash G}\int_H F(hg)\,dh\,d_{H\backslash G}(g).$$

With a discrete subgroup, the inner integral gets replaced by a sum. With z fixed, the function F is simply given by $F(w) = \varphi(z^{-1}w)f(w)$. Cf. [Lan 93], Chapter 12, Theorem 4.3. Since Γ is discrete in G, the spaces G and $\Gamma\backslash G$ are locally isomorphic, so the measures on G and $\Gamma\backslash G$ are "the same" locally. Then Γ may be viewed as having the discrete measure giving points measure 1.

We deal with an extension of Section 5.2 keeping the two variables separate rather than considering conjugations. The conditions QED and LEG of Section 3.1 will be the most important for the rate of decay of a function. Ultimately, we are after G-Gaussians and the results of Section 3.1 on $\Gamma\backslash G$, but the estimate of this section depends only on a weaker condition, so we describe the conditions here for the record.

Let φ be a function on G. We say that φ has **polynomial decay** (with respect to the Hermitian norm defined in Section 2.1) if there exists a positive integer N such that

$$|\varphi(z)| = O(||z||^{-N}) \text{ for } ||z|| \to \infty.$$

Since the Hermitian norm on G is bounded away from 0, this condition is equivalent to the existence of a constant C_N such that

$$|\varphi(z)| \leqq C_N ||z||^{-N} \text{ for all } z \in G.$$

If the condition holds, then we say that φ has polynomial decay of order N. We define **superpolynomial decay** to mean polynomial decay of arbitrarily high order. Note that polynomial decay with respect to the Hermitian norm is the same as linear exponential decay with respect to the polar height, by Chapter 4, Proposition 4.1.1.

We assume throughout that $\varphi \in C(K \backslash G/K)$ *is continuous, even, K-bi-invariant.* We use polar coordinates and the polar Jacobian in terms of the y-coordinate

$$J(y) = \left(\frac{y - y^{-1}}{2}\right)^2,$$

which gives the polar Haar measure $dk_1\, dk_2\, J(y)\, dy/y$. We essentially apply an integral test to get an estimate for the Γ-periodization of $\mathbf{K}_{\varphi}(z, w)$. We follow a method that works in much more general cases (cf. [Har 65], Lemma 37 and [JoL 02], Section 1.2). For $t \geqq 0$, we let

$$G(t) = \text{set of all } z \in G \text{ such that } ||z|| \leqq t \text{ and } A_{G(t)} = A \cap G(t).$$

Lemma 7.1.2. **(a)** *Let* vol *be the Haar volume on G. Then vol(G(t)) has polynomial growth for* $t \geqq 1$. *More precisely, there exists* $c > 0$ *such that*

$$\mathrm{vol}(G(t)) \leqq ct^2 \text{ for } t \geqq 1.$$

(b) *We have*

$$\#(\Gamma_{G(t)}) = O(t^2) \text{ for } t \to \infty.$$

Proof. As to (a), by the polar Haar measure formula, using $||z|| = ||\mathrm{Pol}_A\,(z)||$ and KA^+K open in G (where A^+ is equal to the set of $a \in A$ such that $y_a > 1$), with complement of measure 0, we have

$$\mathrm{vol}(G(t)) = \int_{A^+_{G(t)}} J_{\mathbf{p}}(a) da \text{ up to a constant } = \int_I^t J(y)\frac{dy}{y}.$$

Part (a) is then immediate. (In higher dimension, the volume grows like a power of t.)

For (b), let V be a small symmetric neighborhood of the unit element such that the only element of Γ in VV is I. Then two translates $\gamma_1 V$ and $\gamma_2 V$ by distinct elements of

Γ are disjoint. Hence $\Gamma_{G(t)}V$ has volume $\geqq \#(\Gamma_{G(t)})\text{vol}(V)$. By **HN 2** of Section 4.1, $\Gamma_{G(t)}V \subset \Gamma_{G(ct)}$ with some fixed constant c. Applying (a) concludes the proof.

The previous lemma is going to be applied to show the convergence of certain series, obtained by taking a Γ-trace, i.e., summing over all elements of Γ. We are doing [Har 68], Lemma 9, on SL_2, but under a condition weaker than compact support.

For the next lemma on SL_n, cf. [JoL 02], Chapter 2, Lemma 2.4.

Lemma 7.1.3. *Let φ be a continuous positive function on G with polynomial decay of order $\geqq 6$. Then there is a constant c such that*

$$\mathbf{K}_{\varphi}^{\Gamma}(z,w) = \sum_{\gamma\in\Gamma} \varphi(z^{-1}\gamma w) \leqq c\|z\|^4 \quad \textit{for all } z,w \in G.$$

Proof. Let $G(B-1,B)$ be the annulus consisting of all elements $g \in G$ such that

$$B-1 \leqq \|g\| \leqq B.$$

Replacing φ by its absolute value we may assume $\varphi \geqq 0$. The sum in the lemma is dominated by

$$\sum_{B=1}^{\infty} \sum_{z^{-1}\gamma w\in G(B-1,B)} \varphi(z^{-1}\gamma w).$$

By **HN 3** in Section 4.1, for $g \in G(B)$ we have $\|g^{-1}\| \leqq c_1 B$. Furthermore,

$$z^{-1}\gamma w \in G(B) \Leftrightarrow \gamma \in zG(B)w^{-1}.$$

Let $\gamma_0 \in \Gamma \cap zG(B)w^{-1}$. The map $\gamma \mapsto \gamma\gamma_0^{-1}$ gives an injection

$$zG(B)w^{-1} \hookrightarrow zG(B)G(B)^{-1}z^{-1}.$$

Hence

$$\begin{aligned}
&\#\{\gamma \text{ such that } z^{-1}\gamma w \in G(B)\} \\
&\qquad \leqq \#\{\gamma' \text{ such that } \gamma' \in zG(B)G(B)^{-1}z^{-1}\} \\
&\qquad \leqq \#\{\gamma' \text{ such that } \|\gamma'\| \leqq c_1^2\|z\|^2B^2\} && \text{by } \textbf{HN 2}\text{, Chapter 4} \\
&\qquad \leqq c_2(c_1^2\|z\|^2B^2)^2 && \text{by Lemma 7.1.2.}
\end{aligned}$$

Thus we obtain

$$\sum_{z^{-1}\gamma w\in G(B-1,B)} \varphi(z^{-1}\gamma w) \ll c_2c_1\|z\|^4B^4B^{-N}.$$

The lemma follows at once for $N \geqq 6$ ($5+\varepsilon$ *if you want*).

We may now apply the estimate of Lemma 7.1.3 to the convolution, giving the first example for Theorem 7.1.1.

Proposition 7.1.4. *Let φ be a continuous function on G, with polynomial decay of order $\geqq 6$. Let $\mathbf{K}_\varphi(z,w) = \varphi(z^{-1}w)$. With c as in Lemma 7.1.3, for every $f \in L^1(\Gamma\backslash G)$, we have*

$$|(\mathbf{K}_\varphi * f)(z)| \leqq c\|z\|^4\|f\|_1.$$

If $f \in L^2(\Gamma\backslash G)$, then

$$|(\mathbf{K}_\varphi * f)(z)| \leqq c_1\|z\|^4\|f\|_2 \text{ with } c_1 = c.\mathrm{vol}(\Gamma\backslash G)^{1/2}.$$

Proof. We get

$$\begin{aligned} |(\mathbf{K}_\varphi * f)(z)| &\leqq \int_G |\varphi(Z^{-1}w)f(w)|dw \\ &\leqq \int_{\Gamma\backslash G} \sum_{\gamma\in\Gamma} |\varphi(z^{-1}\gamma w)f(w)|dw \leqq c\|z\|^4\|f\|_1 \end{aligned}$$

by Lemma 7.1.3, thus proving the first inequality. If $f \in L^2$, then we get the second inequality directly from Schwarz and the fact that $\Gamma\backslash G$ has finite measure.

If z ranges over a compact set, then the factor $\|z\|^4$ is bounded, and we get the following:

Proposition 7.1.5. *Let φ be a continuous function on G, with polynomial decay of order $\geqq 6$. Let Ω be a compact subset of G. There is a constant C_φ such that for all $z \in \Omega$ and $w \in$ G we have*

$$|\mathbf{K}_\varphi^\Gamma(z,w)| \leqq C_\varphi(\Gamma,\Omega).$$

or alternatively, $\mathbf{K}_\varphi^\Gamma$ is bounded on $\Omega\times G$ and $G\times\Omega$ (by symmetry). Furthermore, for $f \in L^2(\Gamma\backslash G)$ and $z \in \Omega$,

$$|(\mathbf{K}_\varphi^\Gamma * f)(s)| \leqq C_\varphi(\Gamma,\Omega)vol(\mathscr{F})^{1/2}\|f\|_2.$$

Of course, the constant also depends on φ, which is given a priori, so we usually don't mention it in the constant. Similarly, we might not mention Γ, but we shall deal with various subsets of Γ in the next section, and for emphasis we keep these subsets in the notation.

We recall the complexified heat kernel (complexifying t to τ) from Section 2.3. The estimate of Proposition 7.1.4 of course applies in this case, for the record:

Proposition 7.1.6. *The series*

$$\mathbf{K}_\tau^\Gamma(z,w) = \sum_{\gamma\in\Gamma} \mathbf{K}_\tau(\gamma z,w)$$

converges uniformly for z in a compact subset of G and τ in a compact subset of the right half-plane. The function $\tau \mapsto \mathbf{K}_\tau^\Gamma(z,w)$ is holomorphic in τ in this half-plane.

The above estimates for $\mathbf{g}_t^\Gamma$ *and* $\mathbf{K}_t^\Gamma(z,w)$ *hold also for* $\mathbf{K}_\tau^\Gamma(z,w)$. *In particular, uniformly for* τ *in a compact set,*

$$|\mathbf{K}_\tau^\Gamma(z,w)| \ll \|z\|^4.$$

If $f \in L^1(\Gamma\backslash G/K)$, *then*

$$\mathbf{K}_\tau^\Gamma * f(z) \ll \|z\|^4 \|f\|_1.$$

Proof. Immediate from Proposition 7.1.4 and Section 2.3 (7.2.1), giving the formula for the absolute value of the complexified Gaussian.

Remark. The above estimates give sufficient conditions for convolution of the periodized heat kernel with some function to converge absolutely. The symmetry of the formula shows that the heat kernel is symmetric, and hence symmetric as an integral operator, with respect to the scalar product integration over $\Gamma\backslash G$. This symmetry will be used for instance in Chapter 8 to get the positivity of eigenvalues under appropriate conditions.

7.2 Heat Convolution and Eigenfunctions on Γ\G/K

We first define a distance on $\Gamma\backslash G/K$, We start with the **polar distance function** (Chapter 1, Theorem 1.8.4)

$$d_{\mathbf{p}}(z,w) = d_{G/K}(z,w) = \sigma\left(z^{-1}w\right) \text{ on } G/K \times G/K.$$

We **define**

$$\sigma_{\Gamma\backslash G}(z) = \min_{\gamma\in\Gamma} \sigma(\gamma z) \text{ and } d_{\Gamma\backslash G}(z,w) = \min_{\gamma\in\Gamma} d_{G/K}(\Upsilon_{z,w}) = \min_{\gamma\in\Gamma} d_{G/K}(z,\gamma w).$$

Note that $d_{\Gamma\backslash G}(\gamma z,w) = d_{\Gamma\backslash G}(z,w)$, in other words, $d_{\Gamma\backslash G}$ is defined on $\Gamma\backslash G/K$. Because of the triangle inequality for σ, we get trivially the triangle inequality for $d_{\Gamma\backslash G}$. If $d_{\Gamma\backslash G}(z,w) = 0$, then there is some $\gamma\in\Gamma$ such that $d(\gamma z,\ w) = 0$ so $z,\ w$ are in the same Γ-orbit. Thus $d_{\Gamma\backslash G}$ is a distance function on $\Gamma\backslash G/K$. Note that

$$\sigma_{\Gamma\backslash G}(z) = d_{\Gamma\backslash G}(z, e_{\Gamma\backslash G}).$$

We formulate the integration property more generally than Theorem 7.1.1.

Theorem 7.2.1. *Let* F *be measurable on* $\Gamma\backslash G$. *Let* dg *be any Haar measure on* G. *Let* S *be a measurable subset of* G *that is* Γ*-invariant on the left. Then*

$$\int_{\Gamma\backslash S} \sum_{\gamma\in\Gamma} F(\gamma g)df = \int_S F(g)dg,$$

in the sense that if one side is absolutely convergent, so is the other and the two sides are equal.

Proof. Same as for Theorem 7.1.1.

The three conditions defining **Weierstrass–Dirac** family can thus be formulated by putting the index $\Gamma\backslash G$ to the distance in the conditions of Section 2.1. For a family $\{\psi_c\}$ of K-right invariant functions on $\Gamma\backslash G$ (i.e., functions on $\Gamma\backslash G/K$), these conditions read

WD 1Γ. The function ψ_c are continuous $\geqq 0$.

WD 2Γ. $\displaystyle\int_{\Gamma\backslash G} \psi_c(g)dg = 1.$

WD 3Γ. Given $\delta > 0$, we have $\displaystyle\lim_{c\to 0} \int_{\sigma_{\Gamma\backslash G}(g)\geqq\delta} \psi_c(g)dg = 0.$

Of course, we have the similar formulation for the corresponding function (point-pair invariant) in two variables, using the function $d_{\Gamma\backslash G}(z,w)$ in two variables, right K-invariant in each variable. Thus we take F depending on a parameter z,

$$F_z(w) = F(z,w) = \mathbf{K}_t(z,w),$$

where $\mathbf{K}_t$ is the heat kernel on G/K. Let

$$V_\delta = \{w \in G \text{ such that } \sigma(w) = d(e_G, w) \geqq \delta\}.$$

Then ΓV_δ is left Γ-invariant, and $S = G - \Gamma V_\delta$ also.

Theorem 7.2.2. *The periodized families* $\{\mathbf{g}_t^\Gamma\}$ *and* $\{\mathbf{K}_t^\Gamma\}$ *are WD families on* $\Gamma\backslash G/K$ *for the corresponding Haar measures.*

Proof. We carry out the proof for the Gaussian family in one variable. Each function $\mathbf{g}_t^\Gamma$ is actually > 0, not just $\geqq 0$. The total integral over $\Gamma\backslash G$ is 1 by invoking Theorem 7.2.1 with $S = G$, and the corresponding property for $\{\mathbf{g}_t\}$ on G itself. The property **WD 3Γ** on $\Gamma\backslash G$ also comes from Theorem 7.2.1, by taking $S = G - \Gamma V_\delta$, and noting that

$$\int_{\Gamma\backslash S} \mathbf{g}_t^\Gamma(w)dw \leqq \int_{G-V_\delta} \mathbf{g}_t(w)dw = \int_{\sigma(w)\geqq\delta} \mathbf{g}_t(w)dw,$$

and using **WD 3** for $\{\mathbf{g}_t\}$, Chapter 2, Theorem 2.1.2.

We shall call $\{\mathbf{K}_t^\Gamma\}$ the **heat kernel** on $\Gamma\backslash G/K$. To justify this terminology, we still need to prove the heat equation, which we do in Section 7.3. But first, we squeeze more consequences out of the Weierstrass–Dirac properties.

In light of the definitions, Theorems 7.2.1 and 7.2.2, we now see the possibility of proving some theorems of Chapter 3 in the context of $\Gamma\backslash G$. We state those that will be useful in the sequel. First, we recall the universal property of WD families:

For any bounded measurable function f, $\psi_c * f \to f$ as $c \to 0$, and in particular

$$\mathbf{K}_t^\Gamma * f \to f \text{ as } t \to 0,$$

uniformly on every set on which f is uniformly continuous.

However, we also need the L^1 version (which applies to L^2). The other versions stated and proved in Section 2.2 are also valid, with the same proofs, but for simplicity we stick to L^1, which is sufficient for our purposes.

Proposition 7.2.3. *Let $f \in L^1(\Gamma\backslash G)$. Then $\mathbf{K}_t^\Gamma *_{\Gamma\backslash G} f = \mathbf{K}_t *_G f$, or in full,*

$$\int_{\Gamma\backslash G} \mathbf{K}_t^\Gamma(z,w)f(w)dw = \int_G \mathbf{K}_t(z,w)f(w)dw.$$

The integrals are absolutely convergent, uniformly for z in a compact set and $0 < t \leqq t_0$.

Proof. The first assertion is a special case of Theorem 7.2.1, with

$$F(w) = \mathbf{K}_t(z,w)f(w).$$

If z is in a compact set, then Proposition 7.1.5 guarantees that $\mathbf{K}_t$ is uniformly bounded.

From Proposition 7.2.3, some properties on G apply to prove properties on $\Gamma\backslash G$.

Theorem 7.2.4. L^1-Dirac property. *Let $f \in L^1(\Gamma\backslash G/K)$. Then for $t \to 0$,*

$$\mathbf{K}_t^\Gamma * f \to f \text{ in } L^1(\Gamma\backslash G).$$

Proof. Let $\varphi \in C_c(\Gamma\backslash G/K)$ be L^1-close to f. We have

$$\|\mathbf{K}_t^\Gamma * f - f\|_1 \leqq \|\mathbf{K}_t^\Gamma * (f-\varphi)\|_1 + \|\mathbf{K}_t^\Gamma * \varphi - \varphi\|_1 + \|\varphi - f\|_1.$$

The first term on the right written out in full is

$$\leqq \int_{\Gamma\backslash G}\int_{\Gamma\backslash G} \mathbf{K}_t^\Gamma(z,w)|(f-\varphi)(w)|\,dwdz.$$

But the integral with the reverse order of integration,

$$\int_{\Gamma\backslash G}\left[\int_{\Gamma\backslash G} \mathbf{K}_t^\Gamma(z,w)dz\right]|(f-\varphi)(w)|dw = \|f-\varphi\|_1,$$

is absolutely convergent because the dz integral is 1 by **DIR 2Γ**. Therefore with the reverse order of integration, the integral is $< \varepsilon$ if φ is ε-close to f in L^1. Hence finally the first term on the right by Fubini is $< \varepsilon$. The other terms are immediately determined to be $< \varepsilon$ for t sufficiently small. This concludes the proof.

Proposition 7.2.5. *For $f \in L^1(\Gamma\backslash G/K)$ we have*

$$\mathbf{K}_t^\Gamma * (\mathbf{K}_{t'}^\Gamma * f) = \mathbf{K}_{t+t'}^\Gamma * f.$$

Proof. We have

$$\mathbf{K}_t^\Gamma * (\mathbf{K}_{t'}^\Gamma * f)(z) = \int_{\Gamma\backslash G} \mathbf{K}_t^\Gamma(z,w) \left[\int_G \mathbf{K}_{t'}(w,g) f(g) dg \right] dw_{\Gamma\backslash G}$$

[by Propostion 7.2.3]

$$= \int_G \int_G \mathbf{K}_t(z,w) K_{t'}(w,g) f(g)\, dw\, dg$$

[by Proposition 7.2.3 and Fubini]

$$= (\mathbf{K}_{t+t'} * f)(z) \qquad \text{[by Chapter 2, Theorem 1.4].}$$

$$= \left(\mathbf{K}_{t+t'}^\Gamma *_{\Gamma\backslash G} f\right)(z) \qquad \text{[by Proposition 7.2.3].}$$

This proves Proposition 7.2.5.

Proposition 7.2.6. *The heat kernel on $\Gamma\backslash G$ satisfies the semigroup property*

$$\mathbf{K}_t^\Gamma * \mathbf{K}_{t'}^\Gamma = \mathbf{K}_{t+t'}^\Gamma.$$

Proof. More of the same:

$$\mathbf{K}_t^\Gamma * (\mathbf{K}_{t'}^\Gamma * f)(z) = \int_{\Gamma\backslash G} \mathbf{K}_t^\Gamma(z,w) \left[\int_{\Gamma\backslash G} \mathbf{K}_{t'}^\Gamma(w,g) f(g) dg \right] dw$$

$$= \left(\left(\mathbf{K}_t^\Gamma * K_{t'}^\Gamma \right) * f \right)(z) \qquad \text{[by Fubini].}$$

Hence $\mathbf{K}_t^\Gamma * \mathbf{K}_{t'}^\Gamma$ and $\mathbf{K}_{t+t'}^\Gamma$ have the same value for their convolutions with every $f \in C_c(\Gamma\backslash G/K)$, whence they are equal, thus proving Proposition 7.2.6.

Recall that the finiteness of the volume of $\Gamma\backslash G$ implies that

$$L^2(\Gamma\backslash G) \subset L^1(\Gamma\backslash G).$$

Theorem 7.2.7. *Let $f \in L^1(\Gamma\backslash G/K)$. Suppose that there is one value $t_0 > 0$ such that*

$$\mathbf{K}_{t_0}^\Gamma * f = 0.$$

Then $f = 0$. In words, for each t, the heat convolution operator on $L^1(\Gamma\backslash G/K)$ and so $L^2(\Gamma\backslash G/K)$ is injective, i.e., *its null space is $\{0\}$.*

Proof. By Proposition 7.2.5, we know that for all $t > 0$,

$$\mathbf{K}^{\Gamma}_{t_0+t} * f = 0.$$

By Proposition 7.1.6 we know that $(\mathbf{K}^{\Gamma}_{\tau} * f)(z)$ is holomorphic in τ on the right half-plane. For fixed $z \in \Gamma\backslash G$, the function $(\mathbf{K}^{\Gamma}_{\tau} * f)(z)$ is 0 on a line segment, hence it is identically 0 for all τ. Now we can use the WD property Theorem 7.2.4 to conclude the proof.

Of course, Theorem 7.2.7 is the Γ-result corresponding to Chapter 2, Theorem 2.3.2. For simplicity we stated it here in a form sufficient for the application to Chapter 11 and the eigenfunction expansion. Readers can substitute the more general version if they wish.

By a **heat eigenfunction** f (on $\Gamma\backslash G/K$) we shall mean an eigenfunction $\neq 0$ of convolution $\mathbf{K}^{\Gamma}_t *$ for all t, so

$$\mathbf{K}^{\Gamma}_t * f = \eta(t) f,$$

with a scalar $\eta(t) \neq 0$. We suppose $f \in \mathrm{LEG}_{\sigma}(\Gamma\backslash G/K)$ (LEG and (Γ, K)-invariant) or $f \in L^1(\Gamma\backslash G)$. Then the map $t \mapsto \eta(t)$ is continuous, and

$$\mathbf{K}^{\Gamma}_{t+t'} * f = \mathbf{K}^{\Gamma}_{t'} * \mathbf{K}^{\Gamma}_t * f = \eta(t)\eta(t') f = \eta(t+t') f.$$

Hence $\eta(t+t') = \eta(t)\eta(t')$. Since $t \mapsto \eta(t)$ is continuous, there exists $\lambda \in \mathbf{C}$ such that

$$\eta(t) = e^{-\lambda t}.$$

Theorem 7.2.8. *Let $f \in \mathrm{LEG}_{\sigma}(\Gamma\backslash G/K)$ or $f \in L^1(\Gamma\backslash G/K)$, $f \neq 0$. Suppose f is a heat eigenfunction, with eigenvalue $\eta(t) \neq 0$. Then f is an eigenfunction of Casimir, with eigenvalue $-\lambda$, and $\eta(t) = e^{-\lambda t}$.*

Proof. This comes out again from Theorem 7.1.1 and Chapter 3, Theorems 3.4.1 and 3.4.2.

Theorem 7.2.9. *(a) Let $f \in L^2(\Gamma\backslash G/K)$, $f \neq 0$, be a heat eigenfunction with eigenvalue $\eta(t)$ for $\mathbf{K}^{\Gamma}_t *$. Then $\eta(t)$ is real for all t, so the corresponding eigenvalue λ for ω is real.*
(b) Suppose f is bounded. Then $f \in B^{\infty}$ and $\lambda > 0$, i.e., $\eta < 1$.

Proof. The reality comes from the usual argument for Hermitian linear maps, taking the scalar product $\langle Lf, f\rangle$, using the fact that $\mathbf{K}^{\Gamma}_t$ is real positive and symmetric. Suppose in addition that f is bounded. That $f \in B^{\infty}$ comes by lifting the convolution integral from $\Gamma\backslash G$ to G and invoking Chapter 3, Theorem 3.3.2. That $\lambda \geqq 0$ comes from Chapter 3, Theorem 3.4.3. That $\lambda > 0$ will be proved in Theorem 3.6.

In the case of the upper half-plane ($\mathbf{H}^2 = G/K$, $G = \mathrm{SL}_2(\mathbf{R})$, $K = K(\mathbf{R})$), Maass [Maa 49] originally considered **automorphic forms** on $\mathbf{H}^2$ to be functions invariant under $\mathrm{SL}_2(\mathbf{Z})$ and eigenfunctions of the Laplacian. Thus such eigenfunctions are often called **Maass forms**. In our presentation, we put the emphasis on the heat kernel, but Theorem 7.2.8 shows that one ends up with the same functions if there is some control over the growth.

7.3 Casimir on $\Gamma\backslash G/K$

Let ω be the Casimir operator on G. Since Γ is discrete in G, ω induces a differential operator on $\Gamma\backslash G$, looking like ω locally. In particular, a function on $\Gamma\backslash G$ is an eigenfunction of Casimir if and only if its lift to G is an eigenfunction, and then they have the same eigenvalue.

Theorem 7.3.1. *The functions* $(t,z) \mapsto \mathbf{g}_t^\Gamma(z)$ *and* $(t,z,w) \mapsto \mathbf{K}_t^\Gamma(z,w)$ *satisfy the heat equation.*

Proof. We just give the proof for $\mathbf{g}_t^\Gamma$. It suffices to show that for every $f \in C_c^\infty(\Gamma\backslash G)$, right K-invariant, we have

$$\int_G \partial_t \mathbf{g}_t^\Gamma(w) f(w) dw = \int_{\Gamma\backslash G} (\omega \mathbf{g}_t^\Gamma)(w) f(w) dw.$$

The routine is

$$\begin{aligned}
\int_{\Gamma\backslash G} \partial_t \mathbf{g}_t^\Gamma(w) f(w) dw &= \int_G \partial_t \mathbf{g}_t(w) f(w) dw && \text{[by Theorem 7.1.1]} \\
&= \int_G (\omega g_t)(w) f(w) dw && \text{[by Chapter 2, Theorem 7.5.1]} \\
&= \int_G \mathbf{g}_t(w)(\omega f)(w) dw && \text{[by Chapter 3, Lemma 3.2.1]} \\
&= \int_{\Gamma\backslash G} \mathbf{g}_t^\Gamma(w)(\omega f)(w) dw && \text{[by Theorem 7.1.1]} \\
&= \int_{\Gamma\backslash G} (\omega \mathbf{g}_t^\Gamma)(w) f(w) dw,
\end{aligned}$$

which is valid because f has compact support, and the integral locally is like the integral over G. This concludes the proof. An alternative way would be to ω-differentiate under the sum sign (summing over $\gamma \in \Gamma$). The distribution argument avoids this.

Proposition 7.3.2. *Let f be measurable on $\Gamma\backslash G/K$, and t given. Suppose that the convolution integral $\mathbf{K}_t^\Gamma *_{\Gamma\backslash G} f$ is absolutely convergent. Then f is an eigenfunction of this convolution if and only if its lift to G is an eigenfunction of $\mathbf{K}_t *$, and the two have the same eigenvalue.*

Proof. This is immediate from Theorem 7.2.1.

Theorem 7.3.3. *Let $f \in \mathrm{LEG}^\infty(\Gamma\backslash G/K)$. Taking convolution on $\Gamma\backslash G$, we have for all t,*

$$\omega(\mathbf{K}_t^\Gamma * f) = (\omega \mathbf{K}_t^\Gamma) * f = \mathbf{K}_t^\Gamma * \omega f.$$

Proof. We lift back to G by Theorem 7.2.1, then use Chapter 3, Theorem 3.3.3 and Theorem 3.2.5, and finally come back to $\Gamma\backslash G$ to conclude the proof.

We need the adjointness property of Casimir on $\Gamma\backslash G$.

Proposition 7.3.4.

$$\int_{\Gamma\backslash G} (\omega f_1)\, f_2 d\mu_{\Gamma\backslash G} = \int_{\Gamma\backslash G} f_1 \omega f_2 d\mu_{\Gamma\backslash G}.$$

So far, we need it only in the special case $f_2 = 1$. In general, the following are serviceable conditions: f_1 and f_2 are bounded of order 2 (i.e., up to second derivatives), etc.

The proof can be reduced to the case without gamma, namely with a cutoff function $\beta_N \in C_c^\infty(\Gamma\backslash G)$, $\beta_N \to 1$, written as $\beta_N = h_N^\Gamma$ with $h_N \in C_c^\infty(G)$. Then $\beta_N f_2$ tends to f_2, and

$$\begin{aligned}\int_{\Gamma\backslash G} (\omega f_1)\beta_N f_2 d\mu_{\Gamma\backslash G} &= \int_{\Gamma\backslash G} (\omega f_1)(h_N f_2)^\Gamma d\mu_{\Gamma\backslash G} \\ &= \int_G (\omega f_1)(h_N f_2)\, d\mu_{\Gamma\backslash G} \\ &= \int_G f_1 \omega(h_N f_2) d\mu_G\end{aligned}$$

and then winding back to $\Gamma\backslash G$. Then take the limit for $N \to \infty$.

It's shorter to express all this with the constant function $f_2 = 1$. The next result is the gammafication of Chapter 3, Corollary 3.2.4.

Lemma 7.3.5. *Let f_1 be a C^2 function on $\Gamma\backslash G/K$, and suppose its invariant derivatives up to order 2 restricted to the fundamental domain are bounded. Then*

$$\int_{\Gamma\backslash G} \omega(f_1) d\mu = 0.$$

In Section 3.2 we gave a formula

$$\omega(f^2) = 2(\tilde{Z}_1 f)^2 + 4(\tilde{Z}_2 f)^2 + 4(\tilde{Z}_3 f)^2 + 2f\omega f.$$

Next comes the gammafication of Chapter 3, Proposition 3.2.7.

Theorem 7.3.6. *Let $f_1 \in B(\Gamma\backslash G/K)$. Suppose that the invariant derivatives $\tilde{Z}_j f_1$ and $\tilde{Z}_j^2 f_1 (j = 1,2,3)$ are bounded on the fundamental domain $\mathscr{F}$. If $\omega f_1 = 0$, then f_1 is constant.*

Proof. By Lemma 4.2 and the formula for $\omega(f^2)$, we get 0 equal to the integral over $\mathscr{F}$ of the sum of the three squares in the above formula. Without loss of generality we may suppose that f is real, since the real and imaginary parts of f are annihilated by ω. Then the squares are $\geqq 0$, and the integral being 0 implies that each derivative is 0. By assumption of K-invariance, the derivatives $\tilde{Z}f$ are also 0 for $Z \in k$. Hence there is a basis of g for which all the derivatives with respect to the basis elements are 0, so f is constant.

7.4 Measure-Theoretic Estimate for Convolution on $\Gamma\backslash G$

In Chapter 2, Proposition 2.2.8, we gave measure-theoretic estimates for convolution on a locally compact group G. Here we give the corresponding estimate on $\Gamma\backslash G$. Again the statements are quite general, and apply to a unimodular group and a closed subgroup. We use the same notation Γ as in the application, and suppose the subgroup discrete, but one just replaces the sum by an integral over the subgroup to fit the general case. As in Chapter 2, we use convolution in the form

$$(h\#_G f)(z) = \int_G h(z^{-1}w)f(w)dw.$$

Proposition 7.4.1. *Let G be locally compact, unimodular, with discrete subgroup Γ. Let $h \in L^1(G)$, $\mathbf{K}_h(z,w) = h(z^{-1}w)$, and*

$$\mathbf{K}_h^\Gamma(z,w) = \sum_{\gamma\in\Gamma} h(z^{-1}\gamma w) = \sum_{\gamma\in\Gamma} \mathbf{K}_h(z,\gamma w).$$

Suppose $f \in L^p(\Gamma\backslash G)$ with $p \geqq 1$. Define

$$\mathbf{K}_{h^*\Gamma\backslash G}^\Gamma f)(z) = \int_{\Gamma\backslash G} \mathbf{K}_h^\Gamma(z,w)f(w)dw.$$

Then $\mathbf{K}_h^\Gamma *_{\Gamma\backslash G} f = h\#_G f$ is in $L^p(G)$, and

$$\|h\#f\|_{p,G} = \|\mathbf{K}_h^\Gamma * f\|_{p,\Gamma\backslash G} \leqq \|h\|_{1,G}\|f\|_{p,\Gamma\backslash G}.$$

Proof. Note that the lift of f to G is not in $L^p(G)$, so we cannot quote Chapter 2 Proposition 2.2.7; we have to redo the proof. By Chapter 2, Proposition 2.2.7, we have

$$\int_{\Gamma\backslash G} |\mathbf{K}_h^\Gamma(z,w)||f(w)|dw = \int_{\Gamma\backslash G} |\mathbf{K}_h^\Gamma(z,w)|^{1/p}|f(w)||\mathbf{K}_h^\Gamma(z,w)|^{1/q}dw$$

$$\leqq \left[\int_{\Gamma\backslash G} |\mathbf{K}_h^\Gamma(z,w)||f(w)|^p dw\right]^{1/p} \left[\int_{\Gamma\backslash G} |\mathbf{K}_h^\Gamma(z,w)|dw\right]^{1/q}. \quad (7.3)$$

We can lift the integral on the right to G by Theorem 7.2.1, and get

$$\left[\int_{\Gamma\backslash G} |K_h^\Gamma(z,w)|dw\right]^{1/q} = \left[\int_G |h(z^{-1}w)|dw\right]^{1/q} = \|h\|_{1,G}^{1/q}. \tag{7.4}$$

Raising the right side of (7.3) to the pth power and integrating over dz yields

$$\|\mathbf{K}_h^\Gamma * f\|_{p,\Gamma\backslash G}^p \leqq \int_{\Gamma\backslash G}\int_{\Gamma\backslash G} |\mathbf{K}_h^\Gamma(z,w)||f(w)|^p\, dw\, dz \cdot \|h\|_{1,G}^{p/q} \quad \text{[by Theorem 7.2.1]}$$

$$= \int_{\Gamma\backslash G}\int_G |h(z^{-1}w)|\,|f(w)|^p\, dz\, dw. \|h\|_{1,G}^{p/q}.$$

Now on the inner integral over G (not $\Gamma\backslash G$) we can use the invariance of Haar measure to yield the further equality

$$= \|h\|_{1,G} \int_{\Gamma\backslash G} \|f(w)\|^p dw \|h\|_{1,G}^{p/q}$$

$$= \|h\|_{1,G}^p \|f\|_{p,\Gamma\backslash G}^p \quad [\text{because} 1 + p/q = p].$$

This concludes the proof.

We may rephrase the estimate of Proposition 7.4.1 by saying that $L^1(G)$ acts on $L^p(\Gamma\backslash G)$ by convolution, and that the operator norm of the convolution $\mathscr{C}_h^\Gamma$ is bounded by the L^1-norm of h, or in symbols,

$$\|\mathscr{C}_h^\Gamma\|_{\text{op}} \leqq \|h\|_{1,G}.$$

It is convenient to put here another simple property of convolution. We need a definition whose significance will be apparent in the next chapter. We continue here with a general G, locally compact, unimodular with discrete subgroup Γ, and another closed unimodular subgroup U such that $\Gamma_U\backslash U$ is compact. A function $f \in L^2(\Gamma\backslash G)$ will be called (Γ, U)-**cuspidal** if for all $g \in G$ we have

$$\int_{\Gamma_U\backslash U} f(ug)du = 0.$$

Lemma 7.4.2. *With notation and terminology as above, let $h \in L^1(G)$ and*

$$\mathbf{K}_h(z,w) = h(z^{-1}w).$$

If $f \in L^2(\Gamma\backslash G)$ is (Γ, U)-cuspidal then so is $\mathbf{K}_h^\Gamma * f$.

Proof. By Theorem 7.2.1, we get

$$\begin{aligned}\int_{\Gamma\backslash G} \mathbf{K}_h^\Gamma(uz,w)f(w)dw &= \int_{\Gamma\backslash G} \sum_{\gamma\in\Gamma} \mathbf{K}_h(uz,\gamma w)f(\gamma w)dw \\ &= \int_G \mathbf{K}_h(uz,w)f(w)dw \\ &= \int_G \mathbf{K}_h(z,u^{-1}w)f(w)dw \\ &= \int_G \mathbf{K}_h(z,w)f(uw)dw.\end{aligned}$$

Then integrating over $\Gamma_U\backslash U$,

$$\int_{\Gamma_U\backslash U} \left(\mathbf{K}_h^\Gamma *_{\Gamma\backslash G} f\right)(uz)du = \int_{\Gamma_U\backslash U}\int_G \mathbf{K}_h(z,w)f(uw)\,dw\,du.$$

Interchanging the order of integration and using the cuspidality of f shows that the right side is 0. This proves the lemma.

7.5 Asymptotic Behavior of $\mathbf{K}_t^\Gamma$ for $t \to \infty$

Most of the time, we are interested in the behavior of the heat kernel for $t \to 0$, in connection with the Weierstrass–Dirac property. In a significant case, we are interested in the behavior as $t \to \infty$. *We let* $\mathrm{Vol}(\Gamma\backslash G)$ *be the volume of* $\Gamma\backslash G$ *with respect to which* $\mathbf{K}_t$ *is Dirac (so for us, polar, respectively Iwasawa).*

Theorem 7.5.1. *For fixed z, uniformly for* $w \in G/K$, *we have*

$$\lim_{t\to\infty} \mathbf{K}_t^\Gamma(z,w) = \mathrm{Vol}(\Gamma\backslash G)^{-1}.$$

We state at once the corollary that is relevant for the fixed-point eigenvalue question.

Theorem 7.5.2. *Let* $f \in L^1(\Gamma\backslash G)$. *If f is a heat eigenfunction with eigenvalue* 1, *that is,* $\mathbf{K}_t^\Gamma * f = f$ *for all t, then f is constant (the mean value of f).*

Proof. In the convolution integral limit

$$\lim_{t\to\infty} \int_{\Gamma\backslash G} \mathbf{K}_t^\Gamma(z,w)f(w)dw$$

we can take the limit under the integral, apply the theorem, and obtain

$$\frac{1}{\mathrm{Vol}(\Gamma\backslash G)} \int_{\Gamma\backslash G} f(w)dw = f(z),$$

which proves the result.

The question arises, at which level is Theorem 7.5.1 provable, i.e., is there a cheap proof, or does one have to go through a much more extensive structure? Of course, we already have a proof of Theorem 7.5.2 via Casimir, so we can go ahead to the rest of the book, until the eigenfunction expansion, Theorem 11.4.7 of Chapter 11, to prove the limit in Theorem 7.5.1 above. Note that if the limit exists, then it has to be as stated, as one sees by integrating both sides over $\Gamma\backslash G$.

Chapter 8
Heat Kernel Convolution on $L^2_{\mathrm{cusp}}(\Gamma\backslash G/K)$

The noncompactness of $\Gamma\backslash G$ is seen clearly on the fundamental domain, which goes to infinity in a vertical piece. We need to isolate certain phenomena concerning functions that may diverge in this piece, called a cusp. Note that the length function in the cusp decreases like $1/y$ times the Euclidean length, so what looks like a Euclidean tube actually has a width that shrinks toward 0 going up the cusp.

A function on $\Gamma\backslash G$ is in particular Γ_U-invariant, and hence has a Fourier series (if at all reasonable, certainly an L^2-function). We say that the function is **cuspidal** if the constant term of the Fourier series is 0, that is,

$$\int_{\Gamma_U\backslash U} f(ug)du = 0 \text{ for all } g \in G.$$

It turns out that whatever divergence one doesn't like is due to the concerned function being noncuspidal. We shall make this idea precise in what follows.

Let $L^2_{\mathrm{cusp}}(\Gamma\backslash G/K)$ be the cuspidal subspace of $L^2(\Gamma\backslash G/K)$. By Chapter 7, Lemma 7.4.2, we know that convolution with the heat kernel maps this space into itself. We shall prove that it is a compact operator. We review some basic criteria for such operators in Section 8.1, and then apply them to the concrete situation to get the part of the heat kernel expansion that comes from this convolution operation on the cuspidal space.

We also determine a Hilbert basis for the cuspidal subspace that consists of eigenvectors both for the heat convolution and for the Casimir operator. The corresponding expansion of the heat Gaussian gives rise to one component of the theta relation we are after, namely most of the discrete part of this theta relation. In our concrete case, the only other discrete part consists of the constants.

In the subsequent sections, we use systematically the Iwasawa decomposition

$$z = u_z a_z k_z \text{ or } u(z)a(z)k(z),$$

with

$$u_z = u(x_z) = \begin{pmatrix} 1 & x_z \\ 0 & 1 \end{pmatrix} \text{ and } a_z = \begin{pmatrix} y_z^{1/2} & 0 \\ 0 & y_z^{-1/2} \end{pmatrix},$$

so $y_z = a_z^\alpha$ with the character that we denote by α.

We recall that Γ_∞ is the upper triangular subgroup of Γ.

In Section 8.2, we give an estimate for the $(\Gamma - \Gamma_\infty)$-periodization on a Siegel set, taken concretely to be the subset of $\mathcal{F}$ above Y, namely

$$\mathcal{F}_{\geqq Y} = \mathcal{F} - \mathcal{F}_{<Y} = \{z \in \mathcal{F} \text{ with } y_z \geqq Y\},$$

with $Y > 1$.

In Section 8.3, we give an estimate for the Γ_∞ or Γ_U periodization, using Fourier series estimates.

In Section 8.4 we get the main estimate, which shows that convolution with the heat kernel on $\Gamma\backslash G/K$ is a compact operator on the cuspidal subspace of $L^2(\Gamma\backslash G/K)$. We note that in the Γ-cocompact case, the ensuing eigenspace decomposition is due to Tamagawa in a general setting [Tam 60]. Gangolli [Gan 68] used this decomposition to get at his trace formula in this cocompact case.

Gelfand–Piatetskii-Shapiro [GePS 63] proved the compactness of the convolution operator with $\varphi \in C_c^\infty(G)$, followed by Langlands [Lgld 66], [Lgld 76], Section 8.3. Godement [God 66], [God 67] followed Langlands except for an elegant use of the Poisson formula in the fundamental estimate for the arithmetic case. This proof was the one used in [Lan 75/85]. Borel–Garland [BoG 83] showed that for any function $\varphi \in L^1(G)$, the convolution operator is compact on the discrete part of $L^2(\Gamma\backslash G)$, in the general context of semisimple Lie groups. Actually, the cuspidal part of this theorem is an immediate consequence of [GePS 63]. Indeed, an arbitrary function in L^1 can be L^1-approximated by a sequence $\{\varphi_n\}$ in C_c^∞, and one has the elementary measure-theoretic inequality

$$\|\varphi * f\|_2 \leqq \|\varphi\|_1 \|f\|_2 \text{ for } \varphi \in L^1(G), f \in L^2(\Gamma\backslash G).$$

Thus the convolution operator norm is bounded by the L^1-norm, so the L^1-approximation is also an operator approximation, whence the limit operator is compact.

In this chapter, we give a still different proof exhibiting more directly the use of the cuspidality hypothesis.

8.1 General Criteria for Compactness

We start summarily with some general results of functional analysis, as in [Lan 75/85], Sections 1.3 and 12.4. For the basic definitions and properties of compact operators, see for instance [Lan 93], Chapter 17. If T is a continuous linear operator, we let $|T|_{\text{op}}$ be its operator norm.

Theorem 8.1.1. *Let* $(X, \mathscr{M}_X, dx)$ *and* $(Y, \mathscr{M}_Y, dy)$ *be measured spaces. Assume that* $L^2(X)$, $L^2(Y)$ *have countable orthogonal bases. Let* $Q \in L^2(X \times Y, dx \otimes dy)$. *Then the operator* T_Q *defined on* $f \in L^2(Y)$ *by*

$$T_Q f(x) = \int_Y Q(x,y) f(y) dy$$

is a bounded operator from $L^2(Y)$ *to* $L^2(X)$, *and is compact. We have*

$$|T_Q|_{\text{op}} \leqq \|Q\|_2.$$

Proof. Let $f \in L^2(Y)$. For almost all x our assumption implies that the function Q_x such that $Q_x(y) = Q(x,y)$ is also in $L^2(Y)$. Hence the product fQ_x is in $L^1(Y)$. By Schwarz,

$$|T_Q f(x)|^2 = \left| \int_Y Q(x,y) f(y) dy \right|^2 \leqq \|Q_x\|_2^2 \|f\|_2^2.$$

Integrating gives

$$\|T_Q f\|_2^2 = \int |T_Q f(x)|^2 dx \leqq \|f\|_2^2 \int \int |Q(x,y)|^2 dy dx$$
$$\leqq \|Q\|_2^2 \|f\|_2^2.$$

This proves that $|T_Q|_{\text{op}} \leqq \|Q\|_2$, so Q is a bounded operator with the stated norm bound.

As for compactness, let $\{\varphi_i\}$, $\{\psi_j\}$ be orthonormal bases for $L^2(X)$, $L^2(Y)$ respectively. Then $\{\varphi_i \otimes \psi_j\}$ is an orthonormal basis for $L^2(X \times Y)$. Let

$$Q = \sum c_{ij} \varphi_i \oplus \psi_j$$

be the L^2-series expansion for Q. For each positive integer n, let

$$Q_n = \sum_{i,j=1}^{n} \varphi_i \oplus \psi_j.$$

Then the corresponding operator T_{Q_n} has finite-dimensional image (spanned by the functions $\varphi_1, \ldots, \varphi_n$), and so is compact.

By the first part of the proof, we know that

$$|T_Q - T_{Q_n}|_{\text{op}} \leqq \|Q_n - Q\|_2 \to 0 \text{ as } n \to \infty.$$

Hence the operators T_{Q_n} converge in operator norm to T_Q, which is therefore also compact. This proves Theorem 8.1.1.

Theorem 8.1.2. *Let* X *be a locally compact space with a finite positive measure* dx. *Let* H *be a closed subspace of* $L^2(X, dx)$. *Let*

$$T : H \to \text{BC}(X)$$

be a continuous linear map of H into the Banach space of bounded continuous functions on X, with the sup norm, so there exists $C > 0$ such that for all $f \in H$,

$$\|Tf\|_\infty \leqq C\|f\|_2.$$

Then $T\colon H \to L^2(X)$ is equal to T_Q for some $Q \in L^2(X \times X)$, and is therefore compact (Theorem 8.1.1).

Proof. For each $x \in X$, the functional

$$f \mapsto (Tf)(x)$$

is continuous linear on H. Hence there exists a function $q_x \in H$ such that for all $f \in H$ we have

$$(Tf)(x)\langle f, q_x\rangle = \int f(y)\overline{q_x(y)}dy.$$

Since Tf is a continuous function on X, this also shows that $x \mapsto q_x$ as a map from X into $H \subset L^2(X)$ is weakly continuous, and therefore weakly measurable.

Furthermore, its image is bounded in $L^2(X)$ because

$$\|q_x\|_2^2 = (Tq_x)(x) \leqq C\|q_x\|_2.$$

The theorem then follows from the next lemma.

Lemma 8.1.3. *Let X, Y be finite measured spaces such that the σ-algebras of measurable sets in Y is generated by a countable subalgebra. Given a weakly measurable map*

$$q : X \to L^2(Y)$$

whose image is L^2-bounded, there exists $Q \in L^2(X \times Y)$ such that for almost all $x \in X$ we have

$$q_x(y) = Q(x,y)$$

for all $y \notin S_x$, where S_x is a set of measure 0 in Y, depending on x.

Proof. Consider the space of step functions g on $X \times Y$ with respect to "rectangles," i.e., products of measurable sets in X, Y respectively. The map

$$g \mapsto \int\int g(x,y)q_x(y)\,dy\,dx$$

is L^2-continuous because by the Schwarz inequality,

$$\left|\int\int g(x,y)q_x(y)\,dy\,dx\right|^2 \leqq \|g\|_2^2 \int\int |q_x(y)|^2\,dy\,dx.$$

Hence there exists $Q \in L^2(X \times Y)$ such that for all characteristic functions φ, ψ of measurable sets in X, Y respectively, we have

$$\int \varphi(x)\left[\int \psi(y)\overline{q_x(y)}dy\right]dx = \int \varphi(x)\left[\int \psi(y)\overline{Q(x,y)}dy\right]dx.$$

For each ψ there exists a null set Z_ψ in X such that if $x \notin Z_\psi$, then

$$\int \psi(y)\overline{q_x(y)}dy = \int \psi(y)\overline{Q(x,y)}dy.$$

Using countably many ψ's, we get this relation for all x outside a set Z of measure 0 in X. Hence for $x \in Z$, we get $q_x(y) = Q(x,y)$ for all $y \notin S_x$, where S_x is a set of measure 0 in Y. This proves the lemma.

Actually we used the measurability of the scalar product, and for the few lines needed to justify this, see [Lan 75/85], p. 233.

8.2 Estimates for the $(\Gamma - \Gamma_\infty)$-Periodization

For various subsets Γ' of Γ, we shall consider the Γ'-periodization defined by

$$\mathbf{K}_\varphi^{\Gamma'}(z,w) = \sum_{\gamma\in\Gamma'} \varphi(z^{-1}\gamma w).$$

If Γ' is closed under taking the inverse, then $\mathbf{K}_\varphi^{\Gamma'}$ is also symmetric.

The subsets here of concern to us will be Γ, Γ_∞, Γ_U, and $\Gamma - \Gamma_\infty$. Thus we are especially interested in subsets relevant to cuspidal behavior. Except for Γ, these subsets are not $\mathbf{c}(\Gamma)$-invariant. Thus we go beyond previous methods by using the more refined structure of the fundamental domain and the cusp.

This section deals with estimates for z in the fundamental domain $\mathscr{F}$ of Chapter 6, and y_z going to infinity. For $Y > 1$, let as before

$$\mathscr{F}_{\geqq Y} = \{z \in \mathscr{F} \text{ such that } y_z \geqq Y\}.$$

Lemma 8.2.1. *Uniformly for z, $w \in \mathscr{F}_{\geqq Y}$ and $\gamma \in \Gamma$, we have*

$$\|z^{-1}\gamma w\| \gg_Y \ll \|a_z^{-1}\gamma a_w\|.$$

The Y in the middle means that the implied constants in the inequalities depend only on Y and not on z, w, γ.

Proof. We use Iwasawa coordinates $z = u_z a_z k_z$. Since the Hermitian norm is K-bi-invariant, the inquality to be proved amounts to

$$\|a_z^{-1}u_z^{-1}\gamma u_w a_w\| \gg_Y \ll \|a_z^{-1}\gamma a_w\|.$$

The left side can be written

$$\|a_z^{-1}u_z^{-1}a_z a_z^{-1}\gamma\, a_w a_w^{-1}u_w a_w\|.$$

But $a^{-1}u(x)a = u(x/y_a)$, so for $y_a \geqq Y$, the terms with conjugations of u-components are bounded, as well as their inverses. They can thus be replaced by suitable constants, thus yielding the desired inequality involving only $a_z^{-1}\gamma a_w$, proving the lemma.

So far, we have not met Γ_∞, which comes into the picture in the next lemma.

Lemma 8.2.2. *Given $Y > 1$, there exists a constant C_Y'' and a positive integer M'' such that for all z, $w \in \mathscr{F}_{\geqq Y}$ and $\gamma \in \Gamma - \Gamma_\infty$ we have for all $T \geqq 1$,*

$$\#\{\gamma \in \Gamma - \Gamma_\infty \text{ such that } \|a_z^{-1}\gamma a_w\| \leqq T\} \leqq C_Y'' T^{M''}.$$

For instance $M'' = 14$ will do.

Proof. Write

$$\gamma = \begin{pmatrix} m_1 & m_2 \\ m_3 & m_4 \end{pmatrix} \text{ with } m_i \in o, \text{ so } 1 \leqq |m_i| \text{ if } m_i \neq 0.$$

By hypothesis, $m_3 \neq 0$. By matrix multiplication,

$$a_z^{-1}\gamma a_w = \begin{pmatrix} y_z^{-1/2} m_1 y_w^{1/2} & y_z^{-1/2} m_2 y_w^{-1/2} \\ y_z^{1/2} m_3 y_w^{1/2} & y_z^{1/2} m_4 y_w^{-1/2} \end{pmatrix}.$$

From $\|a_z^{-1}\gamma a_w\| \leqq T$, we obtain

$$1 \leqq |m_3| \leqq T/(y_z y_w)^{1/2} \leqq \frac{1}{Y}T.$$

Then multiplying $y_z^{1/2} m_3 y_w^{1/2}$ by each of the other components, we get the bounds

$$\begin{aligned} |m_1 m_3| &\leqq \frac{1}{y_w}T^2, &&\text{so} \quad |m_1| \leqq \frac{1}{Y}T^2; \\ |m_2 m_3| &\leqq T^2, &&\text{so} \quad |m_2| \leqq T^2; \\ |m_4 m_3| &\leqq \frac{1}{y_z}T^2, &&\text{so} \quad |m_4| \leqq \frac{1}{Y}T^2. \end{aligned}$$

The bound as stated in the lemma is then immediate, because the number of lattice points in a disk of radius R is $\ll R^2$.

By Proposition 4.1.2 of Chapter 4, the Iwasawa component and the polar component have the same order of magnitude in a Siegel set. Thus the above estimate can be used to estimate functions that are K-bi-invariant, and depend only on the polar component of an element in a set $\mathscr{F}_{\geqq Y}$.

Proposition 8.2.3. *Let $Y > 1$. Let φ be a basic Gaussian. There exists a constant $C_\varphi(\Gamma - \Gamma_\infty, \mathscr{F}_{\geqq Y})$ such that for all z, $w \in \mathscr{F}_{\geqq Y}$ we have*

$$\mathbf{K}_\varphi^{\Gamma-\Gamma_\infty}(z,w) = \sum_{\gamma \in \Gamma - \Gamma_\infty} \varphi(z^{-1}\gamma w) \leqq C_\varphi(\Gamma - \Gamma_\infty, \mathscr{F}_{\geqq Y}).$$

Proof. The number of $\gamma \in \Gamma - \Gamma_\infty$ such that for a given integer T we have

$$T \leqq \|z^{-1}\gamma w\| \leqq T+1$$

has the order of magnitude as stated in Lemma 8.2.2. Hence we get the estimate

$$\begin{aligned}\sum_{\gamma \in \Gamma - \Gamma_\infty} \varphi(z^{-1}\gamma w) &\ll \sum_{T=1}^{\infty} \sum_{\|z^{-1}\gamma w\| \leqq T} \varphi(z^{-1}\gamma w) \\ &\ll \sum_{T=1}^{\infty} e^{-c'(\log T)^2}\end{aligned}$$

with some constant c', thereby proving the proposition.

8.3 Fourier Series for the Γ''_U, Γ_∞-Periodizations of Gaussians

8.3.1 Preliminaries: The Γ''_U and Γ_∞-Periodizations

We shall also need an estimate for the Γ''_U or Γ_∞-trace when z, $w \in \mathscr{F}_{\geqq Y}$. Let φ be a basic Gaussian and consider the Γ_U-trace

$$\begin{aligned}\mathbf{K}_\varphi^{\Gamma_U}(z,w) &= \sum_{\gamma \in \Gamma_U} \varphi(z^{-1}\gamma w) = \sum_{m \in \mathfrak{o}} \varphi(z^{-1}u(m)w) \\ &= \sum_{m \in \mathfrak{o}} \varphi(a_z^{-1}u(x_w - x_z + m)a_w),\end{aligned} \tag{8.1}$$

using $z = u(x_z)a_z k_z$, and similarly for w. Abbreviate

$$\varphi_{y_z \cdot y_w}(x) = \varphi(a_z^{-1}u(x)a_w).$$

Then the periodization of $\mathbf{K}_\varphi(z,w)$ can be expressed in terms of the one-variable periodization

$$\mathbf{K}_\varphi^{\Gamma_U}(z,w) = \sum_{m \in \mathfrak{o}} \varphi_{y_z, y_w}(x_w - x_z + m) = \varphi_{y_z, y_w}^{\Gamma_u}(x_w - x_z). \tag{8.2}$$

Let R be the reflection matrix $\text{diag}(\mathbf{i}, -\mathbf{i})$, $R(x+\mathbf{j}y) = -x+\mathbf{j}y$. Write $u_m = u(m)$. One computes trivially

$$\varphi(z^{-1}Ru_m w) = \varphi(a_z^{-1}u_z u_m u_w a_w) = \varphi(a_z^{-1}u(x_w + x_z + m)a_w), \tag{8.3}$$

by $a_z^{-1}u_z^{-1}R = a_z^{-1}RR^{-1}u_z^{-1}R = \pm Ra_z^{-1}u_z$ because R, a_z commute, and $R \in \mathrm{K}$ so gets killed on the left by the K-bi-invariance of φ. Since $\pm\Gamma_U\backslash\Gamma_\infty$ is represented by $\pm R$, we get the formula

$$\mathbf{K}_\varphi^{\Gamma_\infty}(z,w) = 2\varphi_{y_z,y_w}^{\Gamma_U}(x_w - x_z) + 2\varphi_{y_z,y_w}^{\Gamma_U}(x_w + x_z). \tag{8.4}$$

8.3.2 The Fourier Series

The function $\varphi_{y_z,y_w}^{\Gamma_U}$ is periodic, with period lattice $o = \mathbf{Z}[\mathbf{i}]$. Hence it has a Fourier series

$$\varphi_{y_z,y_w}^{\Gamma_U}(x) = \sum_{m\in o} F_m(y_z,y_w)e^{2\pi\mathbf{i}\langle x,m\rangle}, \tag{8.5}$$

where $\langle x,m\rangle = x_1m_1 + x_2m_2$ is the standard dot product, and the Fourier coefficient F_m is given by the usual integral

$$\begin{aligned} F_m(y_z,y_w) &= \int_{o\backslash\mathbf{C}} \varphi_{y_z,y_w}^{\Gamma_U}(x)e^{-2\pi\mathbf{i}\langle x,m\rangle}dx \\ &= \int_{o\backslash\mathbf{C}} \sum_{\ell\in o} \varphi_{y_z,y_w}(x+\ell)e^{-2\pi\mathbf{i}\langle x,m\rangle}dx \\ &= \int_{\mathbf{C}} \varphi_{y_z,y_w}(x)e^{-2\pi\mathbf{i}\langle x,m\rangle}dx. \end{aligned} \tag{8.6}$$

We summarize these formulas in the expression

$$\mathbf{K}_\varphi^{\Gamma_U}(z,w) = \sum_{m\in o} \varphi(z^{-1}u(m)w) = \sum_{m\in o} F_m(y_z,y_w)e^{2\pi i\langle x_w - x_z, m\rangle}. \tag{8.7}$$

The **constant term** of this Fourier series is

$$F_0(y_z,y_w) = \int_{\mathbf{C}} \varphi_{y_z,y_w}(x)dx = \int_{\mathbf{C}} \varphi(a_z^{-1}u(x)a_w)dx.$$

In light of (8.4), the constant term for the Fourier series of $\mathbf{K}_\varphi^{\Gamma_\infty}(z,w)$ is

$$4F_0(y_z,y_w).$$

Proposition 8.3.1. *Let φ be a basic Gaussian. Let $Y > 1$. Then there exists a constant $C(\Gamma_U, \mathscr{F}_{\geqq Y})$ such that for all z, $w \in \mathscr{F}$ we have*

$$\left|\mathbf{K}_\varphi^{\Gamma_U}(z,w) - F_0(y_z,y_w)\right| \leqq C_\varphi(\Gamma_U,\mathscr{F}_{\geqq Y})$$

and

$$\left|\mathbf{K}_\varphi^{\Gamma_\infty}(z,w) - 4F_0(y_z,y_w)\right| \leqq 4C_\varphi(\Gamma_U,\mathscr{F}_{\geqq Y})$$

Proposition 8.3.1 is an immediate consequence of Proposition 8.3.2 below, involving estimates for all the nonzero Fourier coefficients, with appropriate uniformities.

Proposition 8.3.2. *Given a positive integer N, and the basic Gaussian φ, there exists a constant C'_N such that for all z, $w \in \mathscr{F}$, we have the estimate of the Fourier coefficients with $m \neq 0$,*

$$|F_m(y_z,y_w)| \leqq C'_N |m|^{-N}.$$

So for N sufficiently large ($\geqq 3$) there is a constant C''_N such that

$$\left|\sum_{m\neq 0} F_m(y_z,y_m)e^{2\pi\mathbf{i}\langle x_w - x_z, m\rangle}\right| \leqq C''_N,$$

with $C''_N = C'_N \sum_{m\neq 0} |m|^{-N}$.

Proof. We integrate by parts the formula (8.6) for the Fourier coefficients, putting $\partial_1 = \partial_{x_1}$, $\partial_2 = \partial_{x_2}$, to get (up to ± 1)

$$\begin{aligned}
F_m(y_z,y_w) &= \int_{\mathbf{R}^2} \varphi_{y_z,y_w}(x) e^{-2\pi\mathbf{i}(x_1m_1+x_2m_2)}dx_1\,dx_2 \\
&= \int_{\mathbf{R}^2} \varphi_{y_z,y_w}(x)\partial_1^{j_1}\partial_2^{j}2\frac{1}{(2\pi\mathbf{i}m_1)^{j_1}}\frac{1}{(2\pi\mathbf{i}m_2)^{j_2}}e^{-2\pi\mathbf{i}(x_1m_1+x_2m_2)}dx_1\,dx_2 \\
&= \frac{1}{(2\pi\mathbf{i}m_1)^{j_1}(2\pi\mathbf{i}m_2)^{j_2}}\int_{\mathbf{R}^2} \partial_1^{j_1}\partial_2^{j_2}\varphi_{y_z,y_w}(x)e^{-2\pi\mathbf{i}(x_1m_1+x_2m_2)}dx_1dx_2.
\end{aligned}$$

We now need a lemma giving us uniformity.

Lemma 8.3.3. *Fix j_1, j_2. Then*

$$\partial_1^{j_1}\partial_2^{j_2}\varphi_{y_z,y_w} \in L^1(\mathbf{R}^2)(=L^1(\mathbf{C}))$$

uniformly for y_z, $y_w \geqq 1$, say, i.e., there is a constant $C(j_1,\ j_2,\varphi)$ such that

$$\|\partial_1^{j_1}\partial_2^{j_2\varphi_{y_z,y_w}}\|_1 \leqq C(j_1,j_2,\varphi).$$

Proof. The result follows directly from Chapter 3, Proposition 3.1.4(c) and Lemma 3.1.1.

8.4 The Convolution Cuspidal Estimate

We put together the previous estimates for a special type of function. We define a function f on $\Gamma\backslash G$ to be **cuspidal** if

$$\int_{\Gamma_U\backslash U} f(uz)du = 0 \text{ for all } z \in G.$$

The subspace of $\mathrm{L}^2(\Gamma\backslash G/K)$ consisting of cuspidal functions will be denoted by

$$L^2_{\text{cusp}}(\Gamma\backslash G/K).$$

Theorem 8.4.1. *Let φ be a basic Gaussian, $G = \mathrm{SL}_2(\mathbf{C})$, $\Gamma = \mathrm{SL}_2(\mathbf{Z}[\mathbf{i}])$. Let $\mathbf{K}_\varphi$ be the associated point pair invariant. Then there exists a constant C such that for all $f \in L^2_{\text{cusp}}(\Gamma\backslash G/K)$ we have*

$$\|\mathbf{K}^\Gamma_\varphi * f\|_\infty \leqq C\|f\|_2.$$

Proof. Fix $Y > 1$. We have

$$\begin{aligned}
\|\mathbf{K}^\Gamma_\varphi * f\|_\infty &\leqq \sup_{z\in\mathscr{F}} \left|\int_{\mathscr{F}} \mathbf{K}^\Gamma_\varphi(z,w)f(w)d\mu(w)\right| \\
&\leqq \sup_{z\leqq\mathscr{F}_{\leqq Y}} + \sup_{z\in\mathscr{F}_{\geqq Y}} \left|\int_{\mathscr{F}} \mathbf{K}^\Gamma_\varphi(z,w)f(w)d\mu(w)\right| \\
&\leqq C_1 \mathrm{vol}(\mathscr{F})^{1/2}\|f\|_2 + \sup_{z\in\mathscr{F}_{\geqq Y}} \left|\int_{\mathscr{F}} \mathbf{K}^\Gamma_\varphi(z,w)f(w)d\mu(w)\right|
\end{aligned}$$

by Chapter 7, Proposition 7.1.5. Because $\mathscr{F}_{\leqq Y}$ is compact, $C_1 = C_\varphi(\Gamma,\mathscr{F}_{\leqq Y})$. There remains to estimate the second integral:

$$\int_{\mathscr{F}} \mathbf{K}^\Gamma_\varphi f = \int_{\mathscr{F}_{\leqq Y}} \mathbf{K}^\Gamma_\varphi f + \int_{\mathscr{F}_{\geqq Y}} \mathbf{K}^\Gamma_\varphi f.$$

Chapter 7, Proposition 7.1.5, again estimates the first integral on the right by $C_1\mathrm{vol}(\mathscr{F})^{1/2}\|f\|_2$.

We estimate the second integral on the right by decomposing $\mathbf{K}^\Gamma_\varphi$ as the sum

$$\mathbf{K}^\Gamma_\varphi = \mathbf{K}^{\Gamma-\Gamma_\infty}_\varphi + \left(\mathbf{K}^{\Gamma_\infty}_\varphi - 4F_0\right) + 4F_0.$$

We then have to estimate each term of the sum

$$\int_{\mathscr{F}_{\geqq Y}} \mathbf{K}^\Gamma_\varphi f = \int_{\mathscr{F}_{\geqq Y}} \mathbf{K}^{\Gamma-\Gamma_\infty}_\varphi f + \int_{\mathscr{F}_{\geqq Y}} (\mathbf{K}^{\Gamma_\infty}_\varphi - 4F_0)f + \int_{\mathscr{F}_{\geqq Y}} 4F_0 f.$$

The last integral $\int F_0 f$ is 0, that is,

$$\int_{\mathscr{F}_{\geqq Y}} F_0(y_z, y_w) f(w) d\mu(w) = 0,$$

since in terms of the coordinates, it is a repeated integral

$$\int_Y^\infty \int_{\mathscr{F}_U} = \int_Y^\infty \int_{\Gamma_\infty \backslash U} = \int_Y^\infty \frac{1}{2} \int_{\Gamma_U \backslash U},$$

whose vanishing is immediate by the cuspidality assumption on f, i.e., the partial integral of f over $\Gamma_U \backslash U$ is 0.

The integral $\int \mathbf{K}_\varphi^{\Gamma - \Gamma_\infty} f$ is estimated by $C_2 \mathrm{vol}(\mathscr{F})^{1/2} \|f\|_2$, using Proposition 8.2.3, so that $C_2 = C_\varphi(\Gamma - \Gamma_\infty, \mathscr{F}_{\geqq Y})$.

The integral $\int (\mathbf{K}_\varphi^{\Gamma_\infty} - 4F_0) f$ is estimated by $C_3 \mathrm{vol}(\mathscr{F})^{1/2} \|f\|_2$, using Proposition 8.3.1 with $C_3 = 4C_\varphi(\Gamma_U, \mathscr{F}_{\geqq Y})$.

This concludes the proof. Note that we obtain an explicit determination with

$$C = (2C_1 + C_2 + C_3) \mathrm{vol}(\mathscr{F})^{-1/2}.$$

8.5 Application to the Heat Kernel

All the work has been done. We put it together for the payoff.

Theorem 8.5.1. *(a) For each $t > 0$, the convolution $f \mapsto \mathbf{K}_t^\Gamma * f$ maps*

$$L^2_{\mathrm{cus}}(\Gamma \backslash G/k) \to \mathrm{BC}(\Gamma \backslash G/K)$$

into the space of bounded continuous functions.

(b) It is a bounded operator for the sup norm on $\mathrm{BC}(\Gamma \backslash G/K)$ and the L^2-norm on L^2_{cus}.

(c) The image is contained in $L^2_{\mathrm{cus}}(\Gamma \backslash G/K)$, and the convolution map is a compact operator denoted by

$$\mathscr{C}^\Gamma_{t,\mathrm{cus}} : L^2_{\mathrm{cus}}(\Gamma \backslash G/K) \to L^2_{\mathrm{cus}}(\Gamma \backslash G/K)$$

with the L^2-norm on both sides.

(d) For each $t > 0$ this operator is injective.

Proof. First, Theorem 8.4.1 shows that $\mathbf{K}_t^\Gamma * f$ is bounded, so in L^2 since $\Gamma \backslash G$ has finite measure. The continuity of $\mathbf{K}_t^\Gamma * f$ in z comes from Chapter 7, Proposition 7.2.3, and is not a peculiarity of cuspidality.

Since $\Gamma\backslash G$ has finite measure, a bounded measurable function on $\Gamma\backslash G$ is in L^2. Also the L^2-norm is bounded by a constant times the sup norm. Again by Theorem 8.4.1, this implies that convolution with $\mathbf{K}_t^\Gamma$ is continuous for the L^2 norm on both sides. Theorem 8.4.1 and Theorem 8.1.2 show that the convolution operator is compact. That the convolution maps the cuspidal space into the cuspidal space was proved in a more general context in Chapter 5, Lemma 4.2. The injectivity of heat convolution was proved more generally in Chapter 2, Theorem 2.3.2. This concludes the proof.

In this section we deal only with L^2_{cus}, so for simplicity of notation, we omit the subscript cus and write simply $\mathscr{C}_t^\Gamma$ instead of $\mathscr{C}_{t,\text{cus}}^\Gamma$. The family of operators $\{\mathscr{C}_t^\Gamma\}$ will be called the **heat cuspidal family**, and each $\mathscr{C}_t^\Gamma$ is called a **heat cuspidal operator**.

We now deduce the decomposition of $L^2_{\text{cus}}(\Gamma\backslash G/K)$ into eigenspaces, assuming the basic theory of compact operators on Hilbert space H, e.g., [Lan 93], Chapter 17. Let $\mathscr{C}$ be a compact operator of H into itself. Then there is an orthogonal direct sum

$$H = \text{Null}(\mathscr{C}) \oplus \overline{\bigoplus} H_j,$$

where the H_j are finite-dimensional subspaces that are stable under $\mathscr{C}$ and for each j there is a complex number $\eta_j \neq 0$ such that $\mathscr{C} - \eta_j I$ is nilpotent on H_j, the numbers η_j are distinct, say with decreasing absolute value, tending to 0 as $j \to \infty$. The space $\text{Null}(\mathscr{C})$ is the kernel of $\mathscr{C}$ in the sense of algebra. If the operator $\mathscr{C}$ is Hermitian symmetric on each H_j, then H_j is an eigenspace for $\mathscr{C}$, i.e., each element is an eigenvector with eigenvalue η_j.

Let

$$H = L^2_{\text{cus}}(\Gamma\backslash G/K)$$

be the cuspidal subspace of $L^2(\Gamma\backslash G/K)$. We already know from Theorem 8.5.1 that the null space of $\mathscr{C}_t^\Gamma$ is $\{0\}$. Let

H_{BC} = subspace of H consisting of the bounded continuous cuspidal functions. By Theorem 8.4.1, $\mathscr{C}_t^\Gamma$ maps H into H_{BC} and thus $\mathscr{C}_t^\Gamma$ maps H_{BC} into itself. In particular, let f be an eigenfunction of $\mathscr{C}_t^\Gamma$ in H, with nonzero eigenvalue η,

$$\mathscr{C}_t^\Gamma f = \eta f,\ \eta \neq 0.$$

Letting $H_j(t)$ be an eigenspace of $\mathscr{C}_t^\Gamma$, we have the following result:

Theorem 8.5.2. *Let $f \in H = L^2_{\text{cus}}(\Gamma\backslash G/K)$ be an eigenfunction of $\mathscr{C}_t^\Gamma$ for some $t > 0$. Then f is bounded $\mathscr{C}^\infty$, so for each j, $H_j(t) \subset H_{\text{BC}^\infty}$.*

Proof. That f is bounded continuous merely summarizes the above remarks. For the $\mathscr{C}^\infty$ we simply lift the function to G and apply Chapter 3, Theorem 3.3.2, to conclude the proof.

Theorem 8.5.3. *For all $t,\ t' > 0$, the eigenspaces of $\mathscr{C}_t^\Gamma$ and $\mathscr{C}_{t'}^\Gamma$, coincide. The Hilbert space $L^2_{\text{cus}}(\Gamma\backslash G/K)$ is the orthogonal Hilbert space sum of the eigenspaces H_j. An eigenfunction ψ of $\mathscr{C}_t^\Gamma$ has a nonzero eigenvalue*

$$\mathbf{K}_t^{\Gamma} * \psi = e^{-\lambda t}\psi,$$

with λ real > 0. Such $-\lambda$ is the eigenvalue of Casimir on ψ, $\omega\psi = -\lambda\psi$.

Proof. Since $\mathbf{K}_t^{\Gamma}(z,w)$ is symmetric in (z,w), and real positive, it follows that the convolution operator is Hermitian positive semidefinite on each eigenspace $H_j(t)$. Hence each eigenvalue $\eta(t)$ is real > 0, and $\mathscr{C}_t^{\Gamma}$ can be diagonalized on each $H_j(t)$, which thus has an orthonormal basis of $\mathscr{C}_t^{\Gamma}$-eigenfunctions with eigenvalue $\eta_j(t)$. For t', $t > 0$ the operators $\mathscr{C}_t^{\Gamma}$, $\mathscr{C}_{t'}^{\Gamma}$ commute, and by the eigenvalue property, they leave invariant each other's eigenspaces. Fixing t, for each t' an eigenspace for $\mathscr{C}_t^{\Gamma}$ decomposes into a direct sum of eigenspaces for the $\mathscr{C}_{t'}^{\Gamma}$, action. After a finite number of steps using a finite number of values for t' we see that there exists an orthonormal basis for $H_j(t)$ consisting of heat eigenvectors (i.e., eigenvectors for all $\mathscr{C}_{t'}^{\Gamma}$, with all $t' > 0$). By Chapter 3, Theorem 3.4.2, we know that the eigenvalues have the form

$$\eta(t) = e^{-\lambda t},$$

where $-\lambda$ is the eigenvalue of Casimir.

By Theorem 8.5.2, we know that ψ is bounded. By Chapter 7, Theorem 7.2.9(b), we know that $\lambda > 0$. We summarize what we have proved so far.

Now we also get the rest of Theorem 8.5.3, namely the common eigenspaces for two values t, t' are precisely those generated by the eigenfunctions ψ_n for which the corresponding λ_n are equal. This concludes the proof of Theorem 8.5.3.

Theorem 8.5.4. *There exists an orthonormal Hilbert basis $\{\psi_n\}$ of $L^2_{\text{cusp}}(\Gamma\backslash G/K)$ consisting of heat semigroup eigenfunctions, with eigenvalues*

$$\eta_n(t) = e^{-\lambda_n t},$$

with $0 < \lambda_n \leqq \lambda_{n+1}$ for all n, and $\lambda_n \to \infty$.

Proof. The last assertion is immediate from the general fact that the eigenvalues of a compact operator tend to 0, so the λ_n can be ordered so that they tend to ∞.

Theorem 8.5.5. *The L^2-function $Q_t(z,w)$ on $\Gamma\backslash G \times \Gamma\backslash G$ representing the operator $\mathscr{C}_t^{\Gamma}$ (convolution with the periodized heat kernel $\mathbf{K}_t^{\Gamma}$) on $L^2_{\text{cusp}}(\Gamma\backslash G/K)$ is given by the series*

$$Q_t(z,w) = \sum_{n=1}^{\infty} e^{-\lambda_n t}\psi_n(z)\overline{\psi_n(w)}.$$

Proof. Apply Theorem 8.1.2 and Theorem 8.5.4.

We shall call Q_t the **heat kernel's cuspidal component**.

Part IV
Fourier–Eisenstein Eigenfunction Expansions

Chapter 9
The Tube Domain for Γ_∞

This chapter considers the fundamental domain for Γ_∞, which is a tube $\mathscr{T}$. The upper part of this tube is the fundamental domain for Γ as in Chapter 6, but the lower part will play a role. We want to know how the tube is related to the Picard domain $\mathscr{F}$, in several contexts, especially in Chapters 10 and 13. The lower part of the tube is tiled, i.e., is the union of Γ-translates of $\mathscr{F}$. Among other things, integration over $\mathscr{F}$ cannot be done explicitly because of the curved part, but integration over the tube can be carried out explicitly. One then has to estimate the difference. Such an estimate can be carried out because of the tiling.

We are much indebted to Eliot Brenner, whose thesis we have followed for the results of Sections 9.2 through 9.7 as well as the general result in Theorem 9.8.2 and its application to Theorem 9.8.1. In this chapter, we do the minimum needed for the tiling theorem. For an extensive additional treatment of boundary behavior of $\mathscr{F}$ and the corresponding tube; see [Bre 05].

We recommend that readers look at the theorems in Section 9.8 first, to see where we are going, namely the $(\Gamma, \mathscr{F})$-tiling of the tube fundamental domain for Γ_∞. However, Sections 9.3 through 9.6 have to do only with the boundary behavior of $\mathscr{F}$, so they are carried out first. In Section 9.2 we give the proof of the main auxiliary result on $\mathbf{H}^2$ because this proof is short and direct, but it does not generalize, so the reader is warned early about the difficulties that have to be surmounted.

9.1 Differential-Geometric Aspects

We started Chapter 6 with some group-theoretic aspects that were sufficient for some purposes. Here we go further with differential-geometric aspects.

The upper half-plane $\mathbf{H}^2$ and upper half-space $\mathbf{H}^3$ form the basic levels of an inductive system, and also $\mathbf{H}^2$ serves inductively to analyze $\mathbf{H}^3$ in certain respects. Of course, $\mathbf{H}^2$ is more familiar to most people because of its occurrence in complex analysis. The length form giving rise to the Riemannian metric is

$$ds^2 = \left(dx_1^2 + dy^2\right)/y^2,$$

with the complex variable $z = x_1 + y\mathbf{i}$. This metric is called **hyperbolic**.

The space $\mathbf{H}^3$ has a Riemannian metric, also called **hyperbolic**, which is G-invariant. We use its normalization such that the length function is represented by the form

$$ds^2 = (dx_1^2 + dx_2^2 + dy^2)/y^2.$$

Compared to the Euclidean distance, the hyperbolic distance, abbreviated $\mathbf{H}$-distance, becomes smaller by a factor of $1/y$ for $y \to \infty$, and becomes larger as $y \to 0$. Two distinct points $p,\ q \in \mathbf{H}^3$ are joined by a unique curve of minimal length, called the **geodesic** between these points. The curve is actually real analytic, and the segment between the points is part of a curve defined on $(-\infty, \infty)$ by a certain real analytic differential equation that readers will find in basic texts on differential geometry. This infinite curve is called a **geodesic**.

Next we introduce more notation stemming from elementary differential geometry. We let $\bar{\mathbf{H}}$ denote the "completion" given in the two cases (**R** and **C**) by

$$\bar{\mathbf{H}}^2 = \mathbf{H}^2 \cup \mathbf{R} \cup \{\infty\} \text{ and } \bar{\mathbf{H}}^3 = \mathbf{H}^3 \cup \mathbf{C} \cup \{\infty\}.$$

For points $p,\ q \in \bar{\mathbf{H}}$ (not both ∞) we let

$$\mathscr{C}_{\mathbf{H}}(p,q) \text{ or } [p,q]_{\mathbf{H}} \text{ or } [p,q)_{\mathbf{H}} \text{ or } (p,q)_{\mathbf{H}} \text{ or } (p,q)_{\mathbf{H}}$$

denote the unique geodesic segment between p and q in $\mathbf{H}$. For example, if $p,\ q \in \mathbf{H}$, then $[p,q]_{\mathbf{H}}$ is the ordinary geodesic segment in $\mathbf{H}$. If $p \in \mathbf{H}$ and $q = \infty$, then $[p,q)_{\mathbf{H}}$ is the half-open Euclidean vertical segment starting at p and tending to ∞. If both $p,\ q \in \mathbf{C}$, then $(p,q)_{\mathbf{H}}$ is the open Euclidean semicircle perpendicular to $\mathbf{C}$ at p and q, but omitting $p,\ q$. Note that in the half-open or open case, the geodesic segment is uniquely determined by its endpoints, as it is of course for points $p,\ q$ in $\mathbf{H}$ itself. We say that a point of $\mathbf{H}$ lies **between** p and q if it lies on $\mathscr{C}_{\mathbf{H}}(p,q)$.

Geodesics play the role of straight lines in Euclidean geometry. Euclidean proofs essentially go over to the hyperbolic case unchanged. For instance, one may define a subset $\mathscr{C}$ to be **H-convex** if given two points in $\mathscr{C}$, the geodesic segment between the two points is contained in $\mathscr{C}$. It is part of the elementary facts that the fundamental domain is $\mathbf{H}$-convex, both in the case of $\mathbf{H}^3$ and the ordinary case of $\mathbf{H}^2$. Let $\mathbf{H}$ denote $\mathbf{H}^2$ or $\mathbf{H}^3$. Let $\mathscr{D}$ be a subset of $\mathbf{H}$. We define the **hyperbolic convex closure** of $\mathscr{D}$, or **H-convex closure** of $\mathscr{D}$, to be the intersection of all $\mathbf{H}$-convex sets containing $\mathscr{D}$. It is the smallest $\mathbf{H}$-convex set containing $\mathscr{D}$. We denote it by $\mathscr{C}_{\mathbf{H}}(\mathscr{D})$. If $p_1, \dots, p_r \in \bar{\mathbf{H}}$, we let

$$\mathscr{C}_{\mathbf{H}}(p_1, \cdots, p_r)$$

be the $\mathbf{H}$-convex closure of the set of geodesic segments $\mathscr{C}_{\mathbf{H}}(p_i, p_j)$. The situation is entirely analogous to that of Euclidean space.

If three points $p_1,\ p_2,\ p_3$ do not lie on a geodesic, then we call their convex closure $\mathscr{C}_{\mathbf{H}}(p_1, p_2, p_3)$ an **H-triangle**.

A surface in $\mathbf{H}^3$ is called **totally geodesic** if it is closed, and for any two points in the surface, the geodesic passing through the points is contained in the surface. In practice, we deal with complete surfaces in $\mathbf{H}^3$.

As Wu pointed out to us, the easiest way to show that a surface is totally geodesic is to show that it is the fixed-point set of a metric automorphism (**autometry**) of $\mathbf{H}^3$.

Indeed, let g be an autometry of $\mathbf{H}^3$, and Fix(g) its set of fixed points. Let $p \neq q \in$ Fix(g). If any point of the geodesic through p and q is not fixed, then applying g gives another geodesic between p and q, contradiction. The same property and argument of course also apply to geodesic curves. We deal with such curves and surfaces below separately on $\mathbf{H}^2$ and $\mathbf{H}^3$.

Let $\mathbf{H}$ denote $\mathbf{H}^2$ or $\mathbf{H}^3$. Let $\mathscr{D}$ be a subset of $\mathbf{H}$. If $\mathscr{D}$ is contained in a unique geodesic, then we denote this geodesic by $\mathbf{H}^1(\mathscr{D})$. If $\mathscr{D}$ is contained in a unique totally geodesic surface, we denote this surface by $\mathbf{H}^2(\mathscr{D})$. We describe such cases for $\mathbf{H}^2$ and $\mathbf{H}^3$ below. We note for the record the following **trivial properties**:

Fix $i = 1, 2$. Let $\mathscr{D}$, $\mathscr{E}$ be subsets of $\mathbf{H}^3$ such that $\mathbf{H}^i(\mathscr{D})$, $\mathbf{H}^i(\mathscr{E})$ exist.

TP 1. $\mathscr{D} \subset \mathbf{H}^i(\mathscr{E}) \iff \mathbf{H}^i(\mathscr{D}) = \mathbf{H}^i(\mathscr{E})$;
TP 2. $g\mathbf{H}^i(\mathscr{D}) = \mathbf{H}^i(g\mathscr{D})$ for all autometries g of $\mathbf{H}^3$;
TP 3. $g\mathscr{D} \subset \mathbf{H}^i(\mathscr{E}) \iff g\mathbf{H}^i(\mathscr{D}) = \mathbf{H}^i(\mathscr{E}) \iff g^{-1}\mathscr{E} \subset \mathbf{H}^i(\mathscr{D})$.

We use these with $i = 2$ in Section 9.4.

We shall deal with isomorphisms and automorphisms for $\mathbf{H}^2$, $\mathbf{H}^3$, and other such metric spaces, which will be assumed to be metric-preserving unless otherwise specified. We let Aut(X) denote the group of metric automorphisms of a metric space X. When X is also a manifold (as will be the case) we of course assume also that the automorphisms are C^∞ automorphisms.

9.2 The Tube of $\mathscr{F}_{\mathbf{R}}$ and its Boundary Relation with $\partial\mathscr{F}_{\mathbf{R}}$

Using the fixed-point-set criterion mentioned in Section 9.1, and the uniqueness of geodesics through two points, we get easily that

Geodesics on $\mathbf{H}^2$ are of two types. The first is a vertical Euclidean half-line

$$p + \mathbf{R}^+\mathbf{i} \textit{ with } p \in \mathbf{R}.$$

The second is a Euclidean semicircle in $\mathbf{H}^2$, perpendicular to $\mathbf{R}$ in two distinct points (the "endpoints").

The group $G_{\mathbf{R}} = \mathrm{SL}_2(\mathbf{R})$ is a group of autometries of $\mathbf{H}^2$.

We denote by $\mathscr{F}_{\mathbf{R}}$ the standard fundamental domain of $\Gamma_{\mathbf{Z}} = \mathrm{SL}_2(\mathbf{Z})$ acting on $\mathbf{H}^2$, as in the figure below:

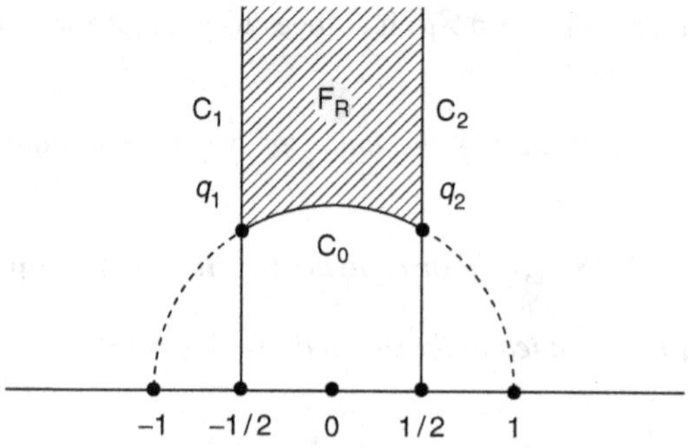

Let q_1, q_2 be the two corners of $\mathscr{F}_{\mathbf{R}}$, that is,

$$q_1 = -\frac{1}{2} + \frac{\sqrt{3}}{2}\mathbf{i} \text{ and } q_2 = \frac{1}{2} + \frac{\sqrt{3}}{2}\mathbf{i}.$$

Then the three sides of $\mathscr{F}_{\mathbf{R}}$ are geodesic segments:

$$\mathscr{C}_\theta = [q_1, q_2]_{\mathbf{H}},\ \mathscr{C}_1 = [q_1, \infty)_{\mathbf{H}},\ \mathscr{C}_2 = [q_2, \infty]_{\mathbf{H}}.$$

It follows by inspection that $\mathscr{F}_{\mathbf{R}}$ is the convex closure

$$\mathscr{F}_{\mathbf{R}} = \mathscr{C}_{\mathbf{H}}(q_1, q_2, \infty).$$

The region $\mathscr{S}_{\mathbf{R}}$ below the fundamental domain is illustrated in the figure below.

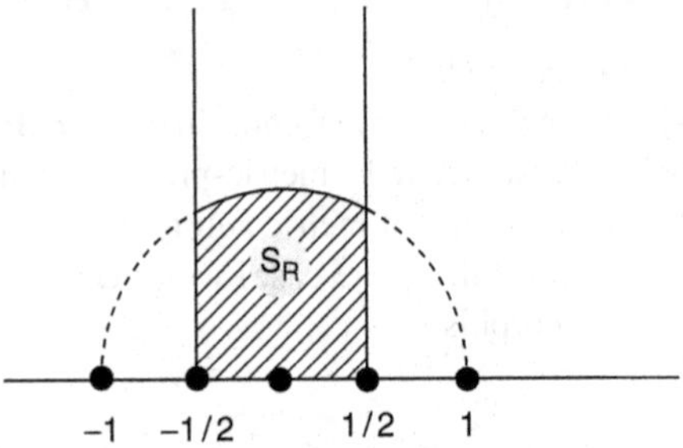

Thus $\mathscr{S}_{\mathbf{R}}$ is the set of points (x_1, y) with $|x_1| \leqq 1/2$ and $0 < y^2 \leqq (1 - x_1^2)$ below the unit circle). We define the **tube**

$$\mathscr{T}_{\mathbf{R}} = \mathscr{S}_{\mathbf{R}} \cup \mathscr{F}_{\mathbf{R}} = [-1/2, 1/2] + \mathbf{R}^+\mathbf{i}.$$

Note that $\mathscr{T}_{\mathbf{R}}$ is a fundamental domain for the action of $\Gamma_{\mathbf{Z},U}$ (the group of unipotent integral matrices over $\mathbf{Z}$). The boundary of $\mathscr{S}_{\mathbf{R}}$ consists of three geodesic segments:

$[q_1, -1/2)_{\mathbf{H}}$, which we denote by ℓ (the left segment);
$[q_1, q_2]_{\mathbf{H}}$, which is a common boundary segment of $\mathscr{S}_{\mathbf{R}}$ and $\mathscr{R}_{\mathbf{R}}$;
$[q_2, 1/2)_{\mathbf{H}} = T\ell$.

Recall (Section 6.2) that $T = \begin{pmatrix} 1 & 1 \\ 0 & 1 \end{pmatrix}$ and $S = \begin{pmatrix} 0 & -1 \\ 1 & 0 \end{pmatrix}$.

Thus to prove that $\partial\mathscr{S}_{\mathbf{R}} \subset \Gamma\partial\mathscr{F}_{\mathbf{R}}$, it suffices to prove it for ℓ. We prove a more precise statement as follows.

Theorem 9.2.1. *The segment ℓ is the union of two $\Gamma_{\mathbf{Z}}$-translates of segments of $\partial\mathscr{F}_{\mathbf{R}}$. Specifically,*

$$\ell = \mathrm{ST}[q_1, q_2]_{\mathbf{H}} \cup ST^2S[q_2, \infty)_{\mathbf{H}} \textit{ or equivalently } S\ell = T[q_1, q_2]_{\mathbf{H}} \cup T^2S[q_2\infty)_{\mathbf{H}}.$$

Hence on $\mathbf{H}$, $\partial\mathscr{S}_{\mathbf{R}}$ is contained in the union of five $\Gamma_{\mathbf{Z}}$-translates of $\partial\mathscr{F}_{\mathbf{R}}$.

Proof. The equivalence comes from the fact that $S^2 = \text{id}$ on $\mathbf{H}$. The assertion is immediately verified by checking the relation via its endpoints, namely

$$S\ell = [q_2, 2)_{\mathbf{H}} = [q_2, q_2+1]_{\mathbf{H}} \cup [q_2+1, 2)_{\mathbf{H}} = T[q_1, q_2]_{\mathbf{H}} \cup T^2 S[q_2 \infty)_{\mathbf{H}}$$

as desired.

Remark. The possibility of using only a finite number of translations of $\mathscr{F}$-boundaries to cover the boundary of $\mathscr{S}_{\mathbf{R}}$ is very special. Already on $\mathrm{SL}_2(\mathbf{C})$, we shall see that one needs infinitely many. On $\mathbf{H}^2$, Theorem 9.2.1 is proved very simply and directly. On $\mathbf{H}^3$, its analogue will require more elaborate arguments, namely those leading to Theorem 9.7.1.

Theorem 9.2.1 suffices to carry out the proof of Theorem 9.8.1 on $\mathbf{H}^2$. It is actually stronger than what is required for this purpose.

9.3 The $\mathscr{F}$-Normalizer of Γ

We shall need a certain subgroup of the normalizer $\mathrm{Nor}(\Gamma)$ in the group of all autometries of $\mathbf{H}^3$. What we need is valid in much greater generality, so this section is self-contained at the foundational level.

Let X be a metric space. Let Γ be a discrete group of autometries, and let $\mathscr{F}$ be a fundamental domain for Γ.

We define

$$\mathrm{Nor}_{\mathscr{F}}(\Gamma) = \mathscr{F} - \textbf{normalizer} \text{ of } \Gamma \text{ in } \mathrm{Aut}(X)$$

to be the set of elements $g \in \mathrm{Aut}(X)$ such that:

Nor$_{\mathscr{F}}$1. $g \in \mathrm{Nor}(\Gamma)$ (g normalizes Γ);
Nor$_{\mathscr{F}}$2. There exists an element $\eta_g \in \Gamma$ such that $g\mathscr{F} = \eta_g \mathscr{F}$.

It is immediate that $\mathrm{Nor}_{\mathscr{F}}(\Gamma)$ is a subgroup of Aut (X) and $\Gamma \subset \mathrm{Nor}_{\mathscr{F}}(\Gamma) \subset \mathrm{Nor}(\Gamma)$.

Theorem 9.3.1. *Let X, $\Gamma, \mathscr{F}$ be as above. Assume the following two conditions:*

FD 3. *$\mathscr{F}$ = closure of its interior;*
FD 4. *$\mathrm{Int}(\mathscr{F})$ is connected, pathwise if you want.*

Let $g \in \mathrm{Nor}_{\mathscr{F}}(\Gamma)$. If $g\mathscr{F} \cap \mathrm{Int}(\mathscr{F})$ is not empty, then $g\mathscr{F} = \mathscr{F}$.

Proof. First we need a lemma.

Lemma 9.3.2. *Let Γ, $\mathscr{F}$, g be as in Theorem 9.3.1. Then*

$$g\Gamma\partial\mathscr{F} = \Gamma\partial\mathscr{F}. \tag{9.1}$$

Proof. Let $\gamma \in \Gamma$ and $z \in \partial\mathscr{F}$, and write

$$g\gamma z = g\gamma g^{-1} gz,$$

Since g is a topological automorphism, it follows that $g\Gamma\partial\mathscr{F} \subset \Gamma\partial\mathscr{F}$, whence the lemma follows by applying this inclusion to g^{-1}.

Back to the theorem, we show first that if $g\mathscr{F} \cap \mathrm{Int}(\mathscr{F})$ is not empty, then $g\mathscr{F} \subset \mathscr{F}$. If not, by the closure and the connectedness assumptions there exists $w \in \mathrm{Int}(\mathscr{F})$ such that $gw \in \partial\mathscr{F}$, or equivalently $\mathrm{Int}(\mathscr{F})$ intersects $\partial(g^{-1}\mathscr{F})$, and therefore intersects $g^{-1}\Gamma\partial\mathscr{F}$, hence $\Gamma\partial\mathscr{F}$ by Lemma 9.3.2. This contradicts the basic property of a fundamental domain, and concludes the proof of $g\mathscr{F} \subset \mathscr{F}$. We must then have $g\mathscr{F} = \mathscr{F}$ because functorially, $g\mathscr{F}$ is a fundamental domain for $\mathbf{c}(g)\Gamma = \Gamma$, and both $\mathscr{F}$, $g\mathscr{F}$ are the closures of their interiors. This concludes the proof.

Remark. The theorem is often applied by changing g and $\mathscr{F}$ by applying elements of Γ. Note that for $\gamma, \eta \in \Gamma$, the element $\gamma g \eta$ is also in $\mathrm{Nor}_{\mathscr{F}}(\Gamma)$.

We list two general statements for later use. The first is obvious.

Lemma 9.3.3. *Let X, Γ, $\mathscr{F}$ be as in Theorem 9.3.1. Then for any open set U of X, there exists $\gamma \in \Gamma$ such that $\gamma\,\mathrm{Int}(\mathscr{F})$ intersects U.*

Lemma 9.3.4. *Let X, Γ, $\mathscr{F}$ be as in Theorem 9.3.1. Let $\tilde{\Gamma}$ be a discrete subgroup of Aut (X) containing Γ as a normal subgroup. Let Δ be a subgroup of $\tilde{\Gamma}$ that is a section of $\tilde{\Gamma}/\Gamma$ and whose elements preserve $\mathscr{F}$. Let $\tilde{\mathscr{F}}$ be a fundamental domain for the action of Δ on $\mathscr{F}$. Then $\tilde{\mathscr{F}}$ is a fundamental domain for the action of $\tilde{\Gamma}$ on X.*

Proof. Let $x \in X$. There exists $\gamma \in \Gamma$ and $z \in \mathscr{F}$ such that $x = \gamma z$. By hypothesis there exists $\tilde{z} \in \tilde{\mathscr{F}}$ and $\delta \in \Delta$ such that $z = \delta\tilde{z}$. This proves the first property **FD 1** (Chapter 6). For the second property **FD 2**, let $\tilde{\gamma} \in \tilde{\Gamma}$ be such that $\tilde{\gamma}\tilde{\mathscr{F}}$ intersects $\mathrm{Int}(\tilde{\mathscr{F}})$. There exists $\delta \in \Delta$ such that $\tilde{\gamma} = \gamma\delta$. Property **FD 2** for $(\Gamma, \mathscr{F})$ implies $\gamma = \mathrm{id}$. Then $\delta\tilde{\mathscr{F}}$ intersects $\mathrm{Int}(\tilde{\mathscr{F}})$. So **FD 2** for $(\Delta, \tilde{\mathscr{F}})$ gives $\delta = \mathrm{id}$, which completes the proof.

9.4 Totally Geodesic Surface in $\mathbf{H}^3$

We shall describe geodesics and totally geodesic surfaces in $\mathbf{H}^3$. Let $\Gamma = \Gamma_o$ be the discrete subgroup $\mathrm{SL}_2(o)$ of $\mathrm{SL}_2(\mathbf{C})$. We return to $\mathscr{F}$ being the Picard fundamental domain of Chapter 6. If we deal simultaneously with $\mathscr{F}_{\mathbf{R}}$, we may write $\mathscr{F}_{\mathbf{C}}$ if needed to make the meaning clear.

Given a point $p \in \mathbf{H}^3$, the geodesics through p are either the vertical half-line $x_p + \mathbf{R}^+\mathbf{j}$, or an open Euclidean semicircle passing through p, and perpendicular to the $\mathbf{C}$-plane of points $x_1 + x_2\mathbf{i}$ $(x_1, x_2 \in \mathbf{R})$.

The group $G = \mathrm{SL}_2(\mathbf{C})$ is a group of isometries of $\mathbf{H}^3$. We make explicit the geodesic surfaces using the fixed-point criterion. Note that some of those we use are not in G (they may be orientation-reversing). For example, the Euclidean planes

$$\mathbf{R} + \mathbf{R}^+\mathbf{j} \text{ and } \mathbf{R}\mathbf{i} + \mathbf{R}^+\mathbf{j}$$

are totally geodesic. The second is fixed under $x_1 \mapsto -x_1$, and the first is fixed under the isometry $x_2 \mapsto -x_2$, that is, under

$$R^* : x_1 + x_2\mathbf{i} + y\mathbf{j} \mapsto x_1 - x_2\mathbf{i} + y\mathbf{j}.$$

Translation by any element $b \in \mathbf{C}$ is an isometry, so the half-planes

$$b + \mathbf{R} + \mathbf{R}^+\mathbf{j} \text{ and } b + \mathbf{R}\mathbf{i} + \mathbf{R}^+\mathbf{j}$$

are totally geodesic.

9.4.1 The Half-Plane $\mathbf{H}^2_\mathbf{j}$

The upper half-plane $\mathbf{H}^2$ has a natural isomorphic image in $\mathbf{H}^3$, replacing $\mathbf{i}$ by $\mathbf{j}$. Indeed, the map

$$x_1 + y\mathbf{i} \mapsto x_1 + y\mathbf{j} \ (x_1 \in \mathbf{R}, y > 0)$$

is a metric embedding of $\mathbf{H}^2$ in $\mathbf{H}^3$, whose image will be denoted by

$$\mathbf{H}^2_\mathbf{j} = \mathbf{R} + \mathbf{R}^+\mathbf{j}.$$

This embedding commutes with the action of $\Gamma_\mathbf{Z}$, acting on $\mathbf{H}^2$ and also on $\mathbf{H}^3$ as a subgroup of $\Gamma = \mathrm{SL}_2(o)$. On $\mathbf{H}^3$, the first figure of Section 9.2 represents the front end of a 3-dimensional tube. We let $\mathscr{F}_{\mathbf{R},\mathbf{j}}$ be the image of $\mathscr{F}_\mathbf{R}$ under that embedding.

It is verified at once that the element $R = \mathrm{diag}(\mathbf{i}, -\mathbf{i})$ stabilizes $\mathbf{H}^2_\mathbf{j}$, that is, $R(\mathbf{H}^2_\mathbf{j}) = \mathbf{H}^2_\mathbf{j}$, and gives thereby an autometry of $\mathbf{H}^2_\mathbf{j}$. Note that R is orientation-reversing, and not in $G_\mathbf{R}$. In addition, note that R normalizes $\Gamma_\mathbf{Z}$ (as a subgroup of Γ). To see this, it suffices to see it on generators, and we have by direct verification

$$\mathbf{c}(R)S = S \text{ and } \mathbf{c}(R)T_1 = T_1^{-1}. \tag{9.2}$$

***$\tilde{R}$* and the half plane $\tilde{\mathrm{R}}\mathbf{H}^2_\mathbf{j}$**. Let $\tilde{\mathrm{R}}$ be the matrix

$$\tilde{\mathrm{R}} = \begin{pmatrix} \zeta_8 & 0 \\ 0 & \zeta_8^{-1} \end{pmatrix},$$

where $\zeta_8 = \mathrm{e}^{\mathrm{i}\pi/4}$ is the primitive 8th root of unity. Another half-plane $\mathbf{H}$-metrically isomorphic to $\mathbf{H}^2_\mathbf{j}$ is

$$\tilde{\mathrm{R}}\mathbf{H}^2_\mathbf{j} \left\{x + y\mathbf{j} \in \mathbf{H}^3 | \mathrm{Re}(x) = 0\right\} = \mathbf{R}\mathbf{i} + \mathbf{R}^+\mathbf{j},$$

i.e., the quaternions in $\mathbf{H}^3$ with real part 0 and positive $\mathbf{j}$-component. The group corresponding to $\Gamma_\mathbf{Z}$ under the isomorphism is the conjugate group

$$\mathbf{c}(\tilde{\mathrm{R}})\Gamma_\mathbf{Z}.$$

whose fundamental domain is

$$\tilde{\mathrm{R}}\mathscr{F}_{\mathbf{R},\mathbf{j}} = \mathscr{C}_{\mathbf{H}}(-\frac{1}{2}\mathbf{i}+\frac{\sqrt{3}}{2}\mathbf{j}, \frac{1}{2}\mathbf{i}+\frac{\sqrt{3}}{2}\mathbf{j}, \infty).$$

We tabulate the conjugations, which show that $\tilde{\mathrm{R}}$ *normalizes* Γ:

$$\begin{aligned} \mathbf{c}(\tilde{R})T_1 &= T_{\mathbf{i}}, \quad \mathbf{c}(\tilde{R})T_{\mathbf{i}} = T_{-\mathbf{i}}, \\ \mathbf{c}(\tilde{R})R &= R, \quad \mathbf{c}(\tilde{R})S = RS. \end{aligned} \tag{9.3}$$

In general, the vertical upper half-plane above a line in $\mathbf{C}$ through 0 and with angle θ measured counterclockwise from the real axis is a geodesic surface denoted by $\mathbf{H}^2(\theta)$. For example, $\mathrm{Fix}(R^*) = \mathbf{H}^2(0)$. Let

$$R_\theta = \begin{pmatrix} e^{i\theta/2} & 0 \\ 0 & e^{-\mathrm{i}\theta/2} \end{pmatrix}.$$

Then $R_\theta(\mathbf{H}^2(0)) = \mathbf{H}^2(\theta)$, so

$$\mathrm{Fix}(\mathbf{c}(R_\theta)R^*) = \mathbf{H}^2(\theta).$$

We have now concluded the list of the upper half-planes in $\mathbf{H}^3$ occurring most often in applications. Next we describe the analogue of upper semicircles for geodesic surfaces, namely upper hemispheres.

Let $\mathbf{S}_r^+(p)$ be the Euclidean upper hemisphere centered at $p \in \mathbf{C}$, of radius r. Then $\mathbf{S}_1^+(0)$ is totally geodesic, fixed under the isometry

$$S^* : z \mapsto (\bar{z})^{-1} = z/|z|^2.$$

Therefore the translation $\mathbf{S}_1^+(p)$ is totally geodesic. Putting a factor r^2 on the map shows that $\mathbf{S}_r^+(0)$ is totally geodesic. Or put another way, let

$$a(r) = \begin{pmatrix} r^{\frac{1}{2}} & 0 \\ 0 & r^{\frac{1}{2}} \end{pmatrix}.$$

Then $a(r)\mathbf{S}_1^+(0) = \mathbf{S}_{\mathrm{r}}^+(0)$, and

$$\mathrm{Fix}(\mathbf{c}(a(r))S^*) = \mathbf{S}_{\mathrm{r}}^+(0).$$

By translation with $p \in \mathbf{C}$, it follows that $\mathbf{S}_r^+(p)$ is totally geodesic.

It can be easily verified that the only totally geodesic surfaces in $\mathbf{H}^3$ are those described above (vertical planes, hemispheres) and their $\mathbf{C}$-translations.

We end the section with some comments about closures.

By the **closure** of a totally geodesic surface in $\overline{\mathbf{H}^3}$, written $\overline{\mathbf{H}^2(\mathscr{E})}$, we mean the closure in the usual compactification of $\mathbf{H}^3$, homeomorphic to the closed disc. The basic examples are:

$\overline{\mathrm{H}^2(\theta)}$, which is the union of ∞ and the upper half plane with the line below it in $\mathbf{C}$;

$\overline{\mathbf{S}_r^+(0)}$, which is the union of the upper hemisphere with the equator in $\mathbf{C}$.

Every element $g \in \mathrm{Aut}(\mathbf{H}^3)$ has a natural extension to a topological automorphism $\bar{g}$ of $\overline{\mathbf{H}^3}$. We have

$$\mathrm{Fix}(\bar{g}) = \overline{\mathrm{Fix}(g)}.$$

9.5 Some Boundary Behavior of $\mathscr{F}$ in $\mathscr{H}^3$ Under Γ

We first label various parts of the boundary of $\mathscr{F}$ and then give the theorem on $\mathscr{H}^3$ analogous to Theorem 9.2.1 on $\mathscr{H}^2$.

Let $\mathscr{C}$ be a subset of $\mathbf{H}^3$. If $\mathscr{C}$ is contained in a unique totally geodesic surface, we denote this surface by $\mathbf{H}^2(\mathscr{C})$. If $\mathscr{C}$ is contained in a unique geodesic, we denote the geodesic by $\mathbf{H}^1(\mathscr{C})$. We list some examples arising from $\mathscr{F}$ and its boundary. See the comments at the end of Section 9.1.

We label four points of $\mathbf{H}^3$, the $\mathbf{H}$-vertices of $\mathscr{F}$, going clockwise as follows:

$$\begin{aligned} p_1 &= -\frac{1}{2} + \frac{\sqrt{3}}{2}\mathbf{j}, \\ p_2 &= -\frac{1}{2} + \frac{1}{2}\mathbf{i} + \frac{1}{\sqrt{2}}\mathbf{j}, \\ p_3 &= \frac{1}{2} + \frac{1}{2}\mathbf{i} + \frac{1}{\sqrt{2}}\mathbf{j}, \\ p_4 &= \frac{1}{2} + \frac{\sqrt{3}}{2}\mathbf{j}. \end{aligned}$$

Note that p_1 and p_4 are the points q_1 and q_2 of Section 9.2. Here on $\mathbf{H}^3$ we have

$$\mathscr{F} = \mathscr{C}_{\mathbf{H}}(p_1, p_2, p_3, p_4, \infty).$$

Note also that the front geodesic segment $[p_1, p_4]_{\mathbf{H}}$ is higher than the back segment $[p_2, p_3]_{\mathbf{H}}$.

9.5.1 *The Faces $\mathscr{B}_i$ of $\mathscr{F}$ and their Boundaries*

The boundary of $\mathscr{F}$ consists of the base $\mathscr{B}_0$ (part of a Euclidean sphere), which we also call the **floor** of $\mathscr{F}$, and the four sides, which we call the **walls** of $\mathscr{F}$. We denote these sides by $\mathscr{B}_i (i = 1, \ldots, 4)$, going clockwise, starting with the line vertically above $-1/2$. We have

$$\mathscr{B}_0 = \mathscr{C}_{\mathbf{H}}(p_1, p_2, p_3, p_4) \quad \text{and} \quad \mathbf{H}^2(\mathscr{B}_0) = \mathbf{S}_1^+(0),$$

where $\mathbf{S}_r^+(p)$ is the Euclidean upper hemisphere of radius r centered at p. We call $\mathscr{B}_0, \mathscr{B}_1, \ldots, \mathscr{B}_4$ the **faces** of $\mathscr{F}$.

For each $i = 1, 2, 3, 4$ the wall $\mathscr{B}_i$ is the **H**-convex closure

$$\mathscr{B}_i = \mathscr{C}_{\mathbf{H}}(p_i, p_{i+1}, \infty).$$

(If $i = 4$, then $i + 1$ means 0.) The wall $\mathscr{B}_i$ lies in a unique totally geodesic surface $\mathbf{H}^2(\mathscr{B}_i)$, which is also a Euclidean half-space, for instance

$$\mathbf{H}^2_{\mathbf{j}} = \mathbf{H}^2(\mathscr{B}_4), \quad \mathbf{H}^2(\mathscr{B}_1) = -\frac{1}{2} + \mathbf{Ri} + \mathbf{R}^+\mathbf{j}, \text{ etc.}$$

Let $z_1, z_2, z_3 \in \bar{\mathbf{H}}^3$. If $\mathbf{H}^2(z_1, z_2, z_3)$ exists (i.e., there is a unique geodesic surface containing these three points), we call $\mathscr{C}_{\mathbf{H}}(z_1, z_2, z_3)$ an

9.5.2 *H-triangle*

For example, the walls $\mathscr{B}_i$ $(i = 1, \ldots, 4)$ are **H**-triangles. The floor $\mathscr{B}_0$ is an **H**-4-gon, and is a union of two **H**-triangles in the obvious way, just as for a quadrilateral in a Euclidean plane, for instance the **H**-triangles

$$\mathscr{C}_{\mathbf{H}}(p_1, p_2, p_3) \text{ and } \mathscr{C}_{\mathbf{H}}(p_2, p_3, p_4).$$

The boundary of the floor $\mathscr{B}_0$ consists of four segments

$$\partial \mathscr{B}_0 = \mathscr{C}_{0,1} \cup \mathscr{C}_{0,2} \cup \mathscr{C}_{0,3} \cup \mathscr{C}_{0,4},$$

going in the clockwise direction. We have $\mathscr{C}_{0,4} = [p_2, p_4]_{\mathbf{H}}$ and $\mathscr{C}_{0,2} = [p_2, p_3]_{\mathbf{H}}$.

By inspection, the boundary of $\mathscr{B}_i$ for $i \neq 0$ consists of three geodesic segments, two infinite and one finite, intersecting only at their endpoints, namely,

$$\partial \mathscr{B}_i = \mathscr{C}_{i,0} \cup \mathscr{C}_{i,1} \cup \mathscr{C}_{i,2},$$

where $\mathscr{C}_{i,1}$, $\mathscr{C}_{i,2}$ are the vertical segments and $\mathscr{C}_{i,0}$ is the finite segment common to the boundary of the floor of $\mathscr{F}$. For example,

$$\mathscr{C}_{1,1} = \mathscr{C}_{\mathbf{H}}(p_1, \infty), \ \mathscr{C}_{1,2} = \mathscr{C}_{\mathbf{H}}(p_2, \infty), \ \mathscr{C}_{1,0} = \mathscr{C}_{\mathbf{H}}(p_1, p_2).$$

Note that for each j, we have $\mathscr{C}_{0,j} = \mathscr{C}_{j,0}$. In addition, certain infinite vertical segments coincide. For example,

$$\mathscr{C}_{1,2} = \mathscr{C}_{2,1} \text{ and } \mathscr{C}_{2,2} = \mathscr{C}_{3,1},$$

and so on, moving clockwise around $\mathscr{F}$. We have for $i = 0, \ldots 4$,

$$\mathbf{H}^2(\mathscr{B}_i) = \mathbf{H}^2(\partial \mathscr{B}_i).$$

The simplest case is that of $i = 4$. So we have an isometry

$$\mathbf{H}^2(\mathscr{B}_4) = \mathbf{H}^2_{\mathbf{j}} \approx \mathbf{H}^2.$$

Note that $\mathscr{B}_4$ corresponds to the fundamental domain $\mathscr{F}_{\mathbf{R}}$ under the isomorphism. In particular, we have the inclusions

$$\mathbf{H}^2(\mathscr{B}_4) = \Gamma_{\mathbf{Z}}\mathscr{B}_4 \subset \Gamma_{\mathbf{Z}}\partial\mathscr{F} \subset \Gamma\partial_{\mathscr{F}}.$$

The significance of these inclusions will become apparent in Section 9.7.

The isometry $\tilde{R}$ and $\tilde{\Gamma}_{\mathbf{Z}}$, $\mathscr{F}_{\mathbf{R},\mathbf{j}}$ We defined $\tilde{\mathrm{R}}$ in Section 9.4. We now look at its effect on some boundary segments of $\mathscr{F}$. We **define** $m_j =$ **midpointof** $\mathscr{C}_{0,j}$, so in particular,

$$m_2 = \text{midpoint of } \mathscr{C}_{0,2} = \frac{1}{2}\mathbf{i} + \frac{\sqrt{3}}{2}\mathbf{j} = \tilde{R}p_4,$$
$$m_4 = \mathbf{j} = \text{midpoint of } \mathscr{C}_{0,4}. \text{ Note that } \tilde{R}(m_4) = m_4.$$

The values on the right are obtained by inspection, and the fact that midpointing commutes with isometries. We define the segments for $i = 1, 2$:

$$\mathscr{C}'_{0,2} = [p_2, m_2]_{\mathbf{H}} \text{ and } \mathscr{C}''_{0,2} = [m_2, p_3]_{\mathbf{H}}.$$

Then

$$\mathscr{C}_{0,2} = \mathscr{C}'_{0,2} \cup \mathscr{C}''_{0,2}.$$

The two segments on the right intersect at the common endpoint m_2.

The next lemma tabulates the effect of the isometry $\tilde{R}$ on certain boundary segments of $\mathscr{F}$.

Lemma 9.5.1. *We have*

$$\mathscr{C}'_{0,2} = \tilde{R}\mathscr{C}_{0,3} \quad \textit{and} \quad \mathscr{C}''_{0,2} = \tilde{R}^{-1}\mathscr{C}_{0,1}, \tag{9.4}$$

so

$$\mathscr{C}_{0,2} = \tilde{R}\mathscr{C}_{0,3} \cup \tilde{R}^{-1}\mathscr{C}_{0,1}. \tag{9.5}$$

The edge $\mathscr{C}_{0,2}$ is not isometric to any other edge $\mathscr{C}_{0,j}$.

Proof. The formulas of (9.4) are seen by verifying that the endpoints of the segments are the same. The segments are then equal. Then (9.4) shows that $\mathscr{C}_{0,1}$, $\mathscr{C}_{0,3}$ are isometric to proper subsegments of $\mathscr{C}_{0,2}$, so $\mathscr{C}_{0,2}$ cannot be isometric to either of them. To see that $\mathscr{C}_{0,2}$ is not isometric to $\mathscr{C}_{0,4}$, we note that both are geodesic segments between points p_i, p_j on $\mathbf{S}^+(0)$, with $\mathrm{Re}(p_i) = -1/2$ and $\mathrm{Re}(p_j) = 1/2$. Since the endpoints of $\mathscr{C}_{0,2}$ are lower on the hemisphere than those of $\mathscr{C}_{0,4}$, it follows that $\mathscr{C}_{0,2}$ has greater $\mathbf{H}$-length than $\mathscr{C}_{0,4}$. This concludes the proof.

We consider the $\mathbf{H}$-triangle $\mathscr{C}_{\mathbf{H}}(m_4, p_4, \infty)$, which is going to provide an example of Lemma 9.3.4. We define the group generated by $\Gamma_{\mathbf{Z}}$ and R,

$$\tilde{\Gamma_{\mathbf{Z}}} = \Gamma_{\mathbf{Z}}\langle R\rangle = \langle S, T_1, R\rangle.$$

Then $\Gamma_{\mathbf{Z}}$ is a normal subgroup (cf. Section 9.4 (9.2)) of index 2. Let $\Delta = \langle I, R\rangle/\pm 1$. Then Δ is a section of $\tilde{\Gamma}_{\mathbf{Z}}/\Gamma_{\mathbf{Z}}$. Each element of Δ preserves $\mathscr{B}_0$.

For the $\mathbf{H}$-triangle $\mathscr{C}_{\mathbf{H}}(m_4, p_4, \infty)$ we shall use the notation

$$\tilde{\mathscr{F}}_{\mathbf{R},\mathbf{j}} = \mathscr{C}_{\mathbf{H}}(m_4, p_4, \infty)$$

Then by direct verification we have the decomposition of $\mathscr{B}_4$ in two $\mathbf{H}$-triangles

$$\mathscr{B}_4 = \tilde{\mathscr{F}}_{\mathbf{R},\mathbf{j}} \cup R\tilde{\mathscr{F}}_{\mathbf{R},\mathbf{j}}.$$

Lemma 9.5.2. *The* $\mathbf{H}$*-triangle* $\tilde{\mathscr{F}}_{\mathbf{R},\mathbf{j}}$ *is a fundamental domain of* Δ *on* $\mathscr{B}_4$*, and a fundamental domain of* $\tilde{\Gamma}_{\mathbf{Z}}$ *on* $\mathbf{H_j}^2$.

Proof. In the above decomposition of $\mathscr{B}_4$, the intersection of the two $\mathbf{H}$-triangles on the right is the common edge $[m_4, \infty)_{\mathbf{H}}$. Since $\mathscr{B}_4$ is the image of $\mathscr{F}_{\mathbf{R}}$ in $\mathbf{H}^2_{\mathbf{j}}$ under $\mathbf{i} \mapsto \mathbf{j}$, it is a fundamental domain of $\Gamma_{\mathbf{Z}}$ on $\mathbf{H}^2_{\mathbf{j}}$. We can then apply Lemma 9.5.2 to conclude the proof.

Next we apply $\tilde{R}$. We note that

$$\tilde{R}\tilde{\mathscr{F}}_{\mathbf{R},\mathbf{j}} = \mathscr{C}_{\mathbf{H}}(m_4, \tilde{R}p_4, \infty), \tag{9.6}$$

$$\tilde{R}\mathbf{H}^2_{\mathbf{j}} = \mathbf{H}^2(\tilde{R}\tilde{\mathscr{F}}_{\mathbf{R},\mathbf{j}}). \tag{9.7}$$

These equalities are immediate from the functoriality of the definitions, together with $\tilde{R}(m_4) = m_4$ for the first equality, and the trivial properties at the end of Section 9.1. We note that $\tilde{R}\tilde{\mathscr{F}}_{\mathbf{R},\mathbf{j}}$ is the right half of $\mathscr{B}_4$ rotated counterclockwise by $\pi/2$.

Lemma 9.5.3. *We have* $\mathrm{Int}(\tilde{\mathrm{R}}\tilde{\mathscr{F}}_{\mathbf{R},\mathbf{j}}) \subset \mathrm{Int}(\mathscr{F})$.

Proof. By inspection.

Lemma 9.5.4. *The set* $\tilde{\mathrm{R}}\tilde{\mathscr{F}}_{\mathbf{R},\mathbf{j}}$ *is a fundamental domain for* $\mathbf{c}(\tilde{R})\tilde{\Gamma}_{\mathbf{Z}}$ *on* $\tilde{R}\mathbf{H}^2_{\mathbf{j}}$.

Proof. Immediate from the functoriality of the definitions.

9.5.3 Isometries of $\mathscr{F}$

We apply the preceding lemmas to isometric images of $\mathscr{F}$.

Lemma 9.5.5. *Let* $g \in \mathrm{Aut}(\mathbf{H}^3)$*. Suppose* $g\mathscr{B}_0 = \mathscr{B}_0$ *but* $g \neq \mathrm{id}$ *on* $\mathscr{B}_0$*. Then*

$$gp_1 = p_4,\ gp_2 = p_3,$$

$$g\mathscr{C}_{0,2} = \mathscr{C}_{0,2},\ g\mathscr{C}_{0,4} = \mathscr{C}_{0,4},\ g\mathscr{C}_{0,1} = \mathscr{C}_{0,3}.$$

Furthermore,

$$\mathrm{Fix}(g) \cap \mathscr{B}_0 = \tilde{R}[p_4, m_4]_{\mathbf{H}}.$$

Proof. First we show that $gp_2 = p_3$. We have $p_2 = \mathscr{C}_{0,1} \cap \mathscr{C}_{0,2}$. The autometry g must map vertices to vertices and edges to edges (of $\mathscr{B}_0$). By Lemma 9.5.1, $g\mathscr{C}_{0,2} = \mathscr{C}_{0,2}$, so $gp_2 = g\mathscr{C}_{0,1} \cap \mathscr{C}_{0,2}$, so $gp_2 = p_2$ or p_3. If $gp_2 = p_2$, then $gp_3 = p_3$, and also $g\mathscr{C}_{0,1} = \mathscr{C}_{0,1}$, $g\mathscr{C}_{0,3} = \mathscr{C}_{0,3}$, so finally g leaves fixed three points that are not on a geodesic, when $g = \mathrm{id}$, contradicting $g \neq \mathrm{id}$ on $\mathscr{B}_0$. Thus we get $gp_2 = p_3$. Furthermore, since $g\mathscr{C}_{0,2} \cap g\mathscr{C}_{0,1} = p_3$, $g\mathscr{C}_{0,2}$ and $g\mathscr{C}_{0,1}$ are the two edges meeting at p_3, so $g\mathscr{C}_{0,1} = \mathscr{C}_{0,3}$. Since $gp_2 = p_3$, g must map the other endpoint p_1 of $\mathscr{C}_{0,1}$ to the other endpoint p_4 of $\mathscr{C}_{0,3}$. Hence finally $gp_4 = p_1$ (the only vertex not yet accounted for).

Since g interchanges the endpoints of $\mathscr{C}_{0,2}$ and $\mathscr{C}_{0,4}$, it fixes the midpoints $\tilde{R}p_4$ and m_4 of these two segments. Hence

$$[\tilde{R}p_4, m_4]_{\mathbf{H}} \subset \mathrm{Fix}(g).$$

Since $g \neq \mathrm{id}$ on $\mathscr{B}_0$, no point of $\mathscr{B}_0$ outside this geodesic segment is fixed by g, so we get the final equality, finishing the proof of the lemma.

Next come similar results with $\mathscr{F}$ instead of $\mathscr{B}_0$.

Lemma 9.5.6. *Let* $g \in \mathrm{Aut}(\mathbf{H}^3)$, $g \neq \mathrm{id}$ *on* $\mathbf{H}^3$. *Assume* $g\mathscr{F} = \mathscr{F}$. *Then:*

(a) $\infty \in \overline{\mathrm{Fix}(g)}$;
(b) $g\mathscr{B}_0 = \mathscr{B}_0$ *and* $g \neq \mathrm{id}$ *on* $\mathscr{B}_0$;
(c) $\mathrm{Fix}(g) \cap \mathscr{F} = \tilde{R}\tilde{\mathscr{F}}_{\mathbf{R},\mathbf{j}}$.

Proof. Since ∞ is the only vertex of $\mathscr{F}$ on the boundary of $\overline{\mathbf{H}^3}$, we must have $g\infty = \infty$, which proves (a).

Since $\mathscr{B}_0$ is the only compact face of $\mathscr{F}$, it follows that g maps $\mathscr{B}_0$ into itself, so acts as an automorphism of $\mathscr{B}_0$ (because $g^{-1}\mathscr{F} = \mathscr{F}$, etc.). Furthermore, if $g = \mathrm{id}$ on $\mathscr{B}_0$, then g fixes a totally geodesic surface and ∞, so $g = \mathrm{id}$ on $\mathbf{H}^3$, a contradiction that proves (b).

We can now apply Lemma 9.5.5. Then m_4, $Rp_4 \in \mathrm{Fix}(g)$, $\infty \in \mathrm{Fix}(g)$. Since

$$\tilde{R}\tilde{\mathscr{F}}_{\mathbf{R},\mathbf{j}} = \mathscr{C}_{\mathbf{H}}(m_4, \tilde{R}p_4, \infty)$$

and the vertices are in $\overline{\mathrm{Fix}(g)}$, we get $\tilde{R}\tilde{\mathscr{F}}_{\mathbf{R},\mathbf{j}} \subset \mathrm{Fix}(g)$. If a point of $\mathscr{F}$ not in $\tilde{R}\tilde{\mathscr{F}}_{\mathbf{R},\mathbf{j}}$ were in $\mathrm{Fix}(g)$, then g would fix the totally geodesic surface $\mathbf{H}^2(\tilde{\mathrm{R}}\tilde{\mathscr{F}}_{\mathbf{R},\mathbf{j}})$ and a point outside this surface, which would imply $g = \mathrm{id}$ on $\mathbf{H}^3$, contrary to hypothesis. This proves (c), and concludes the proof of the lemma.

In practice, the lemma is applied to a conjugate $\gamma^{-1}g\gamma$ of g, and we state the result as it is used later.

Corollary 9.5.7. *Let* $g, \gamma \in \mathrm{Aut}(\mathbf{H}^3)$, $g \neq \mathrm{id}$ *on* $\mathbf{H}^3$. *Assume* $g\gamma\mathscr{F} = \gamma\mathscr{F}$. *Then*

(a) $\gamma_\infty \in \overline{\mathrm{Fix}(g)}$;

(b) $g\gamma\mathscr{B}_0 = \gamma\mathscr{B}_0$;
(c) $\mathrm{Fix}(g) \cap \gamma\mathscr{F} = \gamma\tilde{\mathrm{R}}\tilde{\mathscr{F}}_{\mathbf{R},\mathbf{j}}$.

Lemma 9.5.8. *Let* $g \in \mathrm{Nor}_{\mathscr{F}}(\Gamma)$ *be reflection in the totally geodesic surface* $\mathrm{Fix}(g)$. *Let* $\gamma \in \mathrm{Aut}(\mathbf{H}^3)$. *Suppose that* $\mathrm{Fix}(g) \cap \mathrm{Int}(\gamma\mathscr{F})$ *is not empty. Then*

$$\mathrm{Fix}(g) \cap \gamma\mathscr{F} = \gamma\tilde{R}\tilde{\mathscr{F}}_{\mathrm{R,j}}.$$

Proof. We apply Theorem 9.3.1. Then the hypothesis implies $g\gamma\mathscr{F} = \gamma\mathscr{F}$, and Corollary 9.5.7(c) applies to conclude the proof.

9.6 The Group $\tilde{\Gamma}$ and a Basic Boundary Inclusion

Just as we defined $\tilde{\Gamma}_{\mathbf{Z}}$ in Section 9.5, we now define

$$\tilde{\Gamma} = \langle T_1, T_{\mathbf{i}}, R, S, R^*, \mathbf{c}(\tilde{R})R^* \rangle = \Gamma\langle R^*, \mathbf{C}(\tilde{R})R^* \rangle.$$

Thus $\tilde{\Gamma}$ is generated by the generators of Γ and the additional elements R^*, $\mathbf{c}(\tilde{R})R^*$. We record the effects of conjugation by R^*:

$$\begin{array}{ll} \mathbf{c}(R^*)T_1 = T_1^{-1}, & \mathbf{c}(R^*)T_{\mathbf{i}} = T_{\mathbf{i}}^{-1} \\ \mathbf{c}(R^*)S = S, & \mathbf{c}(R^*)R = R^{-1}. \end{array} \tag{9.8}$$

Lemma 9.6.1. $\tilde{\Gamma} \subset \mathrm{Nor}_{\mathscr{F}}(\Gamma)$.

Proof. There are two conditions to verify. As for the normalization condition, it follows from (9.8) above for R^* and Section 9.4 (9.9) for $\tilde{R}$. As to the second condition, it is trivially satisfied for elements of Γ, and it suffices to show that it is satisfied by R^* and $\mathbf{c}(\tilde{R})R^*$. From the definition

$$R^*(x+yj) = \bar{x}+yj, \tag{9.9}$$

one sees by inspection that

$$R^*\mathscr{F} = R\mathscr{F},$$

so the second condition is satisfied by R^*. Furthermore, directly from the definition,

$$\mathbf{c}(\tilde{R})R^*(x+y\mathbf{j}) = -\bar{x}+y\mathbf{j}. \tag{9.10}$$

Again by inspection, we get

$$\mathbf{c}(\tilde{R})R^*\mathscr{F} = \mathscr{F}.$$

This proves the second property, and concludes the proof of the lemma.

For $z = x + y\mathbf{j} \in \mathbf{H}^3$, we may set $x = z_{\mathbf{C}}$. Then we have $\mathrm{Re}(z_{\mathbf{C}}) = \mathrm{Re}(x)$ and $\mathrm{Im}(z_{\mathbf{C}}) = \mathrm{Im}(x)$. The main theorem will be concerned with the planes

$$\mathbf{H}^2(\mathcal{B}_1) = \{\mathrm{Re}(z_{\mathbf{C}}) = -1/2\}, \ \mathbf{H}^2(\mathcal{B}_2) = \{\mathrm{Im}(z_{\mathbf{C}}) = 1/2\},$$
$$\mathbf{H}^2(\mathcal{B}_3) = \{\mathrm{Re}(z_{\mathbf{C}}) = 1/2\}, \quad \mathbf{H}^2(\mathcal{B}_4) = \{\mathrm{Im}(z_{\mathbf{C}}) = 0\}.$$

We let

$$g_1 = \mathbf{c}(\tilde{R})R^*T_{-1}, \ g_2 = R^*T_{\mathbf{i}} \ g_3 = \mathbf{c}(\tilde{R})R^*T_1, \ g_4 = R^*.$$

Lemma 9.6.2. *For each $i = 1,\ldots,4$, the element $g_i \in \tilde{\Gamma}$ acts on $\mathbf{H}^3$ as the reflection in the totally geodesic surface $\mathbf{H}^2(\mathcal{B}_i)$.*

Proof. Inspection.

Theorem 9.6.3. *For $i = 1,\ldots,4$, we have $\mathbf{H}^2(\mathcal{B}_i) \subset \Gamma\partial\mathcal{F}$.*

Proof. Suppose not. Then there exists $\gamma \in \Gamma$ such that $\mathbf{H}^2(\mathcal{B}_{\mathrm{i}}) \cap \mathrm{Int}(\gamma\mathcal{F})$ is not empty for some i. By Lemmas 9.6.1 and 9.6.2 we have $g_i \in \mathrm{Nor}_{\mathcal{F}}(\Gamma)$. By Lemma 9.5.8 and Lemma 9.6.2, we get

$$\gamma\mathcal{F} \cap \mathbf{H}^2(\mathcal{B}_i) = \gamma\mathcal{F} \cap \mathrm{Fix}(g_i) = \gamma\tilde{R}\tilde{F}_{\mathbf{R},\mathbf{j}}.$$

Thus $\gamma\tilde{R}\tilde{\mathcal{F}}_{\mathbf{R},\mathbf{j}} \subset \mathbf{H}^2(\mathcal{B}_{\mathrm{i}})$. By **TP 1** (one of the trivial properties of Section 9.1) we conclude that $\mathbf{H}^2(\mathcal{B}_i) = \mathbf{H}^2(\gamma\tilde{R}\tilde{\mathcal{F}}_{\mathbf{R},\mathbf{j}})$. By Lemma 9.5.4 it follows that $\gamma\tilde{R}\tilde{\mathcal{F}}_{\mathbf{R},\mathbf{j}}$ is a fundamental domain for $\mathbf{c}(\gamma)\mathbf{c}(\tilde{R})\tilde{\Gamma}_{\mathbf{Z}}$ on $\mathbf{H}^2(\mathcal{B}_{\mathrm{i}})$.

Now, $\mathrm{Int}(\mathcal{B}_{\mathrm{i}})$ is a nonempty open subset of $\mathbf{H}^2(\mathcal{B}_i)$. By Lemma 9.3.3, there exists

$$\eta \in \mathbf{c}(\gamma)\mathbf{c}(\tilde{R})\tilde{\Gamma}_{\mathbf{Z}} \text{ such that } \mathrm{Int}(\eta\gamma\tilde{R}\tilde{\mathcal{F}}_{\mathbf{R},\mathbf{j}}) \cap \mathrm{Int}(\mathcal{B}_{\mathrm{i}}) \text{ is not emty.}$$

By Lemma 9.5.3 applied to $\eta\gamma\mathcal{F}$ instead of $\mathcal{F}$ and $\eta\gamma\tilde{R}\tilde{\mathcal{F}}_{\mathbf{R},\mathbf{j}}$ instead of $\tilde{R}\tilde{\mathcal{F}}_{\mathbf{R},\mathbf{j}}$, we have

$$\mathrm{Int}(\eta\gamma\tilde{R}\tilde{\mathcal{F}}_{\mathbf{R},\mathbf{j}}) \subset \mathrm{Int}(\eta\gamma\mathcal{F}).$$

Since $\mathcal{B}_{\mathrm{i}} \subset \partial\mathcal{F}$, the previous inclusions imply that

$$\mathrm{Int}(\eta\gamma\mathcal{F}) \cap \partial\mathcal{F} \text{ is not empty.}$$

Since $\eta\gamma \in \Gamma$, this contradicts the fundamental domain property, thereby proving the theorem.

9.7 The Set $\mathcal{S}$, its Boundary Behavior, and the Tube $\mathcal{T}$

We apply the preceding section to prove a major property of the region $\mathcal{S}$ below the fundamental domain in the tube $\mathcal{T}$, fundamental domain for Γ_∞. In Theorem 9.2.1, we gave a quite precise results on $\mathrm{SL}_2(\mathbf{R})$ and $\mathbf{H}^2$, by a direct formula. Here we shall prove the corresponding result in the complex case, on $\mathbf{H}^3$. Instead of a finite number of translates, we need an infinite number, and the proof is much more elaborate, requiring the machinery set up in Sections 9.5 and 9.6.

From Section 6.6, recall the **tube**

$$\mathcal{T} = \mathcal{F}_U \times \mathbf{R}^+ = \mathcal{F}_U + \mathbf{R}^+\mathbf{j},$$

and the set

$\mathcal{S}$ = subset of $\mathcal{T}$ lying below the unit sphere, that is, the set of $x + \mathbf{j}y$ with
$$x \in \mathcal{F}_U \text{ and } 0 < y \leqq 1 - |x|^2.$$

Thus $\mathcal{T} = \mathcal{F} \cup \mathcal{S}$. The tube is the convex closure

$$\mathcal{T} = \mathcal{C}_{\mathbf{H}}\left(-\frac{1}{2}, -\frac{1}{2}+\frac{1}{2}\mathbf{i}, \ \frac{1}{2}+\frac{1}{2}\mathbf{i}, \frac{1}{2}, \infty\right).$$

So $\mathcal{T}$ is the **H**-convex set with vertices at ∞ and the four points in **C** directly below the corners of $\mathcal{F}$.

The boundary of $\mathcal{S}$ consists of the **ceiling** $\mathcal{A}_0$, which is none other than the floor $\mathcal{B}_0$ of $\mathcal{F}$, and the four sides, which we denote by $\mathcal{A}_i$ $(i = 1, \ldots, 4)$. Each $\mathcal{A}_i$ occurs just below $\mathcal{B}_i$. We have

$$\partial\mathcal{S} = \bigcup_{i=0}^{4} \mathcal{A}_i.$$

Explicitly,

$$\mathcal{A}_1 = \mathcal{C}_H\left(p_1, p_2, -\frac{1}{2}+\frac{1}{2}\mathbf{i}, -\frac{1}{2}\right),$$

and similarly for the other $\mathcal{A}_i$ going clockwise. Each $\mathcal{A}_i$, like $\mathcal{A}_0 = \mathcal{B}_0$, is part of a totally geodesic surface in $\mathbf{H}^3$. In fact, for all $i = 0, \ldots, 4$,

$$\mathbf{H}^2(\partial\mathcal{A}_i) = \mathbf{H}^2(\mathcal{A}_i) = \mathbf{H}^2(\mathcal{B}_i). \tag{9.11}$$

Theorem 9.7.1. *On* $\mathbf{H}^3$, *we have the inclusion*

$$\partial\mathcal{S} \subset \bigcup_{\gamma\in\Gamma} \gamma\partial\mathcal{F}.$$

Proof. This is immediate from Theorem 9.6.3 and equality (9.11).

The inclusion of Theorem 9.7.1 will play a crucial role in the next section.

9.8 Tilings

We start with a definition of tilings under the following **general conditions**.

Definition. *Let X be a locally compact space, and $\mathcal{S}$ a closed subspace. Let Γ be a discrete group acting topologically on x, and let $\mathcal{F}$ be a fundamental domain, which we assume closed.* (Cf. condition **Til 2** below.)

By a $(\Gamma, \mathscr{F})$-**tiling** of $\mathscr{S}$ we mean a family of translations $\{\gamma_j \mathscr{F}\}_{j \in J}$ of $\mathscr{F}$ by a family of elements of Γ, such that if $j \neq j'$ then $\gamma_j \mathscr{F}$ and $\gamma_{j'} \mathscr{F}$ are disjoint except for boundary points, and $\mathscr{S}$ is the union

$$\mathscr{S} = \bigcup_{j \in J} \gamma_j \mathscr{F}.$$

Each $\gamma_j \mathscr{F}$ is called a **tile** (or $(\Gamma, \mathscr{F})$-tile if reference to the data is necessary). Note that **X** itself is tiled when $\{\gamma_j\}$ ranges over representatives of cosets of the subgroup Γ_{id} of Γ that acts as the identity on **X**. In the case of $\mathbf{H}^2$ or $\mathbf{H}^3$, this subgroup is $\pm I$.

For example, the whole space $\mathbf{H}^3$ is tiled by $\Gamma \bmod \pm 1$, directly from the definition of a fundamental domain: if two translations $\gamma \mathscr{F}$, $\gamma' \mathscr{F}$ have an interior point of one of them in common, then they are equal. We abbreviate "Interior" by **Int**. On $\mathbf{H}^3$ or $\mathbf{H}^2$, a tiling of $\mathscr{S}$ together with the fundamental domain $\mathscr{F}$ is a tiling of $\mathscr{T}$. Thus to get a tiling of $\mathscr{T}$ it suffices to get a tiling of $\mathscr{S}$.

Theorem 9.8.1. *Let $X = \mathbf{H}^2$ or $\mathbf{H}^3$, and let $\mathscr{S}$ be the region of the tube $\mathscr{T}$ below the standard fundamental domain $\mathscr{F}$ ($= \mathscr{F}_{\mathbf{R}}$ or $\mathscr{F}_{\mathbf{C}}$ as the case may be). Then $\mathscr{S}$ has a $(\Gamma, \mathscr{F})$-tiling, and the tiles of $\mathscr{S}$ consist of the tiles of X that are contained in $\mathscr{S}$.*

We are indebted to Eliot Brenner for the proof, under the following **tiling conditions** in addition to the general conditions:

Til 1. $\text{Int}(\mathscr{F})$ is pathwise connected.
Til 2. $\mathscr{S}$ and $\mathscr{F}$ are the closures of their interiors.
Til 3. The $(\Gamma, \mathscr{F})$-tiling of **X** is locally finite.
Til 4. $\partial \mathscr{S} \subset \bigcup_{\gamma \in \Gamma} \gamma \partial \mathscr{F} = \bigcup_{\gamma \in \Gamma} \partial(\gamma \mathscr{F})$.

The four conditions are satisfied in the case of $\mathbf{H}^2$ and $\mathbf{H}^3$. Conditions **Til 1** and **Til 2** are trivial. We proved **Til 3** in Chapter 6, Theorem 6.2.4, and **Til 4** in Theorems 9.1.1 and 9.6.1 of the present chapter. It will therefore suffice to prove the following:

Theorem 9.8.2. *Let X, Γ, $\mathscr{F}$, $\mathscr{S}$ satisfy the general conditions and the tiling conditions. Then $\mathscr{S}$ has a $(\Gamma, \mathscr{F})$-tiling, and the tiles of $\mathscr{S}$ consist of the tiles of X that are contained in $\mathscr{S}$.*

Proof. Directly from the definition of a fundamental domain, it suffices to prove that $\mathscr{S}$ is the union of Γ-translates of $\mathscr{F}$. We first deal with interior points and take care of them in a lemma.

Combining the two lemmas concludes the proof of Theorem 9.8.2.

Lemma 9.8.3. *Let $\gamma \in \Gamma$ be such that $\gamma \mathscr{F}$ intersects* $\text{Int}(\mathscr{S})$. *Then $\gamma \mathscr{F} \subset \mathscr{S}$.*

Proof. Let $w' \in \mathscr{F}$ be such that $\gamma w' \in \text{Int}(\mathscr{S})$. By **Til 2** and continuity, without loss of generality, we may assume $w' \in \text{Int}(\mathscr{F})$. Suppose $\gamma \mathscr{F} \not\subset \mathscr{S}$, so there exists $w'' \in \mathscr{F}$ with $\gamma w'' \notin \mathscr{S}$. Without loss of generality, we may assume $w'' \in \text{Int}(\mathscr{F})$

because the complement of $\mathscr{S}$ is open. Since by **Til 1** the interior $\mathrm{Int}(\mathscr{F})$ is pathwise connected, the γ-image of a path joining w', w'' in $\mathrm{Int}(\mathscr{F})$ must intersect $\partial\mathscr{S}$. So there exists $w \in \mathrm{Int}(\mathscr{F})$ such that $\gamma w \in \partial\mathscr{S}$. By **Til 4**, there exist $\gamma_1 \in \Gamma$, $w_1 \in \partial\mathscr{F}$ such that $\gamma w = \gamma_1 w_1$, so $w = \gamma^{-1}\gamma_1 w_1$, which contradicts the defining property of a fundamental domain. This proves that $\gamma\mathscr{F} \subset \mathscr{S}$, as was to be shown.

Lemma 9.8.4. *Let $z \in \partial(\mathscr{S})$. Then there exists $\gamma \in \Gamma$ such that $z \in \gamma(\mathscr{F})$ and $\gamma\mathscr{F}$ intersects* $\mathrm{Int}(\mathscr{S})$.

Proof. By **Til 2**, there exists a sequence $\{z_n\}$ in $\mathrm{Int}(\mathscr{S})$ such that $\lim z_n = z$. For n sufficiently large, the points z_n lie in a small neighborhood of z, and by **Til 3** there exists a translation $\gamma\mathscr{F}$ such that for infinitely many n, the point z_n lies in $\gamma\mathscr{F}$. Since $\gamma\mathscr{F}$ is closed, it follows that $z \in \gamma\mathscr{F}$. This proves the lemma.

9.8.1 Coset Representatives

On $\mathrm{SL}_2(\mathbf{C})$ one coset of $\Gamma_U\backslash\Gamma_\infty$ has a trivial representative $\gamma = I$ for which $\gamma(\mathscr{F}) = \mathscr{F}$. On the other hand, $R(\mathscr{F})$ is a Γ-fundamental domain above the lower rectangle $R(\mathscr{F}_U)$. The two elements $\{I, R\}$ are representatives of the cosets $\pm\Gamma_U\backslash\Gamma_\infty$. Also, $I \bmod \pm 1$ is the unique representative of the unit coset $\Gamma_\infty\backslash\Gamma_\infty$ such that $\gamma(\mathscr{F}) = \mathscr{F}$. The next theorem shows how a tiling of $\mathscr{T}$ corresponds to a family of coset representatives of $\Gamma_\infty\backslash\Gamma$. This amounts to a geometric representation of a splitting of the sequence

$$1 \to \Gamma_\infty \to \Gamma \to \Gamma_\infty\backslash\Gamma.$$

Part (i) of the next theorem justifies the following *definition*:

$\mathrm{Rep}_{\mathscr{S}}$ =family of coset represntatives γ of $\Gamma_\infty\backslash\Gamma$ (determined mod ± 1) such that $\gamma(\mathscr{F}) \subset \mathscr{S}$.

Theorem 9.8.5. **(i)** *Let $\Gamma_\infty\gamma$ be a coset with $\gamma \in \Gamma - \Gamma_\infty$. Then the coset representative γ can be chosen such that $\gamma(\mathscr{F}) \subset \mathscr{S}$, and such a representative is unique* $\bmod \pm 1$.

(ii) *If γ, $\gamma' \in \Gamma$ represent distinct cosets of Γ_∞, then $\gamma(\mathscr{F})$ and $\gamma'(\mathscr{F})$ have at most boundary points in common.*

(iii) *The family* $\mathrm{Rep}_{\mathscr{S}}$ *is the unique family of coset representatives* $\bmod \pm 1$ *for $\Gamma_\infty\backslash(\Gamma - \Gamma_\infty)$ such that*

$$\bigcup_{\gamma \in \mathrm{Rep}_{\mathscr{F}}} \gamma(\mathscr{F}) = \mathscr{S}.$$

Thus the family $\{\gamma\mathscr{F}\}$ ($\gamma \in \mathrm{Rep}_{\mathscr{S}}$) is a tiling of $\mathscr{S}$.

We then let

$$\mathrm{Rep}_{\mathscr{T}} = \mathrm{Rep}_{\mathscr{S}} \mathrm{U}\{I\}, \text{ so that } \bigcup_{\gamma \in \mathrm{Rep}_{\mathscr{T}}} \gamma(\mathscr{F}) = \mathscr{T}.$$

Thus we obtain a tiling of the tube $\mathscr{T}$.

The proof of Theorem 9.8.5 is immediate from Theorem 9.7.1. Formally, the key is in (ii) of the following lemma.

Lemma 9.8.6. *Let* $\gamma \in \Gamma - \Gamma_\infty$.

(i) *If* $\eta, \eta' \in \Gamma_\infty$ *then* $\eta'\gamma\eta \in \Gamma - \Gamma_\infty$.
(ii) *There exists* $\gamma_\infty \in \Gamma_\infty$ *such that* $\gamma_\infty\gamma\mathscr{F} \subset \mathscr{S}$, *and* γ_∞ *is unique* mod ± 1.
(iii) *The elements* γ_∞, $R\gamma_\infty$ *represent distinct cosets of* $\pm\Gamma_U\backslash\Gamma_\infty$.

Proof. (i) is immediate, with any subgroup replacing Γ_∞. For (ii), since $\mathscr{T}$ is a fundamental domain for Γ_∞, given $z \in \mathrm{Int}(\mathscr{F})$ there exists $\gamma_\infty \in \Gamma_\infty$ such that $\gamma_\infty\gamma z \in \mathscr{T}$. Furthermore $\gamma_\infty\gamma z \notin \mathscr{F}$ (otherwise $\gamma_\infty\gamma = \pm I$, $\gamma \in \Gamma_\infty$). Then Theorem 9.8.1 implies (ii). Finally (iii) is trivial from (ii).

The elements $\gamma \in \Gamma$ that are not in Γ_∞ are those of the form

$$\gamma = \begin{pmatrix} a & b \\ c & d \end{pmatrix} \quad \text{with } c \neq 0.$$

We write $c = c_\gamma$ or $c(\gamma)$ if we want to make explicit the connection of c with γ. The next lemma describes a subset of cosets that will be useful in Section 9.9.

Lemma 9.8.7. *Let* $\gamma \in \Gamma - \Gamma_\infty$. *Then the elements of* $\gamma\Gamma_U$ *lie in distinct cosets of* $\Gamma_\infty\backslash(\Gamma - \Gamma_\infty)$.

Proof. Suppose we have an equality $\gamma u_1 = \gamma_\infty\gamma u_2$ with $\gamma_\infty \in \Gamma_\infty$ and $u_1, u_2 \in \Gamma_U$. Then $\gamma u_1 u_2^{-1}\gamma^{-1} = \gamma_\infty$, whence $u_1 = u_2$ since $c(\gamma) \neq 0$, and we can apply the formula

$$c(\gamma u(x)\gamma^{-1}) = -c(\gamma)^2 x$$

from Section 7.4, (7.4.3). This proves the lemma.

9.9 Truncations

The set $\mathscr{S}$ is defined as part of the tube, bounded from above by the spherical inequality. We shall truncate $\mathscr{T}$ and $\mathscr{F}$ in other ways. Let $Y > 1$. We use systematically the notation

$$\mathscr{F}_{<Y}, \mathscr{F}_{\leqq Y}, \mathscr{F}_{>Y}, \mathscr{F}_{\geqq Y} \text{ and similarly with } \mathscr{T}_{<Y}, \mathscr{T}_{\leqq Y}, \mathscr{T}_{>Y}, \mathscr{T}_{\geqq Y},$$

where for instance

$$\mathscr{T}_{>Y} = \{x + \mathbf{j}y \text{ such that } x \in \mathscr{F}_U \text{ and } y > Y\} = \mathscr{F}_U \times (Y, \infty).$$

The other inequality signs may be used instead, and the notation applies mutatis mutandis to $\mathscr{F}$ and $\mathscr{T}$ respectively. We have for instance

$$\mathscr{T}_{\leqq Y} = \mathscr{S} \cup \mathscr{F}_{\leqq Y} = \mathscr{F}_U \times (0, Y].$$

The set $\mathscr{F}_{\leqq Y}$ is called the **truncated fundamental domain** (at Y).

As in the previous section, we let γ be the matrix with components a, b, c, d.

Recall that on the usual upper half-plane $\mathbf{H}^2$, if $c \neq 0$, and $\gamma \in \mathrm{SL}_2(\mathbf{R})$, then the fractional linear map sends ∞ on a/c. Here on $\mathbf{H}^3$, we note that $\gamma \in \Gamma - \Gamma_\infty$ sends $\mathbf{j}\infty$ to a/c also. Furthermore, on $\mathbf{H}^2$ a fractional linear map sends lines to lines or circles. Here we are interested in a specific analogue, namely for fixed $y > 0$, the plane of elements $x + \mathbf{j}y$ with $x \in \mathbf{C}$. The next proposition describes the image of this plane under γ. Denote this plane by $\mathbf{C} + \mathbf{j}y$. Let

$$r_{c,y} = \frac{1}{2|c|^2 y},$$

As always,

$$\gamma = \begin{pmatrix} a & b \\ c & d \end{pmatrix}.$$

We shall consider the **open ball** (for $c \neq 0$):

$$\mathbf{B}_{a,c,Y} = \mathbf{B}_{\gamma,Y} \text{ centered at } (a/c, r_{c,Y}), \text{ of radius } r_{c,Y}.$$

Its boundary is the **sphere of center** $(a/c, r_{c,Y})$ **and radius** $r_{c,Y}$, defined by the equation

$$\left| x' - \frac{a}{c} \right|^2 + \left| y' - r_{c,y} \right|^2 = r_{c,y}^2.$$

It is the sphere tangent to $\mathbf{C}$ at a/c, of radius $r_{c,y}$.

Proposition 9.9.1. *Let $\gamma \in \Gamma - \Gamma_\infty$. The image of the plane $\mathbf{C} + \mathbf{j}y$ under γ is the sphere with center $(a/c, r_{c,y})$ and radius $r_{c,y}$ omitting the point a/c.*

Proof. For general $\gamma = \begin{pmatrix} a & b \\ c & d \end{pmatrix}$, we know that for $z = x + \mathbf{j}y$,

$$y(\gamma(z)) = \frac{y(z)}{|cz+d|^2} \text{ and } x(\gamma(z)) = \frac{(ax+b)(\overline{cx+d}) + a\bar{c}y^2}{|cz+d|^2}$$

Let $y' = y(\gamma(z))$ and $x' = x(\gamma(z))$. We first show that (x', y') lies on the sphere.

We have

$$\begin{aligned} x' - \frac{a}{c} &= c^{-1}|cz+d|^{-2}(c(ax+b)(\overline{cx+d} + ac\bar{c}y^2 - z|cz+d|^2) \\ &= c^{-1}|cz+d|^{-2}((cb-ad)\bar{c}\bar{x} + (cb-ad)\bar{d}) \\ &= c^{-1}|cz+d|^2(-\bar{c}\bar{x} - \bar{d}). \end{aligned}$$

Also

$$y' - r_{c,y} = \frac{2|c|^2y^2 - |cz+d|^2}{2|c|^2y|cz+d|^2} = \frac{|c|^2y^2 - |cx+d|^2}{2|c|^2y|cz+d|^2}.$$

Then

$$\begin{aligned}\left|x' - \frac{a}{c}\right|^2 + \left|y' - r_{c,y}\right|^2 &= \frac{|cx+d|^2}{|c|^2|cz+d|^4} + \frac{(|c|^2y^2 - |cx+d|^2)^2}{(2|c|^2y)^2|cz+d|^4}\\ &= (2|c|^2y)^{-2}|cz+d|^{-4}(4|c|^2y^2|cx+d|^2 + (|c|^2y^2 - |cx+d|^2)^2)\\ &= (2|c|^2y)^{-2}|cz+d|^{-4}(|cz+d|^2)^2 = \frac{1}{(2|c|^2y)^2} = r_{c,y}^2,\end{aligned}$$

which proves that (x',y') lies on the prescribed sphere.

The converse is done similarly, using the inverse

$$\gamma^{-1} = \begin{pmatrix} d & -b \\ -c & a \end{pmatrix}$$

and showing that if (x',y') lies on the sphere, then its image under γ^{-1} has $\mathbf{j}$-coordinate equal to y. This concludes the proof.

Let $Y > 0$. Let $\mathbf{H}^3_{>Y}$ be the set of elements $x + \mathbf{j}y$ with $y > Y$. We call this set the **half-space above height** Y.

Corollary 9.9.2. *Let* $\gamma \in \Gamma - \Gamma_\infty$. *Then* $\gamma\mathbf{H}^3_{>Y} = \mathbf{B}_{\gamma,Y}$.

Proof. As y increases, the radii in Proposition 9.9.1 decrease and the centers decrease, so using all values of $y > Y$, the spheres fill out the interior of the sphere with parameter Y, as asserted.

We now look at the effect of the elements of $\gamma\Gamma_\infty$ on the upper part of the fundamental domain, so we take $Y > 1$, and consider $\mathscr{F}_{>Y}$. Proposition 9.9.1 and Corollary 9.9.2 determined the image under γ of the horizontal plane $\mathbf{C} + \mathbf{j}Y$ (with fixed Y), as well as the image of the region above this plane, i.e., the half-space with $y > Y$.

For any $\gamma \in \mathrm{Rep}_{\mathscr{S}}$, we *define the subset* S_γ *of* $\mathrm{Rep}_{\mathscr{S}}$ by two equivalent conditions. *First,*

$$S_\gamma^{(1)} = \{\gamma\eta\}_{\eta \in \Gamma_{\infty/\pm 1}},$$

choosing η among the two $\pm$ representatives so that $S_\gamma^{(1)} \subset \mathrm{Rep}_{\mathscr{S}}$. *Second,*

$$S_\gamma^{(2)} = \left\{\gamma' \in \mathrm{Rep}_{\mathscr{S}} \text{ such that } \gamma'(\mathbf{j}\infty) = \gamma(\mathbf{j}\infty)\right\}.$$

We now prove the equivalence of the two conditions. If $\gamma' \in S_\gamma^{(1)}$, then $\gamma' = \gamma\eta$ for some $\eta \in \Gamma_\infty$. Clearly, $\eta(\mathbf{j}\infty) = \mathbf{j}\infty$, so then $\gamma'(\mathbf{j}\infty) = \gamma(\mathbf{j}\infty)$; hence $\gamma' \in S_\gamma^{(2)}$, and we have shown that $S_\gamma^{(1)} \subset S_\gamma^{(2)}$.

Conversely, if $\gamma' \in S_\gamma^{(2)}$, then $\gamma'(\mathbf{j}\infty) = \gamma(\mathbf{j}\infty)$, or $(\gamma^{-1}\gamma')(\mathbf{j}\infty) = \mathbf{j}\infty$, so up to $\pm$ representative, we have for some $\eta \in \Gamma_U$ that $\gamma^{-1}\gamma' = \eta$ or $\gamma^{-1}\gamma' = \eta R$. In the first case, $\gamma' = \gamma\eta$, and, in the second, $\gamma' = \gamma\eta R$; in either case, $\gamma' = \gamma\tilde{\eta}$ for some $\tilde{\eta} \in \Gamma_\infty$, so $\gamma' \in S_\gamma^{(1)}$, and we have shown that $S_\gamma^{(2)} \subset S_\gamma^{(1)}$.

Let $\gamma \in \Gamma - \Gamma_\infty$, and write

$$\gamma(\mathbf{j}\infty) = (x_1(a/c).x_2(a/c)),$$

recalling that if

$$\gamma = \begin{pmatrix} a & b \\ c & d \end{pmatrix} \text{ then } \gamma(\mathbf{j}\infty) = a/c.$$

We use the notation

$$\|a/c\|_\infty = \max(|x_1(a/c)|, |x_2(a/c)|).$$

Note that if $\gamma \in \mathrm{Rep}_{\mathscr{S}}$, then $\|a/c\|_\infty \le 1/2$. Indeed, for $\gamma \in \mathrm{Rep}_{\mathscr{S}}$ and $\gamma' \in S_\gamma$, then $\gamma'(\mathbf{j}\infty) \in \mathscr{F}_U$, so

$$-1/2 \le x_1(\gamma'(\mathbf{j}\infty)) \le 1/2 \text{ and } 0 \le x_2(\gamma'(\mathbf{j}\infty)) \le 1/2.$$

The next two theorems deal successively with those γ such that $\gamma(\mathbf{j}\infty)$ lies in the interior of $\mathscr{F}_U$, and those such that $\gamma(\mathbf{j}\infty)$ lies on the boundary.

Theorem 9.9.3. *Let $\gamma \in \mathrm{Rep}_{\mathscr{S}}$ be such that $\gamma(\mathbf{j}\infty) = a/c \in \mathrm{Int}(\mathscr{F}_U)$. Then for Y sufficiently large, we have*

$$\bigcup_{\gamma' \in S_\gamma} \gamma' \mathscr{F}_{\geqq Y} = \mathbf{B}_{\gamma,Y} \subset \mathscr{S}.$$

Proof. From the $S_\gamma^{(1)}$ realization of S_γ, we have

$$\bigcup_{\gamma' \in S_\gamma} \gamma' \mathscr{F}_{\geq Y} = \gamma \mathbf{H}^3_{\geqq Y} = \mathbf{B}_{\gamma,Y},$$

where the last equality follows from Corollary 9.9.2. The sphere around $\mathbf{B}_{\gamma,Y}$ is given in Proposition 9.9.1. Clearly, for $a/c \in \mathrm{int}(\mathscr{F}_U)$, when Y is sufficiently large, $\mathbf{B}_{\gamma,Y} \subset \mathscr{S}$, which proves the theorem.

Next we deal with the case in which $\gamma(\mathbf{j}\infty)$ lies on the boundary, which splits into two subcases: whether or not the point lies at the corners $\pm 1/2$, $\pm 1/2 + \mathbf{i}/2$.

Theorem 9.9.4. *Let $\gamma \in \mathrm{Rep}_{\mathscr{S}}$ be such that $\gamma(\mathbf{j}\infty) = a/c \notin \mathrm{Int}(\mathscr{F}_U)$.*

(i) *If $\gamma(\mathbf{j}\infty) = a/c$ is not a corner, then for Y sufficiently large, we have*

$$\bigcup_{\gamma' \in S_\gamma} \gamma' \mathscr{F}_{\geqq Y} = B_1^{1/2} \cup B_2^{1/2},$$

where the two sets $B_1^{1/2}$ and $B_2^{1/2}$ are half-spheres lying in $\mathscr{S}$, whose union is Γ_∞-equivalent to a sphere with center $(a/c, r_{c,Y})$, radius $r_{c,Y}$ (different γ's for each of the two half-spheres).

(ii) *If $\gamma(\mathbf{j}\infty) = a/c$ is a corner, then for Y sufficiently large, we have*

$$\bigcup_{\gamma' \in S_\gamma} \gamma' \mathscr{F}_{\geqq Y} = Q_1 \cup Q_2 \cup Q_3 \cup Q_4,$$

where the four sets Q_1, $Q_2 \cdot Q_3$, Q_4 are quarter-spheres lying in $\mathscr{S}$, and whose union is Γ_∞-equivalent to a sphere with center $(a/c, r_{c,Y})$ and radius $r_{c,Y}$ (different γ's for each of the four quarter-spheres).

Proof. As in the proof of Theorem 9.9.3,

$$\bigcup_{\gamma' \in S_\gamma} \gamma' \mathscr{F}_{\geqq Y} = \gamma \mathbf{H}^3_{\geqq Y} = \mathbf{B}_{\gamma,Y}.$$

However, since $\gamma(\mathbf{j}\infty)$ lies in the boundary of $\mathscr{F}_U$, the ball $\mathbf{B}_{\gamma,Y}$ will not be entirely contained in $\mathscr{S}$ for any Y. We now consider the two cases, assuming that $\gamma(\mathbf{j}\infty)$ is in the boundary of $\mathscr{F}_U$: first when $\gamma(\mathbf{j}\infty)$ is not one of the corners of the fundamental rectangle, and second when $\gamma(\mathbf{j}\infty)$ is one of the corners.

In the first case, for Y sufficiently large, no point in $\mathbf{B}_{\gamma,Y}$ will have x-coordinate equal to $\pm 1/2$ or $\pm 1/2 + \mathbf{i}/2$. For such Y, $\mathbf{B}_{\gamma,Y} \cap \mathscr{S}$ is equal to a half-sphere, which we call $B_1^{1/2}$. The complement of $B_1^{1/2}$ in $\mathbf{B}_{\gamma,Y}$ is also a half-sphere, call it $\tilde{B}_1^{1/2}$, which lies in some Γ_∞-translate of $\mathscr{S}$ (indeed, the Γ_∞ translate is defined by one of the actions $z \mapsto z \pm 1$, $z \mapsto z \pm \mathbf{i}$). Using the inverse map associated with this translation one shows that $\tilde{B}_1^{1/2}$ is Γ_∞-equivalent to a second half-sphere in $\mathscr{S}$, which we denote by $B_2^{1/2}$.

A similar argument applies in the second case, when $\gamma(\mathbf{j}\infty)$ is a corner. In both cases, a sketch of the geometry clarifies the description. See the first figure in Section 13.5.

Chapter 10
The $\Gamma_U \backslash U$-Fourier Expansion of Eisenstein Series

This chapter gives basic analytic objects we shall work with. We let $G = \mathrm{SL}_2(\mathbf{C})$ as usual, and let K = unitary subgroup. We let $\Gamma = \mathrm{SL}_2(\mathbf{Z}[\mathbf{i}])$ be the discrete subgroup with coefficients in the Gaussian integers. We then study functions on the space $\Gamma\backslash G$, or $\Gamma\backslash G/K$, especially functions on G that are Γ -invariant on the left and K-invariant on the right. The group G has an Iwasawa decomposition $G = UAK$, each element $g \in$ G being expressible uniquely as a product $g = uak$, with

$$u = u(x) = \begin{pmatrix} 1 & x \\ 0 & 1 \end{pmatrix}, \; x \in \mathbf{C}; \quad a = \begin{pmatrix} a_1 & 0 \\ 0 & a_1^{-1} \end{pmatrix}, \; a_1 > 0; \text{ and } k \in K.$$

Then $U \approx \mathbf{C}$ and $\Gamma_U \approx \mathbf{Z}[\mathbf{i}]$. Thus $\Gamma_U/U = U/\Gamma_U$ is isomorphic to the real torus $\mathbf{C}/\mathbf{Z}[\mathbf{i}] = \mathbf{R}^2/\mathbf{Z}^2$. A function on U that is Γ_U-invariant is thus a function on the torus and has a Fourier series expansion in the ordinary sense, from ordinary analysis on the commutative torus. This chapter determines the Fourier series of Eisenstein series, which are eigenfunctions of Casimir (or of convolution of the heat kernel) depending continuously on a parameter s.

These Eisenstein series are used for the continuous part of the eigenfunction expansion on $\Gamma\backslash G/K$. The ordinary Fourier analysis of this chapter will be used in the next chapter for the more extensive result on the noncommutative group G. We now give a summary statement of the material leading up to the expansion formula of Chapter 11.

For the Fourier expansion in the case of imaginary quadratic fields, see Elstrodt–Grunewald–Mennicke [EGM 85] and [EGM 98], Chapter 8.

10.1 Our Goal: The Eigenfunction Expansion

The situation is analogous to that of Fourier series for periodic functions and the Fourier transform (integral) on the noncompact $\mathbf{R}$ (or Euclidean space), but will include here aspects of both, suitably jazzed up to take into account the noncommutativity of the group structure in addition to the noncompactness.

The kernel function replacing the e^{ixy} of the usual Fourier transform will be first defined by a series, the **Eisenstein series**

$$E(s,z) = \sum_{\gamma\in\Gamma_u\backslash\Gamma} y(\gamma z)^s.$$

This series converges only for Re(s) > 2, so we shall have to determine an analytic continuation (related as usual to a functional equation). Given a function f on $\Gamma\backslash G/K$, we say that f is **cuspidal** if the constant term of its $\Gamma_U\backslash U$ Fourier series is 0, that is,

$$\int_{\Gamma_U\backslash U} f(ug)du = 0 \text{ for all } g \in G.$$

The functions E_s (such that $E_s(z) = E(s,z)$) will be considered for $\mathrm{Re}(s) = 1$, so $s = 1+\mathbf{i}r$ with $r \in \mathbf{R}$. The **Eisenstein coefficient** of a function f with respect to E_s is

$$\langle f, E_s\rangle = \int_{\Gamma\backslash G} f(w)\overline{E(s,w)}d\mu(\mathrm{w}) = \int_{\Gamma\backslash G} f(w)E(1-\mathbf{i}r,w)d\mu(\mathrm{w}),$$

with the Iwasawa-quaternionic measure $d\mu(\mathrm{w})$. Note that this scalar product can also be written in terms of a convolution

$$\langle f, E_S\rangle = (E * f)(\bar{s}).$$

Both ways of looking at it will occur naturally.

The main result we are striving for is the expansion formula

$$f(z) = f_{\mathrm{cusp}}(z) + c_{\mathrm{o}} + c_E \int_{-\infty}^{\infty} \langle f, E_{1+\mathbf{i}r}\rangle E_{1+\mathbf{i}r}(z)dr,$$

with suitable constants c_{o}, c_E to be determined explicitly, the function f_{cusp} being cuspidal. Of course, we have to specify which functions. We carry out the proof first for $f \in C_c^\infty$ $(\Gamma\backslash G/K)$, and next for Gaussians that have been Γ -periodized. This extension is in line with our general structural framework, which finally will use the ultimately normalized periodized heat Gaussian.

With f and f_{cusp} in L^2, we have the series expansion of Chapter 8,

$$f_{\mathrm{cusp}}(z) = \sum \langle f, \psi_n\rangle \psi_n(z)$$

in terms of an orthonormal basis of L^2_{cus} $(\Gamma\backslash G/K)$. We will want (and have) enough smoothness and rapid decay to prove that the expansion also holds pointwise. Here, we deal with the usual Fourier coefficients

$$\langle f_{\mathrm{cus}}, \psi_n\rangle = \int_{\Gamma\backslash G} f_{\mathrm{cus}}(\mathrm{w})\overline{\psi_n(\mathrm{w})}d\mu(\mathrm{w}).$$

The reader will note the similarity between the Fourier series and the Eisenstein integral, the latter being a continuous analogue of the former.

We want to apply this construction to the function

$$f_t(z) = \mathbf{K}_t^{\Gamma}(z,z) = \sum_{\gamma\in\Gamma} \mathbf{g}_t(z^{-1}\gamma z),$$

namely the Γ-periodized heat kernel on the diagonal. From Chapter 4, we know that we are getting into the danger zone if we try to integrate this over $\Gamma\backslash G$. However, we obtain the identity

$$\mathbf{K}_t^{\Gamma}(z,z') = \sum_{n=1}^{\infty} e^{-\lambda_n^t}\psi_n(z)\overline{\psi_n(z')} + c_0 + c_E \int_{-\infty}^{\infty} e^{-2(1+r^2)t} E(1-\mathbf{i}r,z)E(1+\mathbf{i}r.z')dr$$

with a suitable orthonormal basis consisting of eigenfunctions for Casimir. Each $-\lambda_n$ is the eigenvalue of ω on ψ_n. Similarly, $-2(1+r^2)$ is the eigenvalue of Casimir on the Eisenstein function $\mathrm{E}_{1+\mathbf{i}r}$. For $z = z'$, the expansion becomes

$$\mathbf{K}_t^{\Gamma}(z,z) = \sum_{n=1}^{\infty} e^{-\lambda_n t}\,|\psi_n(z)|^2 + c_0 + c_E \int_{-\infty}^{\infty} e^{-2(1+r^2)t}|E(1+\mathbf{i}r,z)|^2 dr.$$

At this point, we have to face the next step, which will allow us to regularize the situation so we can integrate over $\Gamma\backslash G$ safely. This will be the subject of the Eisenstein–cuspidal affair in Chapter 12.

Of course, once all this is carried out, the next step consists in extending it to higher dimension, e.g., $\mathrm{SL}_n(\mathbf{C})$. Foundations are provided in [JoL 02], where we define the term that will replace the Eisenstein integral, and give the conjectural eigenfunction expansion for the heat kernel; see **EFEX 1** of Section 4.6, the preceding conjectures, and the comments in Section 4.7.

10.2 Epstein and Eisenstein Series

Let $\mathbf{F} = \mathbf{Q}(\mathbf{i})$ and $o = \mathbf{Z}[\mathbf{i}]$. Then o is the subring of algebraic integers in $\mathbf{F}$, and is a principal ring, with unique factorization. Its units are $\pm 1, \pm\mathbf{i}$. These are the only elements of absolute value 1 in o. We define the **Epstein series**

$$\mathrm{Ep}(s,z) = \sum_{\substack{(c,d)\neq(0,0)\\ c,d\in o}} y(c,d;z)^s \sum \frac{y(z)^s}{|cz+d|^{2s}}.$$

Theorem 10.2.1. *Let $\delta > 0$. Then the Epstein series converges absolutely in the half-plane*

$$\mathrm{Re}(s) \geqq 2+\delta,$$

and defines a bounded function in this half-plane, uniformly for z in a compact set.

Proof. Uniformly under the stated conditions, we have

$$|cz+d|^2 = |cx+d|^2 + |c|^2y^2 \gg \|(c,d)\|^2$$

for a given norm $\|\cdot\|$ on $\mathbf{C}\times\mathbf{C}$, and (c,d) ranging over the lattice $o\times o$. One could take the Euclidean norm of $\mathbf{R}^4$, or the sup norm of the coordinates on $\mathbf{C}\times\mathbf{C}$; it doesn't matter. For a positive integer n, let $A(n,n+1)$ be the annulus of elements $(x,x')\in\mathbf{C}\times\mathbf{C}$ such that

$$n \leqq \|(x,x')\| \leqq n+1.$$

Then the Epstein series is dominated for s real positive by

$$|\mathrm{Ep}(s,z)| \ll \sum_{n=1}^{\infty} \sum_{(c,d)\in A(n,n+1)} \frac{1}{\|(c,d)\|^{2\cdot s}}.$$

The annulus has volume $\ll n^3$, and so the number of lattice points in the annulus is $\ll n^3$. Hence the inner sum on the lattice points in the annulus has the bound

$$\ll n^3 \cdot \frac{1}{n^{2s}}.$$

Thus the double series dominating the Epstein series converges for $s = 2+\varepsilon$, thus proving the theorem.

We let

$$\Gamma = \mathrm{SL}_2(o)$$

be the discrete subgroup analogous to $\mathrm{SL}_2(\mathbf{Z})$. We let $\Gamma_U = \Gamma\cap U$ be the group of elements $u(b)$ with $b\in o$. For $\gamma\in\Gamma$, $\gamma = \begin{pmatrix} a & b \\ c & d\end{pmatrix}$, the map

$$\begin{pmatrix} a & b \\ c & d\end{pmatrix} \mapsto (c,d)$$

induces a surjection of Γ on the set of relatively prime pairs of elements of o. Indeed, we may view $\mathrm{SL}_2(\mathbf{C})$ as acting on the right of horizontal vectors in $\mathbf{C}^2$. Then the isotropy group of the unit vector $e_2 = (0,1)$ is the group U. Hence Γ_U is the isotropy group in Γ, and we get a bijection

$$\Gamma_U\backslash\Gamma \mapsto \text{relatively prime pairs } (c,d) \text{ in } o.$$

We define the **Eisenstein series** essentially as in Kubota [Kub 68] (see the remarks below). There are three variations. One of them is via the action on $\mathbf{H}^3$, namely for complex s and $z\in\mathbf{H}^3$:

$$E(s,z) = \sum_{\gamma\in\Gamma_U\backslash\Gamma} y(\gamma z)^s. \tag{10.1}$$

By Section 10.1, (10.2), we can rewrite the series in the form

$$E(s,z) = \sum_{(c,d)=1} y(c,d;z)^s, \tag{10.2}$$

the sum being taken over relatively prime pairs (c,d) of elements of o. Since the terms occurring in this sum are part of the terms occurring in the Epstein series, we see that Theorem 10.2.1 applies to give the absolute convergence of the Eisenstein series in the half-plane of the Epstein series, with the same uniformity.

We may also write the Eisenstein series with $g \in G$ instead of z in the form

$$E(s,g) = \sum_{\gamma \in \Gamma_U \backslash \Gamma} (\gamma g)_A^{s\alpha}. \tag{10.3}$$

Each term in the series depends only on the coset gK.

Remark 1. These ways of writing are the cleanest when one is using the group theory on G, Γ, and subgroups. Because of the units in o, terms of the series are counted four times. An alternative would be to define the **Eisenstein series** by using Γ_∞ rather than Γ_U, that is

$$E^\infty(s,z) = \sum_{\Gamma_\infty \backslash \Gamma} y(\gamma z)^s.$$

Furthermore, classical normalizations take these units into account by summing only over certain equivalence classes of pairs, say, up to multiplication by a unit. Thus the above series eliminates a factor 1/4 in the definition of Eisensteir series in [Kub 68], [Asa 70], Szmidt [Szm 83], [JoL 99]. With our definition, the factor 1/4 will reappear in due course, i.e., the eigenfunction expansion of Chapter 11, Theorem 11.3.5, et seq.

Remark 2. The literature is split or whether to use the exponent s or $2s$, and even $1+s$; cf. [EGM 85]. We go with s as in [Kub 68], [Szm 83] for the following reason. The characters on A have two natural bases: α or the dual basis α' relative to the natural trace form. In so-called higher-rank r, it would be $\alpha_1, \ldots, \alpha_r$ and $\alpha'_1 \ldots, \alpha'_r$. Then $\alpha' = \alpha/2$ in our case, and in higher rank, $\alpha'_1, \ldots, \alpha'_r$ can also be given explicitly; cf. for instance [JoL 01] Section 1.4. A character can then be expressed as a linear combination of either base, but positivity is exhibited more clearly in terms of the dual basis, being equivalent to the property that in terms of the dual basis, the coefficients are positive. In addition, in the case of a complex number field such as $\mathbf{Q}(\mathbf{i})$, it is natural to take the norm $\mathbf{N}y = y^2$ to fit number fields, so

$$y^s = (\mathbf{N}y)^{s\alpha'} = a^{s\alpha}.$$

Thus s is the coefficient of the character with respect to the dual basis, but there is a cancellation of the factor 2. Our normalization is influenced by higher-dimensional structures, and thus extends in a natural way to the higher-rank case, where the domain of convergence would be a right half-space in terms of the coefficients of the dual basis $\alpha'_1, \ldots, \alpha'_r$.

The sum over the relatively prime pairs (c,d) is the one that will allow us to connect with the Dedekind zeta function of $\mathbf{Q}(\mathbf{i})$. Obviously, taking only such pairs makes things more complicated than dealing with all pairs, but we defined an **Epstein series** for which the sum is taken over all pairs (c,d) with $c,\ d \in o$ and not both $c,\ d = 0$. Let $\zeta_{\mathbf{F}}$ be the Dedekind zeta function of $\mathbf{F}$, so by definition,

$$\zeta_{\mathbf{F}}(s) = \sum \mathbf{N}a^{-s},$$

the sum being taken over all nonzero ideals a of o. In the special present case in which all ideals are principal, an ideal is determined by a generator, up to multiplication by a unit. Then we have the relation between the Eisenstein and Epstein series:

$$\mathrm{Ep}(s,z) = \zeta_{\mathbf{F}}(s)E(s,z) = \sum_{(c,d)\neq(0,0)} y(c,d;z)^s = \sum_{(c,d)\neq(0,0)} \frac{y^s}{|cz+d|^{2s}}. \tag{10.4}$$

This is obvious, because for an ideal $a = (\nu)$ we have $\mathbf{N}a = |\nu|^2$, and

$$y(\nu c, \nu d; z) = |\nu|^{-2} y(c,d;z).$$

Thus one obtains all pairs from the primitive ones by multiplication by a nonzero element of o. Since the absolute value of a unit is 1, using the norm of the ideal introduces no further duplications of terms.

Our normalization of the Eisenstein series is such that our s is the $2s$ from [JoL 99]. Thus the edge of the half-plane of convergence is now at $\mathrm{Re}(s) = 2$, whereas it was $\mathrm{Re}(s) = 1$ in that reference.

Both Epstein and Eisenstein series are Mellin transforms of **theta series**, defined generally as series of type

$$\sum a_j e^{-\lambda_j t}.$$

For Epstein and Eisenstein, the series are

$$\sum e^{-\lambda(c,d,z)t},$$

where

$$\lambda(c,d,z) = |cz+d|^2/y = y(c,d,z)^{-1} = y(\gamma z)^{-1} = \lambda(\gamma, z).$$

In this chapter, the Mellin transform will be used in the proof of the Fourier series expansion of Theorem 10.5.3 below. The need is also arising to investigate more systematically the asymptotic expansions associated with such series. In particular, this would include an analogue of the Hardy–Littlewood "approximate formula" [HaL 21]. cf. Chapter 14, Theorems 14.1.1 and 14.3.1 for Riemann–Dedekind thetas and zetas.

10.3 The K-Bessel Function

In this section, for the convenience of the reader, we reproduce basic properties of the K-Bessel function. We use the integral definition; cf. [Lan 73/87], Section 20.3. Let a, b be real numbers > 0. Define

K1. $$K_s(a,b) = \int_0^\infty e^{-(a^2t+b^2/t)} t^s \frac{dt}{t}.$$

This is like an integral for the gamma function, but is much better, because first, it is more symmetric, involving both t and $1/t$, and second, it converges absolutely for all complex s, because the presence of $1/t$ cures the blowup that occurs for the gamma integral near 0. So it defines a function entire in s.

Note the formula

$$K_S(1,0) = \Gamma(s).$$

Thus the K-Bessel function may be viewed as a family parametrized by (a,b), and a degenerate member of this family is the gamma function. Of course, the integral defining this special member converges only for $\mathrm{Re}(s) > 0$.

Let us use the invariance of the integral under multiplicative translations, and let $t \mapsto \frac{b}{a}t$. We find that

K2. $$K_s(a,b) = \left(\frac{b}{a}\right)^s K_s(ab) = \left(\frac{b}{a}\right)^{2s} K_s(b,a),$$

where for $c > 0$ we define

K3. $$K_s(c) = \int_0^\infty e^{-e(t+1/t)} t^s \frac{dt}{t}.$$

Warning. For our purposes, we use a normalization of the Bessel integral slightly different from the classical one. If, for instance, $K_s^B(c)$ denotes the K-Bessel function that one finds in tables (see Magnus, Oberhettinger, Erdelyi, Whittaker and Watson, etc.), then we have the relation

$$2K_s^B(2c) = K_s(c).$$

The classical normalization gets rid of some factors in the differential equation, and our normalization gets rid of extraneous factors in the above integral.

In general, the integral of **K3** cannot be simplified any further. From it, we note the functional equation

K4. $$K_s(c) = K_{-s}(c),$$

proved by letting $t \mapsto t^{-1}$ and using the invariance of the integral on $\mathbf{R}^+$ by this transformation.

However, for $s = \frac{1}{2}$, the integral collapses to

K5. $$K_{\frac{1}{2}}(c) = \sqrt{\frac{\pi}{c}}e^{-2c},$$

whence

K6. $$K_{\frac{1}{2}}(a,b) = \frac{\sqrt{\pi}}{a}e^{-2ab}.$$

The proof of **K5** is easy, and runs as follows. Let

$$g(x) = K_{\frac{1}{2}}(x) = \int_0^\infty e^{-x(t+1/t)}t^{\frac{1}{2}}\frac{dt}{t}.$$

Let $t \mapsto t/x$. Then

$$g(x) = \frac{1}{\sqrt{x}}\int_0^\infty e^{-(t+x^2/t)}t^{\frac{1}{2}}\frac{dt}{t}.$$

Let $h(x) = \sqrt{x}g(x)$. We can differentiate $h(x)$ under the integral sign to get

$$h'(x) = -2x\int_0^\infty e^{-(t+x^2/t)}t^{-\frac{1}{2}}\frac{dt}{t}.$$

Let $t \mapsto t^{-1}$, use the invariance of the integral under this transformation, and then let $t \mapsto t/x$. We then find that

$$h'(x) = -2h(x),$$

whence

$$h(x) = Ce^{-2x}$$

for some constant C. We can let $x = 0$ in the integral for $h(x)$ (but not in the integral for $g(x)$!) to evaluate C, which comes out as $\Gamma(\frac{1}{2}) = \sqrt{\pi}$. This proves **K5**.

Even though the Bessel integral does not collapse to an exponential in general, it is significant that this integral is estimated by an exponential in the following manner, which is all that matters for some applications.

K7. *Let* $x_0 > 0$ *and* $\sigma_0 \leqq \sigma \leqq \sigma_1$. *There is a number* $C(x_0, \sigma_0, \sigma_1) = C$ *such that if* $x \geqq x_0$, then

$$K_\sigma(x) \leqq Ce^{-2x}.$$

Proof. First note that $t + 1/t \geqq 2$, if $t > 0$. Split up the integral as

$$\int_0^\infty = \int_0^{1/8} + \int_{1/8}^8 + \int_8^\infty.$$

The middle integral obviously gives an estimate of the type Ce^{-2x}. To estimate the first integral, note that if $t \leqq 1/8$, then

$$\frac{1}{t} \geqq 4 + \frac{1}{2t}.$$

Hence

$$\int_0^{1/8} e^{-x(t+1/t)} t^{\sigma} \frac{dt}{t} \leqq e^{-4x} \int_0^{1/8} e^{-x_0(t+1/2t)} t^{\sigma} \frac{dt}{t}$$

which is of the desired type. The integral to infinity is estimated in the same way, to conclude the proof.

10.3.1 Gamma Function Identities

K8. $$\Gamma(s) \int_{-\infty}^{\infty} \frac{1}{(u^2+1)^s} du = \sqrt{\pi}\, \Gamma\left(s - \frac{1}{2}\right) \text{ for } \mathrm{Re}(s) > \frac{1}{2}.$$

Proof. Consider

$$\Gamma(s) \int_{-\infty}^{\infty} \frac{1}{(u^2+1)^s} du = \int_{-\infty}^{\infty} \int_0^{\infty} e^{-1} \frac{1}{(u^2+1)^s} t^s \frac{dt}{t} du.$$

Let $t \mapsto (u^2+1)t$ and use the invariance of the integral with respect to dt/t, relative to multiplicative translations. Then change the variable $u \mapsto u/\sqrt{t}$. The formula **K8** drops out.

The above formula allows us to find the first term of the expansion of the right-hand side of **K8**. There are of course alternative proofs for this (using the functional equation of the gamma function), but it does no harm to get it in the spirit of the present section. Putting $s = 1$ in the integral of **K8** yields the value π because $1/(u^2+1)$ integrates to the arctangent. To get the coefficient of $s-1$ in the expansion, we differentiate under the integral sign with respect to s, and we must evaluate the integral

$$\int_{-\infty}^{\infty} \frac{\log(u^2+1)}{u^2+1} du.$$

To do this we use a trick shown to Lang by Seeley long ago. Let

$$g(x) = \int_0^{\infty} \frac{\log(u^2x^2+1)}{u^2+1} du,$$

so that $g(0) = 0$. Differentiating under the integral sign and using a trivial partial fraction decomposition yields for x > 0,

$$g'(x) = \frac{\pi}{1+x},$$

Hence $g(1) = \pi \log 2$. This gives us

$$\frac{\Gamma(s - \frac{1}{2})}{\Gamma(s)} = \sqrt{\pi}(1 - (s-1)\log 4 + \cdots).$$

Next we determine a Fourier transform. It costs nothing to give the formula for several variables, $x = (x_1, \ldots, x_m)$, $u = (u_1, \ldots, u_m)$ with the dot product $u \cdot x$ between them. So $u^2 = u \cdot u$ and $|u| = (u^2)^{1/2}$. We use the **inversion measure**

$$d_{\mathrm{inv}}u = \frac{du}{(2\pi)^{m/2}}, \quad \text{with } du = du_1 \cdots du_m,$$

for which Fourier inversion holds without any extraneous factor. We recall the elementary fact that the function $f(u) = e^{-u^2/2}$ is self-dual for the Fourier transform. The proof with a slightly different normalization of the Fourier integral will be recalled in Lemma 10.4.2 below. Then for $\mathrm{Re}(s) > m/2$, we have

K9. $$\Gamma(s) \int_{\mathbf{R}^m} \frac{e^{\mathbf{i}u\cdot x}}{(u^2+1)^s} d_{\mathrm{inv}}u = \frac{1}{2^{m/2}}(1, |x|/2).$$

Proof. As in **K8**, write down the integral for $\Gamma(s)$, interchange the order of integration, let $t \mapsto (1+u^2)t$. Then let $u \mapsto u/(2t)^{1/2}$. The inversion measure $d_{\mathrm{inv}}u$ changes to $d_{\mathrm{inv}}u/(2t)^{m/2}$, giving the integral

$$\int\int e^{-t} e^{-u^2/2} e^{\mathbf{i}u\cdot x/2(t)^{1/2}} t^{s-m/2} 2^{-m/2} d_{\mathrm{inv}}u \frac{dt}{t}.$$

Since the function $e^{-u^2/2}$ is self-dual, this last expression is

$$= \frac{1}{2^{m/2}} \int_0^\infty e^{-t} e^{-x^2/4t} t^{s-m/2} d_{\mathrm{inv}}u \frac{dt}{t},$$

which is the stated answer.

Remark. In the expression of **K9** one can substitute $x = 0$. The integrals are absolutely convergent in the prescribed range $\mathrm{Re}(s) > m/2$, and then **K8** is a special case of **K9**.

For the next formula, we let **M** denote the **Mellin transform**, defined by

$$(\mathbf{M}f)(s) = \int_0^\infty f(t) t^s \frac{dt}{t}.$$

Of course, the definition is valid in a region for s when the integral is absolutely convergent. We shall use the notation

$$d^*t\, d^*u = \frac{dt}{t}\frac{du}{u},$$

and all double integrals in the next formula are on $\mathbf{R}^+ \times \mathbf{R}^+$.

K10. $$\mathbf{M}K_z(s) = \frac{1}{2}\Gamma\left(\frac{s-z}{2}\right)\Gamma\left(\frac{s+z}{2}\right) \quad \text{for } |\mathrm{Re}(z)| < \mathrm{Re}(s).$$

Proof.

$$\Gamma\Big(\frac{s-z}{2}\Big)\Gamma\Big(\frac{s+z}{2}\Big) = \int\int e^{-t}t^{(s-z)/2}e^{-u}u^{(s+z)/2}\,d^*t\,d^*u$$

$$\text{by } t \mapsto t/u: = \int\int e^{-(u+t/u)}t^{s/2}u^z t^{-z/2}\,d^*t\,d^*u$$

$$\text{by } t \mapsto t^2: = 2\int\int e^{-(u+t^2/u)}t^s(u/t)^z\,d^*t\,d^*u$$

$$\text{by } u \mapsto ut: = 2\int\int e^{-(tu+t/u)}t^s u^z\,d^*t\,d^*u$$

$$= 2\int K_z(t)t^s\frac{dt}{t}, \text{ QED.}$$

As a special case of **K 10**, we obtain

K11. $\quad \mathbf{M}K_1(2s+1) = \frac{1}{2}\Gamma(s)\Gamma(s+1)$ for $\mathrm{Re}(s) > 0$.

For $x > 0$ and $\mathrm{Re}(s) > -1$, we have

K12. $\quad \Gamma(s+1)K_{s+{}^1/_2}(x) = \frac{\sqrt{\pi}}{\sqrt{x}}e^{-2x}\int\limits_0^\infty e^{-u}u^s\Big(1+\frac{u}{4x}\Big)^s du.$

Proof. We write down the integrals defining $K_{s+{}^1/_2}$ and the gamma function, and we multiply them, using Fubini's theorem. Then we make a multiplicative translation $u \mapsto u/t$, followed by the inversion $t \mapsto t^{-1}$. We find that the left side is equal to

$$\int\limits_0^\infty\int\limits_0^\infty e^{-((x+u)t+x/t)}t^{{}^1/_2}\frac{dt}{t}u^{s+1}\frac{du}{u}$$

$$= \sqrt{\pi}\int\limits_0^\infty \frac{e^{-2\sqrt{x}\sqrt{x+u}}}{\sqrt{x+u}}u^{s+1}\frac{du}{u} \qquad \text{by } \mathbf{K6}$$

$$= \frac{\sqrt{\pi}}{\sqrt{x}}\int\limits_0^\infty \frac{e^{-2x(1+u)^{{}^1/_2}}}{(1+u)^{{}^1/_2}}x^{s+1}u^{s+1}\frac{du}{u} \qquad \text{by letting } u \mapsto xu.$$

Now we put $(1+u)^{{}^1/_2} = 1+v$ so $du = 2(1+v)dv$ or $dv = (1+u)^{-1/2}du/2$. After this change of variables, the factor e^{-2x} comes out of the integral. Then we let $v = u/2x$ and $dv = du/2x$. A trivial algebraic manipulation gives the asserted answer.

10.3.2 Differential and Difference Relations

Next we come to the differential equation for $K_s(x)$. We show how it comes out easily with the integral definition of the K-Bessel function.

K13. $z^2 K_s''(z) + zK_s'(z) - (4z^2 + s^2)K_s(z) = 0.$

Proof. This relation comes from two types of recurrence relations. One of them, obtained by differentiating under the integral sign for $K_s(z)$, is that

K14. $K_s' = -K_{s+1} - K_{s-1}$ and $K_s'' = 2K_s + K_{s+2} + K_{s-2}$.

On the other hand, we obtain new relations from integrating by parts the integral defining $K_s(z)$, which yields

K15. $K_s(z) = \frac{z}{s}K_{s+1}(z) - \frac{z}{s}K_{s-1}(z),$

or also

$$K_{s+1}(z) = \frac{s}{z}K_s(z) + K_{s-1}.$$

Then the relation for K_s'' in **K14** can be rewritten in the form

K16. $$K_s''(z) = 4K_s(z) + \frac{s+1}{z}K_{s+1}(z) - \frac{s-1}{z}K_{s-1}(z).$$

Thus we have expressed K_s' and K_s'' as linear combinations of K_s, K_{s+1}, K_{s-1}.

It is now clear from **K15** that we can eliminate to get a linear relation between K_s, K_s', and K_s'', which turns out to be the one stated in **K13**.

Remark. For the classical normalization in other books and tables, the factor of 4 in the differential equation **K13** disappears.

For more formulas we refer readers to our paper on the Bessel fundamental class [JoL 96].

10.4 Functional Equation of the Dedekind Zeta Function

For general number fields, see [Lan 70/94], Chapter 13. Here we deal with the number field $\mathbf{F} = \mathbf{Q}(\mathbf{i})$. We let $\zeta_{\mathbf{F}}$ be the Dedekind zeta function already encountered in Section 10.2. We define the **fudge factors**

$$\Phi_{\mathbf{F},\infty}(s) = (2\pi)^{-s}\Gamma(s),$$
$$\Phi_{\mathbf{F}}(s) = (\mathbf{D}^{1/2})^s \Phi_{\mathbf{F},\infty}(s) = \pi^{-s}\Gamma(s),$$

where $\mathbf{D} = \mathbf{D}(o) = 4$ is the discriminant. The fudge factor is that of the functional equation for the zeta function $\zeta_{\mathbf{F}}$, which we recall. We denote the **completed Dedekind zeta function** by its classical notation (almost)

$$\Lambda(\zeta_{\mathbf{F}})(s) = \Lambda\zeta_{\mathbf{F}}(s) = \Phi_{\mathbf{F}}(s)\,\zeta_{\mathbf{F}}(s).$$

We view Λ as the **completion operator**, completing the function standing to its right. The next theorem is the lowest-level prototype for giving a meromorphic expression for the functions we shall encounter, exhibiting explicitly the poles, residues, and a functional equation. We define the **theta function** for $t > 0$:

$$\theta(t) = \sum_{c \in o} e^{-\pi |c|^2 t}.$$

Theorem 10.4.1. *We have*

$$4\Lambda\zeta_{\mathbf{F}}(s) = \int_1^\infty (\theta(t) - 1) t^s \frac{dt}{t} + \int_1^\infty (\theta(t) - 1) t^{1-s} \frac{dt}{t} - \left(\frac{1}{s} + \frac{1}{1-s}\right).$$

The integrals trivially converge absolutely for all s, and define entire functions of s. Trivially from the above expression we have the functional equation

$$\Lambda\zeta_{\mathbf{F}}(s) = \Lambda\zeta_{\mathbf{F}}(1-s).$$

We now develop the proof.

We use the Euclidean Lebesgue measure on $\mathbf{C} = \mathbf{R}^2$. In particular, the torus $\mathbf{R}^2/\mathbf{Z}^2$ has volume 1. For $x, x' \in \mathbf{C}$ we use the **standard scalar product**

$$\langle x, x' \rangle = x_1 x_1' + x_2 x_2' = \mathrm{Re}(x\bar{x}').$$

Thus $|x|^2 = x\bar{x}$.

The **Fourier transform** is normalized in a standard way. For a function f on $\mathbf{R}^2 = \mathbf{C}$, we define it by

$$\check{f}(x') = \int f(x) e^{-2\pi \mathbf{i} \langle x, x' \rangle} dx.$$

So $dx = dx_1\, dx_2$. One has **Fourier inversion** for sufficiently smooth functions, decreasing sufficiently fast, for instance functions in the Schwartz space,

$$f^{\vee\vee} = f^- \text{ where } f^-(x) = f(-x).$$

This is routine advanced calculus; cf. for instance [Lan 97], Chapter 14, Theorem 2.1. For weaker conditions, see the comments after Theorem 14.1.1.

For an o-periodic ($\mathbf{Z}^2$-periodic) function f, one has the Fourier series. For $c \in o = \mathbf{Z}^2$, the Fourier coefficients of f are given by

$$a_c = \int_{\mathbf{C} \backslash o} f(x) e^{-2\pi \mathbf{i} \langle x, c \rangle} dx.$$

Thus the Fourier series of f is

$$f(x) = \sum_{c \in o} a_c \chi_c(x) = \sum_{c \in o} a_c e^{2\pi \mathbf{i} \langle x, c \rangle},$$

with the character χ_c such that $\chi_c(x) = e^{2\pi\mathbf{i}\langle x,c\rangle} = e^{2\pi\mathbf{i}\mathrm{Re}(x\bar{c})}$. The characters χ_c with $c \in o$ are precisely the characters on $\mathbf{C}$ that vanish on the lattice $o = \mathbf{Z}^2$. They correspond to those complex numbers x' such that

$$\mathrm{Re}(x\bar{x}') \in \mathbf{z} \text{ for all } x \in o.$$

Thus the lattice o is self-dual for the given scalar product.

Fourier Expansion of periodized function. *Let f be in the Schwartz space, and let*

$$f^{\mathrm{o}}(x) = \sum_{m\in\mathbf{Z}^2} f(x+m)$$

be its $\mathbf{Z}^2$-periodization, so g is periodic, C^∞. Let

$$f^{\mathrm{o}}(x) = \sum a_m e^{-2\pi\mathbf{i}x\cdot m}$$

be the Fourier series. Then the Fourier coefficients a_m are given by

$$a_m = \check{f}(m).$$

Proof. By the simplest version of what we call twisted Fubini on $\mathbf{R}^2/\mathbf{Z}^2$, we have

$$\begin{aligned} a_m &= \int_{\mathbf{R}^2/\mathbf{Z}^2} f^{\mathrm{o}}(x) e^{-2\pi\mathbf{i}x\cdot m} dx \\ &= \int_{\mathbf{R}^2/\mathbf{Z}^2} \sum_{k\in\mathbf{Z}^2} f(x+k) e^{-2\pi\mathbf{i}x\cdot m} dx \\ &= \int_{\mathbf{R}^2/\mathbf{Z}^2} \sum_{k\in\mathbf{Z}^2} f(x+k) e^{-2\pi\mathbf{i}(x+k)\cdot m} dx \\ &= \int_{\mathbf{R}^2} f(x) e^{-2\pi\mathbf{i}x\cdot m} dx \\ &= \check{f}(m). \end{aligned}$$

This proves the formula.

Putting $x = 0$ yields what is called the **Poisson summation formula**

$$\sum_{m\in\mathbf{Z}^2} a_m = f^{\mathrm{o}(0)} = \sum_{m\in\mathbf{Z}^2} f(m) = \sum_{m\in\mathbf{z}^2} \check{f}(m).$$

The formula will be applied especially to the exponential quadratic function

$$f(x) = e^{-\pi|x|^2} = e^{-\pi\langle x,x\rangle}.$$

Lemma 10.4.2. *This function is self-dual, that is, $\check{f} = f$.*

Proof. One may argue on the corresponding function in one $\mathbf{R}$-variable, since the exponent splits in a sum $x_1^2 + x_2^2$. In one variable u, with $f(u) = e^{-\pi u^2}$, let $g = \check{f}$.

Determine $g'(u)$ by first differentiating under the integral sign, then integrate by parts. You get $g'(u) = -2\pi u g(u)$. Hence $g(u) = Ce^{-\pi u^2}$, and $C = g(0) = 1$; qed.

Remark. For the other normalization of the Fourier transform, without π in the exponent, one has to take $dx/\sqrt{2\pi}$ for the measure to get $e^{-x^2/2}$ self-dual.

Lemma 10.4.3. *For $t > 0$, let $f_t(x) = e^{-\pi|x|^2}t$. Then*

$$\check{f}_t(v) = \frac{1}{t}f(v/t^{1/2}) = \frac{1}{t}e^{-\pi|v|^2/t}.$$

Proof. This comes directly from the self-duality of Lemma 10.4.2, and the change of variables $x \mapsto x/t^{1/2}$ in the Fourier inversion integral. We have $1/t$ instead of $1/t^{1/2}$ in front because there is a factor $1/t^{1/2}$ for each of the two real variables of $\mathbf{C} = \mathbf{R}^2$.

Recall the **theta series** for $t > 0$:

$$\theta(t) = \sum_{c \in o} e^{-\pi|c|^2 t} = \sum_{c \in o} f_t(c).$$

Lemma 10.4.4. *We have*

$$\theta(t^{-1}) = t\theta(t).$$

Proof. This is immediate from the Poisson formula and Lemma 10.4.3.

We are now ready for the final steps in the proof of Theorem 10.4.1. By definition,

$$\Lambda\zeta_{\mathbf{F}}(s) = \pi^{-s}\Gamma(s)\zeta_{\mathbf{F}}(s) = \pi^{-s}\Gamma(s)\frac{1}{4}\sum_{\substack{c \in o \\ c \neq 0}} \frac{1}{|c|^2 s}.$$

The factor 1/4 comes from the four units, the fact that every ideal is principal, and a generator is determined up to multiplication by a unit, which has absolute value 1. Then we use the formula for $\mathrm{Re}(s) > 1$:

$$\pi^{-s}\Gamma(s)\frac{1}{|c|^2 s} = \int_0^\infty e^{-\pi|c|^2 t}t^s\frac{dt}{t}.$$

obtained by letting $t \mapsto t/\pi|c|^2$ in the integral. From the definitions, we get

$$4\Lambda\zeta_{\mathbf{F}}(s) = \int_0^\infty (\theta(t) - 1)t^s\frac{dt}{t}.$$

We write

$$\int_0^\infty = \int_1^\infty + \int_0^1.$$

The first integral from 1 to ∞ yields the first term in Theorem 10.4.1. For the second integral, we obtain

$$\begin{aligned}
\int_0^1 (\theta(t)-1)t^s\frac{dt}{t} &= \int_0^1 \left(\frac{1}{t}\theta(1/t)-1\right)t^s\frac{dt}{t} && \text{by Lemma 10.4.4}\\
&= \int_1^\infty (t\theta(t)-1)t^{-s}\frac{dt}{t} && \text{by } t\mapsto 1/t\\
&= \int_1^\infty (t\theta(t)-t)t^{-s}\frac{dt}{t} + \int_1^\infty (t-1)t^{-s}\frac{dt}{t}\\
&= \int_1^\infty (\theta(t)-1)t^{1-s}\frac{dt}{t} - \left(\frac{1}{s}+\frac{1}{1-s}\right),
\end{aligned}$$

which proves Theorem 10.4.1.

10.5 The Bessel–Fourier $\Gamma_U\backslash U$-Expansion of Eisenstein Series

Let x be a variable in $\mathbf{C}$, $x = x_1 + x_2\mathbf{i}$ with x_1, x_2 real. We continue to view $\mathbf{C}$ as $\mathbf{R}^2$, and $\mathbf{C}/o \cong \mathbf{R}^2/\mathbf{Z}^2$ is a torus, having volume 1 with respect to the standard Haar Lebesgue measure. On $\mathbf{C}$ we consider again the positive definite $\mathbf{R}$-bilinear symmetric form

$$(x,x') \mapsto \mathrm{Re}(x\bar{x}') = \langle x,x'\rangle.$$

With respect to this form, the lattice $o = \mathbf{Z}^2$ is self-dual, meaning that the characters on $\mathbf{C}/o$ are precisely the functions χ_c with $c \in o$ such that

$$\chi_c(x) = e^{2\pi\mathbf{i}\langle x,c\rangle} = e^{2\pi\mathbf{i}\mathrm{Re}(x\bar{c})}.$$

Hence a function f on the torus (so on $\mathbf{C}$ and $\mathbf{Z}[\mathbf{i}]$-periodic) has a Fourier series expansion

$$f(x) = \sum_{c\in o} A_c e^{2\pi\mathbf{i}\langle x,c\rangle}.$$

The term A_o with $c = 0$ is called the **constant term** of the Fourier expansion of f, or simply the **constant term** of f. In the group context, when f is also viewed as a function on $\Gamma\backslash G/K$, we specify that it is the $\Gamma_U\backslash U$-**constant term** of f.

We shall determine the Fourier expansion of the Eisenstein series, especially the structure of the constant term, and the other Fourier coefficients, which are quite different. This is where the Bessel function will come in. The same formalism of Fourier transform and Poisson summation as in the previous section will occur. We need some formulas concerning the formalism of Fourier transforms.

Lemma 10.5.1. *Let f be a function in the Schwartz space of* $\mathbf{C} = \mathbf{R}^2$. *Let* $b \in \mathbf{C}$. *Let* $g(v) = f(v+b)$. *Then*

$$\check{g}(w) = e^{2\pi \mathbf{i}\langle b,w\rangle}\check{f}(w).$$

Proof. Immediate by the change of variables $v \mapsto v - b$ in the Fourier transform integral.

In Lemma 10.4.3 we dealt with a multiplicative translation, and in Lemma 10.5.1 we deal with an additive translation. In what follows, we use the same exponential function $f(v) = e^{-\pi|v|^2}$ with both kinds of translations. Let $r > 0$, let $b \in \mathbf{C}$, and let

$$f_{r,b}(v) = e^{-\pi r|b+v|^2}. \tag{10.5}$$

Applying first Lemma 10.5.1 (i.e., making the additive translation in the Fourier transform), and then Lemma 10.4.3 (the multiplicative translation), we get

$$\check{f}_{r,b}(w) = e^{2\pi \mathbf{i}\langle b,w\rangle}\frac{1}{r}e^{-\pi|w|^2/r}. \tag{10.6}$$

Applying the Poisson summation formula to the theta series

$$\theta(r,b+o) = \sum_{d\in o} e^{-\pi r|b+d|^2} = \sum_{d\in o} f_{r,b}(d)$$

yields

$$\theta(r,b+o) = \frac{1}{r}\sum_{d\in o} e^{-\pi|d|^2/r}e^{2\pi \mathbf{i}\langle b,d\rangle}. \tag{10.7}$$

As to the Bessel function, we shall use mostly the simpler formulas of Section 10.3, namely for $a,\ b,\ c > 0$,

$$K_s(a,b) = \int_0^\infty e^{-(a^2t+b^2/t)}t^s\frac{dt}{t} = \left(\frac{b}{a}\right)^s K_s(ab),$$

where

$$K_s(c) = \int_0^\infty e^{-c(t+1/t)}t^s\frac{dt}{t}.$$

We use the trivial facts that K_s is entire in s, $K_s = K_{-s}$, and for Proposition 10.5.2, we shall use the exponential decay of **K7**.

The meromorphic continuation and functional equation of the Dedekind zeta function was the first step in an inductive procedure for higher-level similar functions. We deal here with the next level, and for $z = x + \mathrm{j}y \in \mathbf{H}^3$ we define the **completed Eisenstein series** (see Section 10.4 for $\Phi_{\mathbf{F}}$ and $\zeta_{\mathbf{F}}$):

$$\Lambda E(s,z) = \Phi_{\mathbf{F}}(s)\zeta_{\mathbf{F}}(s)E(s,z) = \Lambda\zeta_{\mathbf{F}}(s)E(s,z).$$

The meromorphic continuation of this function will be carried out by exhibiting explicitly a Fourier expansion. Indeed, the Γ-invariance of the Eisenstein series shows that as a function of $x \in \mathbf{C} = \mathbf{R}^2$ the function is periodic with respect to the lattice $o = \mathbf{Z}[\mathbf{i}]$. Thus there has to be a Fourier expansion. The constant term is structurally different from the other terms. The constant term will involve y and the lower-level ζ-function of Theorem 10.4.1. It reflects the poles, such as they are. The nonconstant terms will have Fourier coefficients that come from the Bessel function. The series formed with these terms is entire in s. Thus we now define the **Bessel–Fourier** series

$$\mathrm{BF}(s,z) = \sum_{cd\neq 0} \left|\frac{d}{c}\right|^s yK_s(\pi\, y|cd|)e^{2\pi\mathbf{i}}\langle x, cd\rangle. \tag{10.8}$$

The sum is taken over all pairs (c,d) with $c,\ d \in o$ and $cd \neq 0$. Note that the Fourier coefficients depend only on the absolute values $|c|$ and $|d|$. In particular, one could use $\langle x, \overline{cd}\rangle$ in the exponent without changing the value of the sum. The Fourier coefficients with $m \in o,\ m \neq 0$, of this Fourier series are thus

$$\sum_{cd=m} \left|\frac{d}{c}\right|^s yK_s(\pi y|cd|) = \sum_{cd=m} yK_s(\sqrt{\pi y}|c|, \sqrt{\pi y}|d|). \tag{10.9}$$

Note that formally,

$$\mathrm{BF}(s,z) = \mathrm{BF}(-s,z), \tag{10.10}$$

$$\mathrm{BF}(s-1,z) = \sum_{cd\neq 0} \left|\frac{d}{c}\right|^{s-1} yK_{s-1}(\pi\, y|cd|)e^{2\pi\mathbf{i}\langle x,cd\rangle} \tag{10.11}$$

$$[\text{by } \mathbf{K2}] = \sum_{cd\neq 0} y \int_0^\infty e^{-(\pi y|c|^2 t + \pi y|d|^2/t)} t^{s-1}\frac{dt}{t} e^{2\pi\mathbf{i}\langle x,cd\rangle}.$$

$$= \sum_{cd\neq 0} yK_{s-1}(\sqrt{\pi y}|c|, \sqrt{\pi y}|d|)e^{2\pi\mathbf{i}\langle x,cd\rangle}.$$

Remark. Readers can compare this formula easily with the general case of number fields if so interested. It is exactly what's going to appear in the next theorem. Up to different normalizations, we shall get Kubota's formula [Kub 68], p. 269; cf. also Asai [Asa 70] and [JoL 99], Theorem 3.2, Zagier [Zag 79], as well as Szmidt [Szm 83] (2.3). Note that although our Eisenstein series E differs from that of Kubota and others such as Szmidt by a factor of 4, there is equality of the normalized functions, that is, for Szmidt's E^* [Szm 83] (2.4), we do have the equality

$$\Lambda E(s) = E^*(s).$$

Proposition 10.5.2. *For each z, the Bessel–Fourier series* $\mathrm{BF}(s,z)$ *defines an entire function of s. Let $y_1 > 0$. Uniformly for z such that $y(z) \geqq y_1$, it is absolutely and uniformly convergent in every vertical strip of finite width; in fact, there exists a constant $q > 0$ such that* $\mathrm{BF}(s-1,z) = 0(e^{-qy})$ *for $y \to \infty$.*

Proof. The number of pairs c,d such that $|cd| = n$ (positive integer) grows polynomially in n. Each factor $|d/c|^s$ is $0\left(n^{\mathrm{Re}(s)}\right)$ for $n \to \infty$. Hence the convergence follows from the exponential decay of formula **K7**, as does the exponential decay of the whole series $\mathrm{BF}(s-1,z)$.

In preparation for the next theorem, we describe how the Epstein and Eisenstein series are Mellin transforms of a theta series. We use the expression (10.8) of Section 10.2, for the Epstein zeta function with $\mathrm{Re}(s) > 2$,

$$\mathrm{Ep}(s,z) = \zeta_{\mathbf{F}}(s)E(s,z) = \sum_{(c,d)\neq(0,0)} y(c,d;z)^s.$$

Let

$$\theta_{\mathrm{EP}}(t) = \sum_{(c,d)\neq(0,0)} e^{-t|cz+d|^2/y(z)}.$$

Then

$$(\mathbf{M}\theta)(s) = \Gamma(s)\mathrm{Ep}(s,z).$$

We have a similar relation for the Eisenstein series.

For the application, we decompose the sum over (c,d) into two sums

$$\sum_{(c,d)\neq(0,0)} = \sum_{c=0,d\neq0} + \sum_{c\neq o}\sum_{d\in o}.$$

The first sum for the Epstein series yields

$$\sum_{c=0,d\neq0} = \sum_{d\neq0} \frac{y^s}{|d|^{2s}} = y^s 4\zeta_{\mathbf{F}}(s). \tag{10.12}$$

On the other hand,

$$\int_0^\infty e^{-\pi t/y(c,d;z)}\, t^s \frac{dt}{t} = y(c,d;z)^s \pi^{-s}\Gamma(s) = \Phi_{\mathbf{F}}(s)y(c,d;z)^s. \tag{10.13}$$

Hence

$$\Phi_{\mathbf{F}}(s)\sum_{c\neq0}\sum_{d\in o} y(c,d;z)^s = \sum_{c\neq0}\sum_{d\in o}\int_0^\infty e^{-\pi t|cz+d|^2/y} t^s \frac{dt}{t}. \tag{10.14}$$

Thus we get a decomposition of the Epstein series into a "constant term" (10.12) and another term (10.14).

We now apply this to the determination of the Fourier series.

Theorem 10.5.3. *The function* $\Lambda E(s,z) = \Lambda E(s,x+\mathbf{j}y)$ *has the* $\Gamma_U\backslash U$*-Fourier expansion*

$$\Lambda E(s,z) = \sum_{m\in o} A_m(\Lambda E)(s,y)e^{2\pi\mathbf{i}\langle x,m\rangle}$$

as follows. The 0*-Fourier coefficient is given by*

$$A_o(\Lambda E)(s,y) = y^s 4\Lambda\zeta_{\mathbf{F}}(s) + y^{2-s}4\Lambda\zeta_{\mathbf{F}}(s-1).$$

For $m \neq 0$*, the Fourier coefficient is given by*

$$A_m(\Lambda E)(s,y) = A_{-m}(\Lambda E)(s,y) = \sum_{cd=m} \left|\frac{d}{c}\right|^{s-1} yK_{s-1}(\pi\, y|cd|).$$

Thus

$$\begin{aligned}\Lambda E(s,z) &= y^s 4\Lambda\zeta_{\mathbf{F}}(s) + y^{2-s}4\Lambda\zeta_{\mathbf{F}}(s-1) + \mathrm{BF}(s-1,z).\\ &= y^s 4\Lambda\zeta_{\mathbf{F}}(s) + y^{2-s}4\Lambda\zeta_{\mathbf{F}}(s-1)\\ &\quad + \sum_{cd\neq 0} \left|\frac{d}{c}\right|^{s-1} yK_{s-1}(\pi y|cd|)e^{2\pi\mathbf{i}\langle x,cd\rangle}.\end{aligned}$$

For E itself, putting $\phi_{11}(s) = \Lambda\zeta_{\mathbf{F}}(s-1)/\Lambda\zeta_{\mathbf{F}}(s)$*, we get*

$$E(s,z) = 4y^s + \phi_{11}(s)4y^{2-s} + \Lambda\zeta_{\mathbf{F}}(s)^{-1}\mathrm{BF}(s^{-1},z).$$

Remark. The notation ϕ_{11} stems from the higher-dimensional case, when the function ϕ_{11} becomes a matrix (ϕ_{ij}) called the **scattering matrix**. For the rest of this book, we abbreviate $\phi_{11} = \phi$, omitting the subscript.

Proof. Since $\Lambda\zeta_{\mathbf{F}}(s) = \Phi_{\mathbf{F}}(s)\zeta_{\mathbf{F}}(s)$, multiplying (10.12) by $\Phi_{\mathbf{F}}(s)$ yields the first term in the formula of the theorem (constant term).

For the next terms, first note that $A_m = A_{-m}$ because there is a bijection between pairs (c,d) with $cd = m$, and $(-c,d)$ with $-cd = -m$. The terms inside the K-function have absolute values around them. Next, we use (10.14) to get

$$\begin{aligned}&\Phi_{\mathbf{F}}(s)\sum_{c\neq 0}\sum_{d\in o} y(c,d;z)^s\\ &= \sum_{c\neq 0}\sum_{d\in o}\int_0^\infty e^{-\pi t(|cx+d|^2+|c|^2y^2)/y} t^s \frac{dt}{t}\\ &= \sum_{c\neq 0}\int_0^\infty e^{-\pi t|c|^2 y}\sum_{d\in o} e^{-\pi\, ty^{-1}|cx+d|^2} t^s\frac{dt}{t}\\ &= \sum_{c\neq 0}\int_0^\infty e^{-\pi t|c|^2 y}\sum_{d\in o} yt^{-1}e^{-\pi y|d|^2/t}e^{2\pi\mathbf{i}\langle cx,d\rangle}t^s\frac{dt}{t}\end{aligned}$$

[by the Poisson summation formula (10.7) with $b = cx$]

$$= \sum_{c\neq 0}\sum_{d\in o} y\int_0^\infty e^{-\pi y|c|^2 t}\, e^{-\pi y|d|^2/t}\, t^{s-1}\frac{dt}{t}e^{2\pi\mathbf{i}\langle cx,d\rangle}.$$

We split the sum over $d \in o$ into one term with $d = 0$ and a sum over $d \neq 0$. The term with $d = 0$ allows us to compute the integral, using a change of variables

$$t \mapsto t/\pi y|c|^2.$$

Then the expression $y^{2-s}4\Lambda\zeta_{\mathbf{F}}(s-1)$ falls out, giving the second term of the formula of the theorem.

There remains the double sum over $c \neq 0$ and $d \neq 0$. This sum can be rewritten in the form

$$\sum_{cd\neq 0} y \int_0^\infty e^{-\pi y|c|^2/t} e^{-\pi y|d|^2/t} t^{s-1} \frac{dt}{t} \cdot e^{2\pi \mathbf{i}\langle x,cd\rangle}.$$

which is precisely the desired Bessel–Fourier series (10.1), thus concluding the proof of the theorem.

10.5.1 The Constant Term

The constant term mentioned at the beginning of the section can be written naively in terms of the x-coordinate ($x \in \mathbf{C}$) as the integral

$$A_0(f) \int_{o\backslash\mathbf{C}} f(x)dx.$$

Viewing the function f on the group G, or on G/K, we have the integral expression

$$A_0(f)(g) = \mathrm{CT}_{\Gamma_U\backslash U}(f)(g) = \int_{\Gamma_U\backslash U} f(ug)du.$$

We may also use the variables (x,y) as coordinates on the space G/K, $G = \mathrm{SL}_2(\mathbf{C})$, with $g(\mathbf{j}) = x + \mathbf{j}y$, so we may also write

$$A_0(f)(g) = A_0(f)(y) = A_0(f,y).$$

Theorem 10.5.3 determines this constant term for the function ΛE_s, with s as parameter, namely the $\Gamma_U\backslash U$-constant term of ΛE (or ΛE_s) is

$$A_0(\Lambda E)(s,y) = 4y^s\Lambda\zeta_{\mathbf{F}}(s) + 4y^{2-s}\Lambda\zeta_{\mathbf{F}}(s-1).$$

Note that the constant term in this case of E depends on zeta functions of lower level than the Eisenstein series, namely gamma and classical Dedekind zeta functions, and is thus meromorphic. Dividing by $\Lambda\zeta_{\mathbf{F}}(s)$ then yields the constant term of E_s itself. Chapter 12 will go deeper into this constant term.

Next come more analytic properties, especially the functional equation.

Theorem 10.5.4. *The function $\Lambda E(s,z)$ is meromorphic in s. Indeed, the constant term $A_o(\Lambda E)(s,y)$ is meromorphic, as shown above by the Dedekind zeta functions, and the Bessel series is entire. The constant term is invariant under $s \mapsto 2-s$ and so is the Bessel series* $\mathrm{BF}(s-1,z)$. *In particular,*

$$\Lambda E(s,z) = \Lambda E(2-s,z).$$

Proof. The first assertion is immediate from Proposition 10.5.2, Theorem 10.5.3, and the above remark that the Dedekind zeta function is meromorphic. Note that the poles can be read off the constant term trivially. Under the change $s \mapsto 2-s$, the two terms of the constant term get permuted by the functional equation of the Dedekind zeta function $\Lambda\zeta_{\mathbf{F}}(w) = \Lambda\zeta_{\mathbf{F}}(1-w)$. The Bessel–Fourier series is invariant under $s \mapsto 2-s$ because $K_w = K_{-w}$ for all complex w, so K_{s-1} is invariant under $s \mapsto 2-s$, and the pair (c,d) goes into (d,c), so the terms of the series get permuted, thus concluding the proof.

Corollary 10.5.5. *The only poles of $\Lambda E(s,z)$ are at $s = 0$, 2. They are among those of $\Lambda\zeta_{\mathbf{F}}(s)$ and $\Lambda\zeta_{\mathbf{F}}(s-1)$, and are simple, determined by the above constant term. The residues are*

$$\operatorname*{res}_{s=2} \Lambda E(s,z) = 1 = 4\operatorname*{res}_{s=1} \Lambda\zeta_{\mathbf{F}}(s) \quad \textit{and} \quad \operatorname*{res}_{s=2} E(s,z) = \frac{\pi^2}{\zeta_{\mathbf{F}}(2)} = \frac{1}{\Lambda\zeta_{\mathbf{F}}(2)}.$$

Thus

$$\Lambda \mathrm{E}(\mathrm{s},\mathrm{z}) = \frac{1}{\mathrm{s}-2} - \frac{1}{\mathrm{s}} + H(\mathrm{s},\mathrm{z}) \textit{ with } \mathrm{H}(\mathrm{s},\mathrm{z}) \textit{ entire in } \mathrm{s}.$$

The Eisenstein function E(s,z) *itself has poles at* $\mathrm{s} = 2$ *and at the zeros of* $\zeta_{\mathbf{F}}$.

Proof. The first assertion comes from the known poles of $\Lambda\zeta_{\mathbf{F}}(w)$ at $w = 0$, 1. Thus $\Lambda\zeta_{\mathbf{F}}(s)$ has poles at 0, 1 and $\Lambda\zeta_{\mathbf{F}}(s-1)$ has poles at 1, 2. The poles at 1 cancel because the residues in Theorem 10.4.1 are minus each other. The residues of the Dedekind zeta function can also be read off Theorem 10.4.1, and give immediately the residues of ΛE.

We are interested not only in the residue of the Eisenstein series but in the constant term of the power series at $s = 2$. The determination of this constant term in the case of $\mathrm{SL}_2(\mathbf{R})$ and $\mathrm{SL}_2(\mathbf{Z})$ is called **Kronecker's limit formula**. The determination in the general case was done by Asai [Asa 70]; cf. [JoL 99]. Here we are interested only in a special case. We have

$$y^{2-s} = 1 - (\log y)(s-2) + O(|s-2|^2)$$

$$\Lambda\zeta_{\mathbf{F}}(s-1) = \operatorname*{res}_{s=1} \Lambda\zeta_{\mathbf{F}}(s)\frac{1}{s-2} + \operatorname*{con}_{s=1} \Lambda\zeta_{\mathbf{F}}(s) + O(|s-2|),$$

where $\operatorname*{con}_{s=1}$ denotes the constant term in a Laurent expansion at $s = 1$.

Theorem 10.5.6. *At $s = 2$, the power series expansion of the Eisenstein series starts with*

$$\begin{aligned}\mathrm{E}(\mathrm{s},\mathrm{z}) &= \frac{1}{s-2} - 4\cdot \log y + 4\cdot \underset{s=1}{\mathrm{con}}\, \Lambda\zeta_{\mathbf{F}}(s) + \mathrm{BF}(1,z) + O(|s-2|)\\ &= \underset{s=1}{\mathrm{res}}\, \Lambda\zeta_{\mathbf{F}}(s)\cdot \frac{4}{s-2} - 4\cdot \underset{s=1}{\mathrm{res}}\, \Lambda\zeta_{\mathbf{F}}(s) \log\, y + 4y^2\Lambda\zeta_{\mathbf{F}}(2)\\ &\quad + 4\cdot \underset{s=1}{\mathrm{con}}\, \Lambda\zeta_{\mathbf{F}}(s) + \mathrm{BF}(1,z) + O(|s-2|).\end{aligned}$$

Proof. Immediate from Theorem 10.5.3.

For the record, we mention one more analytic property to be used later (e.g., Theorem 10.3.1 of Chapter 10), applied to the Casimir operator.

Proposition 10.5.7. *For each z, and C^∞ differential operator D_z, the function $D_z\Lambda E(s,z)$ is analytic in s except possibly for a simple pole at $s = 2$ or 0.*

Proof. This is immediate from the analytic expression of Theorem 10.5.3, differentiating the integral definition of the K-Bessel function $K_s(\dots)$.

We call the function $E(s,z)$ the **Eisenstein function**. When we want to emphasize that s is viewed as a parameter, we use the notation $E_s(z)$. By Theorem 10.5.3, we get a meromorphic expression for $E(s,z)$. As a function of s, we see that $E(s,z)$ has poles at the zeros of the zeta function $\zeta_{\mathbf{F}}(s)$.

10.6 Estimates in Vertical Strips

We shall give polynomial estimates for the growth of the Epstein and Eisenstein series in appropriate vertical strips. For present purposes, we need to work on the line $\mathrm{Re}(s) = 1$, so the strip has to include this line. We need both upper and lower estimates. Since the Eisenstein series is equal to the Epstein series divided by the Dedekind zeta function, we need a lower bound for this function on the vertical lines to ensure convergence of the integral. This need is left unstated by Zagier [Zag 79], who states on p. 315 concerning the spectral decomposition, "We now give a rough indication, ignoring analytic problems, of how the Rankin–Selberg method implies the spectral decomposition formula for $L^2(\Gamma\backslash H)$." We are indebted to Sarnak for the reference to Titchmarsh and for explaining how to fix some of the problems accompanying Zagier's formal arguments by means of Proposition 10.6.1 and Theorem 10.6.2.

We start with the Epstein series, which is at the root of Theorem 10.5.3, namely

$$\mathrm{Ep}(s,z) = \sum_{(c,d)\neq(0,0)} \frac{y^s}{|cz+d|^{2s}} = \zeta_{\mathbf{F}}(s)E(s,z).$$

We shall give additional information to supplement Theorem 10.2.1, by using the functional equation. We recall the completed function

$$(\Lambda E)(s,z) = \pi^{-s}\Gamma(s)\zeta_{\mathbf{F}}(s)E(s,z) = \pi^{-s}\Gamma(s)\mathrm{Ep}(s,z) = \Phi_{\mathbf{F}}(s)\mathrm{Ep}(s,z).$$

We use Theorems 10.4.1 and 10.5.3 systematically, the functional equations being

$$\Phi_{\mathbf{F}}(s)\zeta_{\mathbf{F}}(s) = \Phi_{\mathbf{F}}(1-s)\zeta_{\mathbf{F}}(1-s) \text{ and } (\Lambda E)(s,z) = (\Lambda E)(2-s,z).$$

Proposition 10.6.1. *Outside a neighborhood of the poles ($s = 0$ or 2), in a strip*

$$-\delta \leqq \sigma \leqq 2+\delta, \text{ say, with } \sigma = \mathrm{Re}(s), s = \sigma + \mathbf{i}r,$$

we have uniformly for z in a compact set,

$$|\mathrm{Ep}(s,z)| \ll (1+|r|)^2.$$

Proof. We apply the functional equation with $\sigma = 2+\delta$, so on the sides of the strip,

$$-\delta \leqq \mathrm{Re}(s) \leqq 2+\delta.$$

All the functions involved are of order 1, and one can apply the Phragmén–Lindelöf theorem as usual in analytic number theory; cf. for instance [Lan 70/94], Sections 13.5 and 17.2.

Theorem 10.6.2. *Given ε, uniformly for z in a compact set and $1 \leqq \mathrm{Re}(s) \leqq 10$ (say), s outside a neighborhood of 2, for $s = \sigma + \mathbf{i}r$,*

$$|\mathrm{E}(z,\sigma+\mathbf{i}r)| \ll (1+|r|)^{2+\varepsilon}.$$

The implied constant depends only on the compact set, ε, and the neighborhood. This comes from the Dedekind zeta lower bound

$$|\zeta_{\mathbf{F}}(\sigma+\mathbf{i}r)| \gg |r|^{-\varepsilon} \text{ for } \|r\| \to \infty.$$

Proof. By Proposition 10.6.1, the Dedekind zeta lower bound implies the Eisenstein upper bound. As to that lower bound, we use the classical arguments from Titchmarsh [Tit 51], Sections 3.3 and 3.4. Let $\zeta = \zeta_{\mathbf{F}}$. We have

$$\zeta(s) = \exp \sum_{\mathfrak{p}} \sum_{m=1}^{\infty} \frac{1}{m\mathbf{N}\mathfrak{p}^{ms}},$$

and hence

$$|\zeta(s)| = \exp \sum_{\mathfrak{p}} \sum_{m=1}^{\infty} \frac{\cos(mr\log N\mathfrak{p})}{m\mathbf{N}\mathfrak{p}^{m\sigma}}.$$

It follows that

$$\zeta^3(\sigma)|\zeta(\sigma+\mathbf{i}r)|^4|\zeta(\sigma+2\mathbf{i}r)| = \exp \sum_{\mathfrak{p}} \sum_{m=1}^{\infty} \frac{3+4\cos(mr\log\mathbf{N}\mathfrak{p})+\cos(2mr\log\mathbf{N}\mathfrak{p})}{m\mathbf{N}\mathfrak{p}^{m\sigma}}. \quad (10.15)$$

From the positivity

$$3+4\cos\theta+\cos 2\theta = 2(1+\cos\theta)^2 \geqq 0, \tag{10.16}$$

it follows that

$$\zeta^3(\sigma)\,|\zeta(\sigma+\mathbf{i}r)|^4\,|\zeta(\sigma+2\mathbf{i}r)| \geqq 1 \ \text{ for } \sigma > 1. \tag{10.17}$$

For the convenience of the reader, we first recall how this implies that ζ has no zero on the line $\mathrm{Re}(s) = 1$. Fix $r \neq 0$. Suppose $\zeta(1+\mathbf{i}r) = 0$. Then for $\sigma > 1$,

$$\zeta^3(\sigma) = O(1/(\sigma-1)^3) \ \text{ and } \ \zeta(\sigma+2\mathbf{i}r) = O(1) \ \text{ for } \sigma \to 1,$$

but

$$\zeta(\sigma+\mathbf{i}r) = O(\sigma-1).$$

Hence the left side of (10.17) is $O(\sigma-1)$, contradiction.

We now give the proof proper for Theorem 10.6.2 by superimposing another estimate. Of course, r is no longer fixed. We recall the following from basic analytic number theory:

Lemma 10.6.3. *For* $1 \leqq \sigma \leqq 10$, *say, we have*

$$|\zeta(\sigma+\mathbf{i}r)| = O(|r|^\varepsilon) \ \textit{for } r \to \infty.$$

Proof. See for instance [Lan 70/94], Chapter 13, Theorem 5. Actually, a better estimate with the log is known with different types of arguments, as in Titchmarsh [Tit 51], Chapter 3, Theorem 3.3.5, but we won't need this stronger estimate.

The desired lower bound with $2r$ instead of r now follows from Lemma 10.6.3 and (10.17). This proves Theorem 10.6.2.

It is convenient to insert another estimate depending also on the previous section, especially Theorem 10.5.3.

Proposition 10.6.4. *Let* $s \neq 1$, 2 *and* $s \neq$ *zero of* $\zeta_{\mathbf{F}}$. *Let* $\varepsilon > 0$. *Suppose*

$$-\varepsilon < \mathrm{Re}(s) < 1+\varepsilon.$$

Then for $\sigma = \mathrm{Re}(s)$,

$$E(s,z) = O(y^\sigma + y^{2-\sigma}) \ \textit{for } y = y_z \to \infty.$$

The constant implicit in this estimate is uniform in some neighborhood of each s.

Proof. We use the Fourier expansion of Theorem 10.5.3. By Proposition 10.5.2 (via **K7**, exponential decay of the Bessel function), the Bessel series is bounded, and the two terms of the constant term determine the growth. We divide the constant term in Theorem 10.5.3 by $\Lambda\zeta_{\mathbf{F}}(s)$. We get y^s, giving rise to the estimate y^σ, and also $y^{2-s}\Lambda\zeta_{\mathbf{F}}(s-1)/\Lambda\zeta_{\mathbf{F}}(s)$. Both $\zeta_{\mathbf{F}}(s-1)$ and $\zeta_{\mathbf{F}}(s)$ are bounded away from 0 and ∞

on the vertical line $\text{Re}(s) = \sigma > 2$, because of the Euler product. The quotient of the gamma factor is $\Gamma(s-1)/\Gamma(s) = 1/(s-1)$, which actually tends to 0 as $|s| \to \infty$. The quotient due to the power of π is bounded, so the stated estimate follows.

The estimates of this section will be used in Section 11.3 (among other places) to obtain estimates for a convolution $E * f$ with functions f that are C^∞ with compact support. In Section 11.4 we shall complement the present estimates with more significant ones, to take care of uniformities necessary to apply the formalism to the convolution of the heat kernel (which does not have compact support) with Eisenstein series.

10.7 The Volume–Residue Formula

We now compute the volume of the fundamental domain (Iwasawa measure),

$$\text{Vol}(\mathscr{F}) = \text{Vol}(\Gamma\backslash H^3).$$

following Sarnak [Sar 83]. We proved in Corollary 10.5.5 that

$$\operatorname*{res}_{s=2} E(s,z) = \frac{\pi^2}{\zeta_{\mathbf{F}}(2)}.$$

However, for this section, this is irrelevant in the sense that we establish a direct relation between the volume and the residue in Theorem 10.7.2 below. The technique used in the proof involving the fundamental domain will be used again, so we extract generally relevant observations here.

Let R be the reflection mentioned at the beginning of Section 4.1. Then $\{\pm I, \pm R\}$ is a subgroup of order 4 in Γ. For $x = x_1 + \mathbf{i}x_2 \in \mathbf{C}$, let

$$Y(x) = (1 - x_1^2 - x_2^2)^{1/2}.$$

The fundamental domain $\mathscr{F}$ itself consists of elements $z = (x,y)$ with $x \in \mathscr{F}_U$ and $y \geqq Y(x)$. Then for $\gamma \in \Gamma_U\backslash\Gamma$, $\gamma \notin \Gamma_\infty = \Gamma_U(\pm I, \pm R)$, and x in the fundamental rectangle, we have

$$0 < y(\gamma z) \leqq Y(x).$$

Proposition 10.7.1. *Let $\mathscr{S} = \{(x,y) | x \in \mathscr{F}_U,\ 0 < y \leqq Y(x)\}$ be the region below the sphere and above the fundamental rectangle. There is a set of coset representatives $\text{Rep}_{\mathscr{S}}$ of $\Gamma_\infty\backslash(\Gamma - \Gamma_\infty)$ (the cosets other than the unit coset) mapping $\mathscr{F}$ into $\mathscr{S}$, and such that the family $\{\gamma\mathscr{F}\}$ $(\gamma \in \text{Rep}_{\mathscr{S}})$ tiles $\mathscr{S}$.*

This is essentially a repetition of Chapter 9, Theorem 7.5, for the convenience of the reader.

Next we complement the formula of Section 6.4.

Theorem 10.7.2. *With respect to the Iwasawa measure* $dx_1\, dx_2\, dy/y^3$, *we have independently of* z,

$$\operatorname*{res}_{s=2} E(s,z) = \frac{2}{\mathrm{Vol}(\Gamma\backslash\mathbf{H}^3)} = \frac{1}{\mathrm{Vol}(\Gamma\backslash G)}.$$

Proof. From Theorem 10.5.3 we know that

$$E(s,z) = 4y^s + 4\phi_{11}(s)y^{2-s} + \mathrm{BF}(s-1,z)/\Lambda\zeta_{\mathbf{F}}(s),$$

and the function $\phi_{11}(s)$ has

$$4\operatorname*{res}_{s=2}\phi_{11}(s) = \operatorname*{res}_{s=2} E(s,z).$$

From Section 10.3, **K7**, we know that the Bessel series has exponential decay for $y(z) \to \infty$. Hence $\mathrm{BF}(s-1,z)$ is uniformly integrable over the fundamental domain, restricting s to $\mathrm{Re}(s) > 2$. Let

$$h(s) = \int_{\mathscr{F}} (E(s,z) - 4y^s)\frac{dx_1\, dx_2\, dy}{y^3}. \tag{10.18}$$

Then

$$h(s) = \int_{\mathscr{F}} \left(y^{2-s}\operatorname*{res}_{s=2} E(s,z)\frac{1}{s-2} + H(s,z) \right) d\mu(z), \tag{10.19}$$

with $H(s,z)$ holomorphic in s for $\mathrm{Re}(s) \geqq 2$. Therefore (10.19) yields

$$\operatorname*{res}_{s=2} h(s) = \mathrm{vol}(\mathscr{F}) \cdot \operatorname*{res}_{s=2} E(s,z).$$

On the other hand, the four elements $\gamma = \pm I,\ \pm R$ are such that $y(\gamma z) = y$, so they contribute the term $4y^s$ in the sum defining the Eisenstein series, taken over $\gamma \in \Gamma_U\backslash\Gamma$. Therefore

$$E(s,z) - 4y^s = \sum_{\gamma \neq \pm I, \pm R} y(\gamma z)^s.$$

Hence (10.18) yields

$$h(s) = \int_{\mathscr{F}} \sum_{\gamma \neq \pm I, \pm R} y(\gamma z)^s dx_1\, dx_2\, dy/y^3.$$

We integrate the series expression term by term. Under translation back by an element in Γ_U, the full sum for $E(s,z)$ amounts to an integral over the fundamental domain for Γ_U, counted four times. By Proposition 10.7.1,

$$h(s) = \int_{\mathscr{F}_U} \int_0^{Y(x)} 4y^s \frac{dy}{y^3}\, dx_1\, dx_2 = \int_{\mathscr{F}_U} \frac{4Y(x)^{s-2}}{s-2}\, dx_1\, dx_2.$$

Hence the residue of $h(s)$ at $s = 2$ is $4(1/2) = 2$. This proves the theorem.

10.8 The Integral over $\mathcal{F}$ and Orthogonalities

The shape of the fundamental domain is about as nice as it could be, subject to ordinary calculus. Let $Y > 1$. We define the **truncated fundamental domain** to be

$$\mathcal{F}_{\leqq Y} = \text{set of points } (x,y) \in \mathcal{F} \text{ such that } y \leqq Y \text{ and } x \in \mathcal{F}_U.$$

Note that for $y \geqq 1$, the condition $x_1^2 + x_2^2 + y^2 \geqq 1$ is valid for all x. We may reformulate this in the form

$$\mathcal{F}_{\leqq Y} = \mathcal{F}_{\leqq 1} \cup (\mathcal{F}_U \times [1.,Y]).$$

Thus $\mathcal{F}_{\leqq Y}$ has a natural product structure, except for the constant "curved" part $\mathcal{F}_{\leqq 1}$. Similarly, for $1 < Y_0 < Y$ we have

$$\mathcal{F}_{\leqq Y} = \mathcal{F}_{\leqq Y_O} \cup (\mathcal{F}_U \times [Y_O,Y]).$$

We shall integrate over the fundamental domain instead of $\Gamma\backslash G$ (or $\Gamma\backslash \mathbf{H}^3$). We reduce the integral to an absolutely convergent one. The arguments will refine those of the volume computation by using an extra function f, which we assume bounded measurable.

Let H be the finite group $(\pm I, \pm R)$, where R is as before, defined at the beginning of Section 10.1, giving the map $x \mapsto -x$ when $z = x + \mathbf{j}y$. By Proposition 10.7.1, the region $\mathcal{S} = \{(x,y) | x \in \mathcal{F}_U,\ 0 < y \leqq Y(x)\}$ is tiled, as the union of the γ-translates $\gamma\mathcal{F}$ with $\gamma \in \mathrm{Rep}_{\mathcal{S}}$, so

$$\mathcal{S} = \mathrm{Rep}_{\mathcal{S}}\mathcal{F}.$$

It is the region above the fundamental rectangle (half the square of $\Gamma_U\backslash U$), and below the sphere of radius 1. For $Y_0 > 1$, we then have a decomposition of the truncated tube into a disjoint union (except for boundary points)

$$\mathcal{T}_{(0,Y_0]} = \mathcal{F}_U \times (0,Y_0] = \mathcal{F}_{\leqq Y_0} \cup \mathrm{Rep}_{\mathcal{S}}\mathcal{F}.$$

Note that $\mathcal{F}_{\leqq Y_0}$ is the part of the fundamental domain lying above the sphere and below the constant Y_0.

We recall that a function f on $\Gamma\backslash G$ is called **cuspidal** if

$$\int_{\Gamma_U\backslash U} f(uz)du = 0 \quad \text{for all } z \in \mathrm{G}.$$

We may take functions in $L^2(\Gamma\backslash G)$ since $\Gamma\backslash G$ has finite measure, so $L^2 \subset L^1$. We shall give two orthogonality properties of $L^2_{\text{cusp}}(\Gamma\backslash G)$ (the subspace of cuspidal functions in L^2), with respect to the constant functions and with respect to the Eisenstein series. We start with a general consequence of cuspidality.

Let $\mathcal{T}$ be the tube, $\mathcal{T} = \mathcal{F} \cup \mathcal{S}$. Let $0 < Y_0 \leqq Y$. Let $\mathcal{T}_{[Y_0,Y]}$ be the subset of elements $z \in \mathcal{T}$ such that $Y_0 \leqq y \leqq Y$ if $z = x + \mathbf{j}y$. Let f be cuspidal, and h any continuous function of y. Then

$$\int_{\mathcal{T}_{[Y_0,Y]}} f(z)h(y_z)\,d\mu(z) = 0. \tag{10.20}$$

Proof. Trivially from cuspidality,

$$\int_{Y_0}^{Y} \int_{\Gamma_U\backslash\mathbf{C}} f(x+\mathbf{j}y)h(y)\frac{dx\,dy}{y^3} = 0.$$

Thus there is no problem evaluating such integrals over horizontally truncated parts of the tube. Problems come from the curved part when one is integrating over $\mathcal{F}$. In what follows, we deal with $Y_0 > 1$. Then from (10.18), we get for instance

$$\int_{\mathcal{F}_{\leq Y}} f(z)y^s\,d\mu(z) = \int_{\mathcal{F}_{\leqq Y_0}} f(z)y^s\,d\mu(z). \tag{10.21}$$

Theorem 10.8.1. *Let f be bounded measurable and cuspidal, on $\Gamma\backslash G/K$. Then the following limit exists for all s, and actually*

$$\lim_{y\to\infty} \int_{\mathcal{F}_{\leqq Y}} E(s,z)f(z)\,d\mu(z) = 0.$$

Proof. We separate the part of the integral over $\mathcal{F}$ that causes divergence. So we apply the above decomposition to $\mathcal{F}_{\leqq Y}$ with $Y > Y_0$. We have

$$\begin{aligned}
\int_{\{\leqq y} f(z)E(s,z)\,d\mu(z) &= \int_{\mathcal{F}_{\leqq Y}} f(z) \sum_{\Gamma_U\backslash\Gamma} y(\gamma z)^s d\mu(z) \\
&= 4 \int_{\mathrm{Rep}_{\mathcal{T}}\mathcal{F}_{\leqq Y}} f(z)y^s d\mu(z) \\
&= 4 \int_{\mathcal{F}_{\leqq Y}} f(z)y^s d\mu(z) + 4 \int_{\mathrm{Rep}_{\mathcal{S}}\mathcal{F}_{\leqq Y}} f(z)y^s d\mu(z). \\
\text{[by (2)]} &= 4 \int_{\mathcal{F}_{\leqq Y_0}} f(z)y^s d\mu(z) + 4 \int_{\mathrm{Rep}_{\mathcal{S}}\mathcal{F}_{\leqq Y}} f(z)y^s d\mu(z).
\end{aligned}$$

The second integral on the right in (10.2) *is absolutely convergent for $Y \to \infty$ to the integral over* $\mathrm{Rep}_{\mathcal{S}}\mathcal{F}$, i.e.,

$$\int_{\mathcal{S}} f(z)y^s\,d\mu(z)$$

is absolutely convergent for $\mathrm{Re}(s) > 2$.

Proof. This is immediate, since f is assumed bounded, so $\|f\|_\infty$ can be taken out as a factor in front of the integral, and

$$\int_0^1 \frac{y^{2+\varepsilon}}{y^3} dy < \infty.$$

By (10.20), (10.21), and the absolute convergence just proved, we get

$$\begin{aligned}\lim_{Y\to\infty} \int_{\mathscr{F}_{\leqq Y}} f(z)E(s,z)\,d\mu(z) &= 4\int_{\mathscr{F}_{\leqq Y_0}} f(z)y^s\,d\mu(z) + 4\int_{\mathscr{S}} f(z)y^s\,d\mu(z) \qquad (10.22)\\ &= 4\frac{1}{2}\int_{\Gamma_U\backslash\mathbf{C}}\int_0^{Y_0} f(x+\mathbf{j}y)y^s\frac{dx\,dy}{y^3}\\ &= 0, \text{ by cuspidality,}\end{aligned}$$

thereby proving Theorem 10.8.1.

Theorem 10.8.2. *Let f be cuspidal and bounded on $\Gamma\backslash G/K$. Then*

$$\int_{\Gamma\backslash G} f(z)\,d\mu(z) = 0.$$

Proof. We write with abbreviated notation

$$\begin{aligned}\int_{\mathscr{F}_{\leq Y}} fE &= \int_{\mathscr{F}_{\leqq Y_0}} fE + \int_{\mathscr{F}_{[Y_0,Y]}} f(E-4y^s) + \int_{\mathscr{F}_{[Y_0,Y]}} 4fy^s\\ &= \int_{\mathscr{F}_{\leqq Y_0}} fE + \int_{\mathscr{F}_{[Y_0,Y]}} f(E-4y^s) \text{ [by (10.20)]}.\end{aligned}$$

We let $Y\to\infty$. On the right, we use the boundedness of $E(s,z)-4y^s$ on $\mathscr{F}_{[Y_0,\infty)}$ as in Theorem 10.7.2. On the left, we use Theorem 10.8.1. We get

$$0 = \int_{\mathscr{F}_{\leqq Y_0}} f(z)E(s,z)\,d\mu(z) + \int_{\mathscr{F}-\mathscr{F}_{\leqq Y_0}} f(z)(E(s,z)-4y^s)\,d\mu(z). \qquad (10.23)$$

Let r_2 denote the residue of $E(s,z)$ at $s=2$. Then

$$\lim_{s\to 2}(s-2)E(s,z) = r_2 = \lim_{s\to 2}(s-2)(E(s,z)-4y^s).$$

We multiply (10.23) by $s-2$ and let $s\to 2$, to obtain

$$0 = \int_{\mathscr{F}_{\leqq Y_0}} + \int_{\mathscr{F}-\mathscr{F}_{\leqq Y_0}} f(z)r_2\,d\mu(z) = r_2\int_{\mathscr{F}} f(z)\,d\mu(z),$$

which proves Theorem 10.8.2.

Remark 1. We don't need to know anything about the residue of $E(s,z)$ at 2 besides the fact that it isn't 0.

Remark 2. The above two theorems can be interpreted as orthogonality relations on $\Gamma\backslash G$: Theorem 10.8.1 gives a weak orthogonality between bounded cuspidal functions and Eisenstein series; Theorem 10.8.2 gives orthogonality between such functions and the constant functions.

The next statements are less delicate. They involve only absolute convergence. They will not be used further in this book. We use BC to denote Bounded Continuous.

Theorem 10.8.3. *Let* $f \in \mathrm{BC}(\Gamma\backslash G/K)$, *and suppose* $E_s f \in L^1(\Gamma\backslash G/K)$ *for all* s *with* $\mathrm{Re}(s)$ *in some open interval (2,c), uniformly in compact subsets. Then* f *is cuspidal if and only if it satisfies either one of the following conditions:*

(i) *There exists* $\sigma \in (2,c)$ *such that for all* s *with* $\mathrm{Re}(s) = \sigma$ *we have* $[E_s, f] = 0$.
(ii) *For all* s *with* $\mathrm{Re}(s) \in (2,c)$ *we have* $[E_s, f] = 0$.

Proof. We use the group Fubini formalism, replacing E_s by its defining series:

$$\int_{\Gamma\backslash G} \sum_{\gamma\in\Gamma_U\backslash\Gamma} y(\gamma z)^s f(z)\, d\mu(z) = \int_{\Gamma_U\backslash G} y(z)^s f(z)\, d\mu(z)$$

$$= \int_A a^{s\alpha} \int_{\Gamma_U\backslash U} f(ua) du \delta^{-1}(a)\, da. \tag{10.24}$$

If f is cuspidal, the inner integral is 0, thus proving the implication one way. Note that the scalar product $[E_s, f]$ is analytic in s for $\mathrm{Re}(s) \in (2,c)$, so (i) implies (ii). To prove that (ii) implies (i), we have to use families of test functions as follows. Write $s = \sigma + \mathbf{i}r$, $r = \mathrm{Im}(s)$. Abbreviate

$$\mathrm{Tr}(f)(a) = \int_{\Gamma_U\backslash U} f(ua) du.$$

Assume (i). Let $\varphi = \varphi(r)$ be a Euclidean Gaussian on the line, and integrate the expression of (10.24) for $[E_s, f]$ against φ. Interchanging the order of integration, we get

$$0 = \int_A \mathrm{Tr}(f)(a) \left(\int_{-\infty}^{\infty} a^{(\sigma+\mathbf{i}r)} \varphi(r) dr \right) \delta^{-1}(a) da.$$

The integral over $(-\infty, \infty)$ is a Fourier–Mellin integral, and by picking φ arbitrarily we get arbitrary Gaussians back, for instance yielding a Dirac family on A. Since the integral of $\mathrm{Tr}(f)$ against all elements of such a family is 0, it follows that $\mathrm{Tr}(f) = 0$, as was to be shown.

For a higher-dimensional version of the above theorem, see [JoL 02], Chapter 3, Theorem 3.3.5.

Corollary 10.8.4. *Let* $f \in \mathrm{BC}(\Gamma\backslash G/K)$ *and suppose* $E_s f \in L^1(\Gamma\backslash G/K)$ *for all* $s \neq 2,\ 1 \leqq \mathrm{Re}(s) \leqq 2+\varepsilon$, *uniformly for s in compact sets. Then f is cuspidal if and only if* $[E_s, f] = 0$ *for all s with* $\mathrm{Re}(s) = 1$.

Proof. Immediate from Theorem 10.8.3 and the meromorphic property of the scalar product $[E_s, f]$ in the variable s.

In higher-dimensional versions, the corollary is only a conjecture; see [JoL 02], Chapter 4, Conjecture 4.6.1. The next chapter will use the same group Fubini formalism in a more extensive context.

Chapter 11
Adjointness Formula and the $\Gamma\backslash G$-Eigenfunction Expansion

With an Iwasawa decomposition $G = UAK$ and a discrete subgroup Γ, one gets a basic adjointness formula for the trace with respect to $\Gamma_U\backslash\Gamma$ and the integral over $\Gamma_U\backslash U$; see, for instance, [Lan 73/85] Section 13.1. For a formulation of such adjointness on SL_n, see [JoL 02], Chapter 3. We use the simpler version in the context when one integrates a Γ-invariant function against an Eisenstein series, in which case what comes out is called the Rankin–Selberg method, which relates the above integral with the Mellin transform of the $\Gamma_U\backslash U$-constant term of f. This is the first step in getting a Fourier-like expansion for a function in the noncommutative setting of $\Gamma\backslash G/K = \Gamma\backslash\mathbf{H}^3$ in our case. Such an expansion involves three basic pieces, which partake of both Fourier series and Fourier integrals, as well as Fourier inversion, in both its series and integral forms. Aside from the $\Gamma_U\backslash U$-constant term that we saw in Chapter 5, there is a $\Gamma\backslash G$-**constant term**, defined by the total integral

$$\int_{\Gamma\backslash G} f(g)dg,$$

with respect to a suitably normalized Haar measure on the (right) homogeneous space $\Gamma\backslash G$. Since $\Gamma\backslash G$ has finite measure, the constant functions on $\Gamma\backslash G$ are in L^2, and form a one-dimensional space. There is another space, the cuspidal space, consisting of those functions that have the $\Gamma_U\backslash U$-constant term equal to 0. Finally, there is a space orthogonal to both, and having a dense subspace of functions that have the analogue of a Fourier-like integral expression. For the orthogonalities, see Section 10.8. They are weaker than usual, because they involve an improper integral.

Determining the above spaces is the program of the present chapter. It is an important feature of the expansion that the families of functions (discrete and continuous) giving rise to the expansion are eigenfunctions of heat kernel convolution and of Casimir. Thus we refer to Chapter 3, where we laid the groundwork for such functions on G, and Chapter 8 for Γ-periodized functions. A basic aspect of the situation is the consideration of the continuous family, namely the Eisenstein family, as lying in a space of test functions for heat convolution such that the desired convolution formalism holds with absolutely convergent integrals. The groundwork for this was laid in Chapter 3.

We are of course principally interested in the eigenfunction expansion of the heat kernel, because this is the one that gives rise to our desired theta inversion formula on G/K. We go through C^∞ functions with compact support first. We start with a beautiful idea of Zagier [Zag 79] giving a direct formal access to the expansion. We thank Sarnak for his help in surmounting some of the analytic difficulties that arise in justifying the idea.

What we call the eigenfunction expansion is usually called the spectral decomposition, because a version was originally formulated by Roelcke [Roe 66/67] in the context of functional analysis and its "spectral theorem." See also Selberg [Sel 89], Elstrodt–Grunewald–Mennicke [EGM 98], Sections 6.3 and 6.4, referring systematically to Roelcke, and using the approach via functional analysis. For further historical comments, see Section 11.3 below, especially the end of the Section. Our treatment completely avoids the "spectral theorem," and leads directly to the eigenfunction expansion, the eigenfunctions being smooth and either in L^2 or in the space of linear exponential growth. It is the explicit pointwise eigenfunction expansion aspect that leads into the determination of explicit formulas and theta inversion, and ultimately to new zeta functions by taking the Gauss transform.

11.1 Haar Measure and the Mellin Transform

In the present situation $G = UAK$, the group U is abelian, normal in UA, and K is compact. We give K the Haar measure normalized so that the total measure of K is 1. The group A is isomorphic to the real positive group $\mathbf{R}^+$. We can use the Iwasawa coordinate $y = a_1^2 = a^\alpha$ from Sections 1.1 and 6.1 to get such an isomorphism. Then

$$da = dy/y$$

is a normalization of Haar measure on A. Finally, we have an isomorphism of U with $\mathbf{C} = \mathbf{R}^2$, with real coordinates x_1, x_2. Then $dx_1\, dx_2$ is a normalization of Haar (Lebesgue) measure on $\mathbf{C}$, and in terms of these coordinates, we write Haar measure on U as

$$du = dx_1\, dx_2.$$

Chapter 1, Proposition 1.2.1, shows that for a function f $\in C_c(UA) = C_c(AU)$, a Haar measure is determined by the functional

$$f \mapsto \int_U \int_A f(au)\,da\,du = \int_A \int_U f(au)\,du\,da.$$

On the other hand, because of the conjugation action of A on n (cf. Section 1.2), when we reverse the variables and use the ua decomposition, Haar measure on UA is given by the integral

$$f \mapsto \int_U \int_A f(ua)\delta(a)^{-1} du\, da,$$

where $\delta(a)$ is the determinant of the conjugation by a on n. Since n is an eigenspace of $\mathbf{R}$-dimension 2 for this action, with character α, we get

$$\delta(a) = a^{2\alpha} = y^2.$$

Thus we call $\delta(a)^{-1}da$ the A-**component** or A-**factor** of Haar measure.

Hence we determine a measure μ on UA by

$$d\mu(ua) = \delta(a)^{-1} du\, da = \frac{du\, dy}{y^3} = \frac{dx_1\, dx_2\, dy}{y^3},$$

such that the Haar measure on G is what we call the **Iwasawa measure**

$$dg = a^{-2\alpha} du\, da\, dk = y^{-3} du\, dy\, dk,$$

which fits the quaternionic coordinates perfectly. More fully, for $f \in C_c(G)$, we have

$$\int_G f(g)dg = \int_U \int_A \int_K f(uak)a^{-2\alpha} du\, da\, dk. \tag{11.1}$$

If, as will be the case, f is K-invariant, then

$$\int_G f(g)dg = \int_U \int_0^\infty f(ua_y)y^{-3}\, dy\, du.$$

Let $g(\mathbf{j}) = x + y\mathbf{j}$ and $f(g) = f_{\mathbf{h}}(x + \mathbf{j}y)$. Then we may also write

$$\int_G f(g)dg = \int_{-\infty}^{\infty} \int_{-\infty}^{\infty} \int_0^{\infty} f_{\mathbf{h}}(x + \mathbf{j}y)y^{-3}\, dy\, dx_1\, dx_2. \tag{11.2}$$

So much for Haar measure on G, as related to an invariant measure on $\mathbf{H}^3$.

We recall the Mellin transform. Let φ be a function on A. Its **Mellin transform** $\mathbf{M}\varphi$ is defined by the integral

$$(\mathbf{M}\varphi)(s) = \int_A \varphi(a)a^{s\alpha} da = \int_0^\infty \varphi(a_y)y^s \frac{dy}{y}.$$

Note that α constitutes a $\mathbf{C}$-basis for the characters on A, and that the transform is normalized by picking such a basis and the measure on the real multiplicative group. Thus $\mathbf{M}\varphi$ is a function of one complex variable. One might write

$$\varphi(a) = \varphi_{\mathbf{h}}(y) \text{ or also } \varphi_\alpha(y).$$

We recall that inverting the Mellin transform amounts to inverting a Fourier transform after a change of variables. Indeed, write $s = \sigma + \mathbf{i}r$. For any function φ on $\mathbf{R}^+$ for which the Mellin integral converges absolutely, we have

$$(\mathbf{M}\varphi)(s) = (\mathbf{M}\varphi)(\sigma+\mathbf{i}r) = \int_0^\infty \varphi(y)y^{\sigma+\mathbf{i}r}\frac{dy}{y}.$$

We change variable, letting $y = e^v$. We put

$$\varphi_\sigma(y) = \varphi(y)y^\sigma \text{ and } \psi_\sigma(v) = \varphi(e^v)e^{v\sigma} = \varphi_\sigma(e^v),$$

so that

$$(\mathbf{M}\varphi)(\sigma+\mathbf{i}r) = \hat{\psi}_\sigma(-r) \text{ where } \hat{\psi}_\sigma(r) = \int_{-\infty}^{\infty} \psi_\sigma(v)e^{-v\mathbf{i}r}dv.$$

Under conditions for which Fourier inversion is valid, we get

$$\psi_\sigma^- = \frac{1}{2\pi}\hat{\hat{\psi}}_\sigma \text{ or } \varphi(e^v)e^{v\sigma} = \frac{1}{2\pi}\int_{-\infty}^{\infty}(\mathbf{M}\varphi)(\sigma+\mathbf{i}r)e^{-v\mathbf{i}r}dr,$$

which can be expressed in terms of the multiplicative variable y, namely the **Mellin inversion formula** with a given σ reads

$$\varphi(y) = \frac{1}{2\pi}\int_{\mathrm{Re}(s)=\sigma}(\mathbf{M}\varphi)(s)y^{-s}\,dIm(s).$$

One needs serviceable conditions for the steps in the formal proof to hold. There are standard spaces for this, e.g., the Schwartz space. We formulate one set of standard but less-restrictive conditions for inversion.

Theorem 11.1.1. *Let φ be a function on $\mathbf{R}^+$ and σ real. Let φ_σ, ψ_σ be the functions defined by $\varphi_\sigma(y) = \varphi(y)y^\sigma$ and $\psi_\sigma(\upsilon) = \varphi_\sigma(e^\upsilon)$. If ψ_σ and $\hat{\psi}_\sigma$ are in $L^1(\mathbf{R}^+)$, then they are both bounded continuous; and Mellin inversion holds for φ with this σ, or equivalently, Fourier inversion holds for ψ_σ.*

11.1.1 Appendix on Fourier Inversion

Theorem 11.1.1 amounts to standard criteria of advanced calculus, which we recall for the convenience of the reader. We shall use constantly that if $f \in L^1$ then $\hat{f}$ is trivially uniformly continuous, bounded, with bound

$$\|\hat{f}\|_\infty \leqq \|f\|_1.$$

Theorem 11.1.2. *If f and $\hat{f}$ are in L^1, then Fourier inversion holds, that is,*

$$\hat{\hat{f}} = \frac{1}{2\pi} f^{-}. \tag{11.3}$$

Proof. The Fourier transform was defined with respect to the standard Lebesgue measure, so one has to throw in the $1/\sqrt{2\pi}$ factor for each transform. To avoid this, we let

$$d_1 x = dx/\sqrt{2\pi}.$$

Let $h(x) = ce^{-x^2 c'}$ with constants $c, c' > 0$ chosen such that $\hat{h} = h$. For $r > 0$ let $h_r(x) = h(rx)$, so $\hat{h}_r(x) = (1/r)h(x/r)$ directly from a change of variables in the definition of the Fourier transform. Suppose first f bounded. Then

$$\begin{aligned}\int \hat{f}(x)e^{-ixy}h(rx)d_1x &= \int\int f(t)e^{-itx}e^{-ixy}h_r(x)d_1t\,d_1x \\ &= \int f(t)\hat{h}_r(t+y)d_1t \\ &= \int f(t)\frac{1}{r}h\left(\frac{t+y}{r}\right)d_1t = \int f(ru-y)h(u)d_1u\end{aligned}$$

with the change of variables $u = (t+y)/r$. Letting $r \to 0$, if f is bounded, we get by the dominated convergence theorem

$$\hat{\hat{f}}(y)h(0) = f(-y)\hat{h}(0),$$

which concludes the proof under the additional assumption that f is bounded. But:

Theorem 11.1.3. *If f, $\hat{f} \in L^1$ then f is essentially bounded.*

Proof. We owe the following proof to Peter Jones. First we recall some L^1 simpler facts. For φ, $f \in L^1$ we have

$$(\varphi * f)^{\wedge} = \hat{\varphi}\hat{f}. \tag{11.4}$$

This is trivial by Fubini. Also for any function $\varphi \in L^1$, trivially

$$\|\hat{\varphi}\|_\infty \leqq \|\varphi\|_1. \tag{11.5}$$

Now let $\varphi \in C_c^\infty(\mathbf{R})$. Then $\varphi * f$ is bounded, trivially, by $\|\varphi\|_\infty \|f\|_1$. We need the more serious inequality

$$\|\varphi * f\|_\infty \leqq \|\varphi\|_1 \|\hat{f}\|_1. \tag{11.6}$$

Proof. Since $\varphi * f$ is bounded, in L^1, and by (11.4), $(\varphi * f)^{\wedge} = \hat{\varphi}\hat{f} \in L^1$, we know from the bounded version of the theorem that $(\varphi * f)^{\wedge\wedge} = (\varphi * f)^{-}$. Then by (11.5) twice and (11.4),

$$\|\varphi * f\|_\infty = \|(\varphi * f)^{\wedge\wedge}\|_\infty \leqq \|(\varphi * f)^{\wedge}\|_1 = \|\hat{\varphi}\hat{f}\|_1 \leqq \|\hat{\varphi}\|_\infty \|\hat{f}\|_1 \leqq \|\varphi\|_1 \|\hat{f}\|_1$$

as desired.

To show that $\|f\|_{\infty,\text{ess}} \leqq \|\hat{f}\|_1$, without loss of generality we may assume that $\|\hat{f}\|_1 = 1$. (If $\|\hat{f}\|_1 = 0$ then $f = 0$ a.e.; if not, divide f by the constant $\|\hat{f}\|_1$.) We apply (11.6′) to functions $\{\varphi_{\hat{n}}\}$, which form a Dirac sequence (cf. [Lan 93], Section 8.3, and $\varphi_n \in C_c^\infty(\mathbf{R})$. In particular, the functions φ_n are now such that

$$\|\varphi_n\|_1 = 1, \text{ so (11.6) gives } \|\varphi_R * f\|_\infty \leqq 1. \tag{11.6′}$$

Suppose f is not essentially bounded by 1. Given $C > 1$, there exists a set S of measure $\delta > 0$ such that $|f(x)| \geqq C$ for all $x \in S$, so $\|f\|_{1,S} \geqq C\delta$. By Corollary 3.4 of the above reference, we have

$$\|f - \varphi_n * f\|_1 \to 0 \quad \text{as} \quad n \to \infty. \tag{11.7}$$

Then restricting L^1-norms to S, we get

$$\|f - \varphi_n * f\|_1 \geqq \|f - \varphi_n * f\|_{1,S} \geqq \|f\|_{1,S} - \|\varphi_n * f\|_{1,S}.$$

But

$$\|\varphi_n * f\|_{1,S} \leqq \|\varphi_n * f\|_\infty \delta \leqq \delta \quad \text{[by (11.6″), for all } n\text{]}.$$

Thus we get $\|f - \varphi_n * f\|_1 \geqq (C-1)\delta$, which contradicts (11.7) and proves that f is essentially bounded by 1; qed.

Next we give a standard sufficient condition under which f, $\hat{f}$ are in L^1.

Theorem 11.1.4. *Suppose f differentiable and f, $f' \in L^1$. Then*

$$(\hat{f'})(x) = \mathbf{i}x\hat{f}(x) \textit{ so } \hat{f}(x) = O(1/|x|) \text{ for } x \to \infty. \tag{11.8}$$

If f, f', f'' are in L^1, then $(f'')(x) = (\mathbf{i}x)^2\hat{f}(x)$ and $\hat{f}(x) = O(1/x^2)$. Etc.

Proof. We use systematically that $\|\hat{f}\|_\infty \leqq \|f\|_1$ so $\|(\hat{f'})\|_\infty \leqq \|f'\|_1$. We have

$$(\hat{f'})(y) = \int_{-\infty}^{\infty} f'(x)e^{-\mathbf{i}xy}\,d_1x = \lim_{B\to\infty} \int_{-B}^{B} f'(x)e^{-\mathbf{i}xy}\,d_1x.$$

We thank Coifman for pointing out to us the following elegant trick.

Let φ_B be a C^∞ bump function equal to 1 on $[-B,B]$ and decreasing to 0 outside $[-B-1,B+1]$. Then from the last expression we get

$$(\hat{f'})(y) = \lim_{B\to\infty} \int_{-\infty}^{\infty} f'(x)\varphi_B(x)e^{-\mathbf{i}xy}\,d_1x.$$

We integrate by parts, with $u = \varphi_B(x)e^{-\mathbf{i}xy}$ and $dv = f'(x)dx$, so $v = f(x)$. The uv term is 0, and the other term becomes

$$\lim_{B\to\infty} - \int_{-\infty}^{\infty} f(x)[\varphi_B(x)(-\mathbf{i}y)e^{-\mathbf{i}xy} + \varphi_B'(x)e^{\mathbf{i}xy}]d_1x = -(-\mathbf{i}y)\hat{f}(y),$$

which proves the formula, and hence also the estimate.

Iterating, if f, f', f'' are in L^1, then

$$(\hat{f''})(x) = (\mathbf{i}x)^2 \hat{f}(x).$$

The desired estimate $\hat{f}(x) = O(1/x^2)$ follows. One can continue this by induction.

11.2 Adjointness Formula and the Constant Term

We next go to the quotient space $\Gamma\backslash G$. Note that $\Gamma\backslash\mathbf{H}^3$ is messier because of duplications, the kernel in the representation of G on $\mathbf{H}^3$, and singularities, so we prefer dealing with $\Gamma\backslash G$. We assume that the reader is acquainted with the most elementary properties of invariant measures on homogeneous spaces. In particular, if H is a closed unimodular subgroup of G, then fixing a Haar measure on H determines a unique G-invariant measure on $H\backslash G$ (as right homogeneous space for G), satisfying the **group Fubini theorem**

$$\int_G f(g)dg = \int_{H\backslash G} \left[\int_H f(wg)dw \right] dg.$$

This is the case if H is discrete, with elements having measure 1, so the integral over H is the sum over the elements of H. We meet this case now with $H = \Gamma = \mathrm{SL}_2(\mathbf{Z}[\mathbf{i}])$. Note that locally, G and $\Gamma\backslash G$ are topologically and measure isomorphic.

Since Γ is discrete in G, the Iwasawa Haar measure on G automatically defines a right-invariant measure on $\Gamma\backslash G$, denoted by $d_{\Gamma\backslash G}(g)$, again called the **Iwasawa measure** on $\Gamma\backslash G$. One often writes more simply dg, omitting the index $\Gamma\backslash G$.

Eisenstein series can be defined with more general functions than characters, simply as the $\Gamma_U\backslash\Gamma$-trace of a function, under an assumption that the series converges absolutely. For simplicity, let $\varphi \in C(A)$, let $\pi_A : G \to A$ be the Iwasawa projection on A, and **define**

$$\begin{aligned} E(\varphi, g) = \mathrm{Tr}_{\Gamma_U\backslash\Gamma}(\varphi \circ \pi_A)(g) &= \sum_{\gamma \in \Gamma_U\backslash\Gamma} \varphi((\gamma g)_A) \\ &= \text{abbreviated } \mathrm{Tr}_{\Gamma_U\backslash\Gamma}(\varphi)(g). \end{aligned}$$

One calls $\mathrm{Tr}_{\Gamma_U\backslash\Gamma}(\varphi)$ the **Eisenstein series** of φ or the **φ-Eisenstein series**, or also the **Eisenstein trace** of φ. If $\varphi = \chi$ is a character, we get the Eisenstein series defined previously.

For symmetry purposes, it is also appropriate to define the constant term integral as a **trace**, namely for a function f on $\Gamma\backslash G/K$, and $g = uak$ the Iwasawa decomposition, we write

$$\mathrm{Tr}_{\Gamma_U\backslash U}(f)(g) = \mathrm{Tr}_{\Gamma_U\backslash U}(f)(a) = \int_{\Gamma_U\backslash U} f(ug)dg = \int_{\Gamma_U\backslash U} f(ua)du.$$

We also call this integral the **cuspidal integral** of f.

11.2.1 Adjointness Formula

Let dg be the Iwasawa Haar measure on G, $dg = du\, d\mu_A(a)\, dk$, where as in Section 11.1

$$d\mu_A(a) = \delta(a)^{-1}da.$$

Let $[.,.]$ be the symmetric integral scalar product (integrating the product of two functions without any complex conjugate, over an indicated measure space). Then for $f \in C(\Gamma\backslash G/K)$, $\varphi \in C(A)$ such that

$$\mathrm{Tr}_{\Gamma_U\backslash\Gamma}|\varphi| \cdot |f| \in L^1(\Gamma\backslash G),$$

we have the adjointness formula

$$[\mathrm{Tr}_{\Gamma_U\backslash\Gamma}\varphi, f]_{\Gamma\backslash G} = [\varphi, \mathrm{Tr}_{\Gamma_U\backslash U} f]_{d\mu_A}.$$

Proof. We have

$$\begin{aligned}[\mathrm{Tr}_{\Gamma_U\backslash\Gamma}\varphi, f]_{\Gamma\backslash G} &= \int_{\Gamma\backslash G} \sum_{\gamma\in\Gamma_U\backslash\Gamma} \varphi(\gamma g)f(g)dg \\ &= \int_{\Gamma_U\backslash G} \varphi(g)f(g)dg \\ &= \int_A \int_{\Gamma_U\backslash U} f(ua)\varphi(a)\, \delta(a)^{-1}du\,da,\end{aligned}$$

which proves the formula.

A function f will be called **cuspidal** if its cuspidal integral is 0, that is,

$$\int_{\Gamma_U\backslash U} f(ug)du = 0 \text{ for all } g \in G.$$

From the adjointness formula, we shall get various conditions for cuspidality.

The adjointness was made the starting point of the exposition in [Lan 73/85], Chapter 13. The material in that reference has a different emphasis from the present chapter, which is directed toward more explicit formulas for various expansions of functions. Cf. [Zag 79] and [Szm 83].

We now translate the adjointness formula into the notation of the Iwasawa quaternion coordinates. If we view the above function as a function of the coordinate y via $g(\mathbf{j}) = x + \mathbf{j}y$, then

$$\mathrm{Tr}_{\Gamma_U \backslash \Gamma} \varphi(g) = \sum_{\gamma \in \Gamma_U \backslash \Gamma} \varphi(y(\gamma g)) = \sum_{(c,d)=1} \varphi(y(c,d;z)).$$

For $f \in \mathrm{L}^1(\Gamma_U \backslash G/K)$ we have the integral formula

$$\int_{\Gamma_U \backslash G} f(g) d_{\Gamma_U \backslash G}(g) = \int_0^\infty \int_0^1 \int_0^1 f_{\mathbf{h}}(x + \mathbf{j}y) y^{-3} dx_1 \, dx_2 \, dy. \tag{11.10}$$

If for each $a \in A$ the function $u \mapsto f(ua)$ is in $L^2(\Gamma_U \backslash U)$, then f has a Fourier series (L^2-convergent) in the usual manner. The constant term is given by the cuspidal integral

$$\mathrm{CT}_{\Gamma_U \backslash U}(f)(a) = \int_{\Gamma_U \backslash U} f(ua) du = \int_0^1 \int_0^1 f_{\mathbf{H}}(x + \mathbf{j}y) dx_1 \, dx_2. \tag{11.11}$$

No constant factor appears because the volume of $\mathbf{Z}[\mathbf{i}] \backslash \mathbf{C}$ is 1 for the standard Haar measure on $\mathbf{C} = \mathbf{R}^2$. Thus the U-measure is normalized so that $\Gamma_U \backslash U$ has measure 1. For a function f on $\mathbf{H}^3$, so a function of (x, y), $x \in \mathbf{C}$ and $y > 0$, Γ-invariant on the left, we use the notation for the **constant term** (with respect to $\Gamma_U \backslash U \approx \mathbf{C}/o$):

$$\mathrm{CT}_{\Gamma_U \backslash U}(f)(y) = A_0(f)(y) = A_0(f, y) = \int_{\mathbf{C}/o} f(x + \mathbf{j}y) dx.$$

We shall apply the above to functions f obtained by convolution as follows. Let $E = E(s, g)$ be a function of two variables: s in some open set in $\mathbf{C}$, and $g \in \Gamma \backslash G$. For a function f on $\Gamma \backslash G$, we use the *Iwasawa convolution*

$$(E * f)(s) = (E *_{\Gamma \backslash G} f)(s) = \int_{\Gamma \backslash G} E(s, g) f(g) d\mu_{\mathrm{Iw}}(g)$$

under conditions for which the integral is absolutely convergent. The measure is the *Iwasawa measure* induced on $\Gamma \backslash G$, more convenient than the polar measure for the moment.

We now reproduce the adjointness formula in the present context and repeat the proof in terms of the (x, y) coordinates. We let $E(s, g)$ be the usual Eisenstein series.

Proposition 11.2.1. *Let f be a continuous function on $\Gamma \backslash G/K$. Let s have $\sigma = \mathrm{Re}(s) > 2$, and be such that the convolution integral for $(E * f)(\sigma)$ is absolutely convergent. Then letting $\varphi(y) = A_0(f)(y) y^{-2}$, we have*

$$(E * f)(s) = \int_0^\infty A_0(f)(y) y^{s-2} \frac{dy}{y} = \mathbf{M}\varphi(s) = \mathbf{M}A_0(s - 2).$$

Proof. Directly from the definition of the Eisenstein series,

$$\int_{\Gamma\backslash G} E(s,g)f(g)dg = \int_{\Gamma\backslash G} \sum_{\gamma\in\Gamma_U\backslash\Gamma} y(\gamma g)^s f(g)dg$$

$$= \int_{\Gamma_U\backslash G} f(g)y(g)^s dg$$

$$= \int_0^\infty \int_{\mathbf{C}/\mathfrak{o}} f(x+\mathbf{j}y)y^{s-2}\, dx\, \frac{dy}{y}$$

$$= \int_0^\infty A_0(f,y)y^{s-2}\frac{dy}{y}$$

as was to be shown.

In words, Proposition 11.2.1 says that the convolution of the Eisenstein series with a function f is the Mellin transform of the constant term of f with an appropriate shift. This is called the **Rankin–Selberg method** in the SL_2 industry. It is a special case of a general integral adjointness formula for more general groups G such as SL_n (semisimple Lie groups, etc.). For a treatment of SL_n, see [JoL 02].

11.3 The Eisenstein Coefficient $E * f$ and the Expansion for $f \in C_c^\infty(\Gamma\backslash G/K)$

We shall use estimates from Chapter 8 especially Section 8.6, to apply Proposition 11.2.1.

The analytic continuation of $E*f$ will first be carried out with $f\in C_c^\infty(\Gamma\backslash\mathbf{H}^3)$. We are much indebted to Sarnak, who provided us with the rest of this section, justifying Zagier's beautiful formal arguments [Zag 79] for such f. We are striving toward Theorem 11.3.5, which is a basic result, proved in this section for $f\in C_c^\infty(\Gamma\backslash G/_{\circ}K)$.

From Section 2.2 (6c) we know the eigenvalue of Casimir on the function y^s. Casimir is an invariant differential operator, so each term $y(\gamma z)^s$ is an eigenfunction with this eigenvalue.

We use the notation E_s for the function such that

$$E_s(g) = E(s,g),$$

and similarly for E_{χ_ζ}. As usual, we let $\tau = 2\rho$ be the trace of the Lie (a,n) representation. Letting $[,]$ denote the symmetric integral scalar product (no complex conjugation) with respect to Iwasawa measure, for a test function f, we have

$$[E_s, f]_{\Gamma\backslash G} = (E * f)(s).$$

If f has compact support, it follows trivially that the integral defining the convolution $E * f(s)$ converges absolutely and uniformly for s in a compact set avoiding the poles, and so $E * f(s)$ is meromorphic in the complex variable s. One may also prefer to take the convolution with ΛE_s, which has a pole in s only at $s = 2$. The convolution is with respect to the Iwasawa Haar measure as in Sections 11.1 and 11.2.

The next theorem goes back to Chapter 2, Proposition 2.4.4, and the bilinear form of Section 1.6, **C**-bilinear extension of the trace form on a.

Theorem 11.3.1. *The Eisenstein function E_s is an eigenfunction of Casimir, with eigenvalue*

$$\mathrm{ev}(\omega_G, E_s) = 2s(s-2) \text{ or invariantly } \mathrm{ev}(\omega_G, E_{\chi_\zeta}) = \check{B}_{\mathbf{C}}(\zeta,\zeta) - \check{B}_{\mathbf{C}}(\tau,\zeta).$$

This holds for all s where E does not have a pole. Alternatively, the same thing holds with E_s replaced by ΛE_s (whose only pole is at 2), so for $s \neq 2$.

Proof. We know the analytic expression of Chapter 10, Theorem 10.5.3. Then it's a question of differentiating a series, or DUTIS and convergence. We pick one way. Let $f \in C_c^\infty(\Gamma\backslash G/K)$. Then for $\mathrm{Re}(s) > 2$, we have

$$[\omega E_s, F] = [E_s, \omega f] \text{ and also} = 2s(s-2)[E_s, f].$$

The desired conclusion falls out. But $[E_s, \omega f]$ is analytic (meromorphic) in s, so the relation is also true for all s (not a pole), as was to be shown.

Remark. The Eisenstein function provides a basic example for Chapter 3, Theorem 3.4.1. First E_s is in LEG^∞, and it is a heat eigenfunction

$$\mathbf{K}_t * E_s = \mathbf{K}_{t,\mathrm{Iw}} *_{Iw} E_s = e^{2s(s-2)t} E_s.$$

Furthermore, the heated Eisenstein function

$$E_S^\#(t,z) = e^{2s(s-2)t} E_s(z)$$

satisfies the heat equation for $\mathrm{Re}(s) \geqq 1,\ s \neq 2$, and more generally whenever s is not a pole.

Proposition 11.3.2. *Let $f \in C_c^\infty(\Gamma\backslash G/K)$. The function $E * f(s)$ is holomorphic in the region $\mathrm{Re}(s) \geqq 1$, except for a simple pole at $s = 2$. For any $N > 0$, it satisfies the bound*

$$|(s-2)||E * f(s)| \ll_N (1+|t|)^{-N} \text{ in the strip } 1 \leqq \mathrm{Re}(s) \leqq 10.$$

In other words, $E * f$ has superpolynomial decay in the strip.

Proof. Using Theorem 11.3.1, we get

$$E * f(s) = \int_{\Gamma\backslash G} f(z)\omega^m E(s,z)(2s(s-2))^{-m} d\mu(z)$$
$$= (2s(s-2))^{-m} \int_{\Gamma\backslash G} \omega^m f(z)E(s,z)d\mu(z).$$

By Theorem 10.6.2 of Chapter 10 and the fact that $\omega^m f$ is fixed in $C_c^\infty(\Gamma\backslash G)$, this last expression has the desired decay in the half-plane, proving the proposition.

The above can now be used to justify the next steps in Zagier. Indeed, the constant term $A_0(f,y)$ defined in Section 11.2 for $0 < y < \infty$ is bounded for $y \to 0$, and $A_0(f,y) \to 0$ as $y \to \infty$. Hence Proposition 11.2.1 applies to give for $\mathrm{Re}(s) > 2$,

$$\int_0^\infty \frac{A_0(f,y)}{y^2} y^s \frac{dy}{y} = E * f(s).$$

By Proposition 11.3.2, Mellin inversion is valid for $\mathrm{Re}(s) > 2$, that is, we can apply Theorem 11.1.1, to get

$$\frac{A_0(f,y)}{y^2} = \frac{1}{2\pi \mathbf{i}} \int_{\mathrm{Re}(s)=2+\varepsilon} E * f(s) y^{-s} ds.$$

By Proposition 11.3.2, the integral is sufficiently rapidly convergent that the integration over the vertical line can be shifted to $\mathrm{Re}(s) = 1$, thus yielding the following result:

Lemma 11.3.3. *For $f \in C_c^\infty(\Gamma\backslash \mathbf{H}^3)$ we have*

$$A_0(f,y) = \mathrm{res}_2(E * f) + \frac{1}{2\pi \mathbf{i}} \int_{\mathrm{Re}(s)=1} (E * f)(s) y^{2-s} ds$$
$$= \mathrm{res}_2(E * f) + \frac{1}{2\pi} \int_{-\infty}^{\infty} (E * f)(1 - \mathbf{i}r) y^{1+\mathbf{i}r} dr.$$

The second expression is obtained by putting $s = 1 + \mathbf{i}r$, and then letting $r \mapsto -r$.

Lemma 11.3.4. *We have the relation*

$$\int_{\mathrm{Re}(s)=1} (E * f)(\bar{s}) y^s d\mathrm{Im}(s) = \frac{1}{8} \mathrm{CT}_{\Gamma_U\backslash U} \int_{\mathrm{Re}(s)=1} (E * f)(\bar{s}) E(s,z) d\mathrm{Im}(s)$$

The integral on the right converges absolutely.

Proof. The absolute convergence of the integrals comes from Theorem 10.6.2 of Chapter 10 and Proposition 11.3.2. The constant term $\mathrm{CT}_{\Gamma_U\backslash U}$ is an integral over a compact group (torus), and can be taken inside the integral on the vertical line, so the right side by Chapter 10, Theorem 10.5.1, is equal to

$$\int_{\mathrm{Re}(s)=1} (E*f)(\bar{s})\mathrm{CT}_{\Gamma_U\backslash U}E(s,z)d\mathrm{Im}(s)$$

$$= 4\int_{\mathrm{Re}(s)=1} (E*f)(\bar{s})\left[y^s + y^{2-s}\frac{\Lambda\zeta_{\mathbf{F}}(s-1)}{\Lambda\zeta_{\mathbf{F}}(s)}\right]d\mathrm{Im}(s). \tag{11.12}$$

The right side is a sum of two integrals, and we claim that they are equal. Indeed, for $\mathrm{Re}(s) = 1$ we have $2-s=\bar{s}$, $2-\bar{s}=s$, $s-1=1-\bar{s}$, so using the functional equations

$$\Lambda E(\bar{s}) = \Lambda E(2-\bar{s}) \quad \text{and} \quad \Lambda\zeta_{\mathbf{F}}(1-\bar{s})$$

we get

$$\begin{aligned}(E*f)(\bar{s})y^{2-s}\frac{\Lambda\zeta_{\mathbf{F}}(s-1)}{\Lambda\zeta_{\mathbf{F}}(s)} &= (E*f)(\bar{s})y^{\bar{s}}\frac{\Lambda\zeta_{\mathbf{F}}(s-1)}{\Lambda\zeta_{\mathbf{F}}(s)}\\ &= (E*f)(2-\bar{s})y^{\bar{s}}\frac{\Lambda\zeta_F(2-\bar{s})}{\Lambda\zeta_{\mathbf{F}}(\bar{s})}\frac{\Lambda\zeta_{\mathbf{F}}(s-1)}{\Lambda\zeta_{\mathbf{F}}(s)}\\ &= (E*f)(s)y^{\bar{s}}.\end{aligned}$$

Making the change of variables $s \mapsto \bar{s}$ in the integral concludes the proof of the lemma.

Note that the $\Gamma_U\backslash U$-constant term of a constant function is the function itself, because the volume of $\Gamma_U\backslash U$ is 1. Then from Lemmas 11.3.3 and 11.3.4, we obtain

$$\mathrm{CT}_{\Gamma_U\backslash U}(f)(z) = \mathrm{res}_2(E*f) + \frac{1}{2\pi\cdot 8}\mathrm{CT}_{\Gamma_U\backslash U}\int_{\mathrm{Re}(s)=1}(E*f)(\bar{s})E(s,z)d\mathrm{Im}(s). \tag{11.13}$$

Next we reach a major structural result that we call the **Fourier–Eisenstein expansion** of a function f, or the **eigenfunction expansion**. It is often also called the **spectral expansion**, especially in a context in which methods of spectral functional analysis are used, which we do not. Let f_{cusp} be defined by the difference

$$f_{\mathrm{cusp}}(z) = f(z) - \mathrm{res}_2(E*f) - \frac{1}{16\pi}\int_{\mathrm{Re}(s=1)}(E*f)(\bar{s})E(s,z)d\mathrm{Im}(s).$$

Then f_{cusp} is Γ-invariant, and $\mathrm{CT}_{\Gamma_U\backslash U}(f_{\mathrm{cusp}}) = 0$, i.e., f_{cusp} has $\Gamma_U\backslash U$-constant term equal to 0, under conditions of absolute convergence as follows.

Theorem 11.3.5. *Let $f \in C_c^\infty(\Gamma\backslash G/K)$. With the Iwasawa measure convolution, we have an expansion*

$$f(z) = f_{\mathrm{cusp}}(z) + \mathrm{res}_2(E * f) + \frac{1}{16\pi} \int\limits_{\mathrm{Re}(s)=1} (E * f)(\bar{s})E(s,z)d\mathrm{Im}(s).$$

The function f_{cusp} is cuspidal and bounded.

Proof. We may decompose the Eisenstein integral on the right side as the sum of its constant term taken care of by Lemma 11.3.4, (11.12), (11.13), and the integral

$$\frac{1}{16\pi} \int\limits_{\mathrm{Re}(s)=1} (E * f)(\bar{s})\mathrm{BF}(s-1,z)\Lambda\zeta_{\mathbf{F}}(s)^{-1}d\mathrm{Im}(s).$$

The exponential decay for the Bessel function (Section 10.3, **K7**) shows that the integral giving the complementary part to the constant term is a bounded function on $\Gamma\backslash G$. Since f is bounded, it follows that f_{cusp} is bounded, thus completing the proof of Theorem 11.3.5.

Note that the factor $16 = 4^2$ exactly takes care of the quadruplication in the definition of Eisenstein series, and the fact that there are two of them inside the integral.

Lemma 11.3.6. *The integral*

$$\int\limits_{\mathrm{Re}(s)=1} (E * f)(\bar{s})\Lambda\zeta_{\mathbf{F}}(s)^{-1}d\mathrm{Im}(s)$$

is absolutely convergent.

Proof. This comes from Theorem 10.6.2 of Chapter 10, namely the lower bound for the zeta function, and the superpolynomial decay of $E * f$ in Proposition 11.3.2.

Remark. In Zagier [Zag 79], on pp. 314–315, an assumption is made only on the decay of f at infinity. But this assumption is insufficient to make the relevant integrals converge, that is, to justify Mellin–Fourier inversion, because the constant term is bounded only near 0. Furthermore, to have inversion one must have both some smoothness and some decay conditions, as in the theorems of the appendix of Section 11.1. Peter Jones told us that there is a structural phenomenon involved in the behavior of the constant term near 0, because of the following fact.

> *Let f be continuous, bounded, not identically 0, and Γ-invariant. Then, say on the upper half-plane,*
>
> $$\int\limits_{\mathbf{R}/\mathbf{Z}} f(x + \mathbf{i}\varepsilon)dx$$
>
> *has a nonzero limit as ε goes to 0.*

Thus to apply Mellin inversion and to shift the line of integration from 2 to 1, something had to be done to avoid systematically divergent integrals. Note that

Kubota [Kub 73] p. 47 does specify that the function "is always assumed to decrease sufficiently rapidly both as $y \to 0$ and $y \to \infty \ldots$"

In addition, on pp. 315–317, the Eisenstein series are integrated over the vertical line $\text{Re}(s) = 1$, after being divided by Riemann's zeta function. This requires a lower bound on the line for this function, and an estimate of $E * f$, here provided by Sarnak's combination of Proposition 11.3.2 and Theorem 10.6.2 from Chapter 10.

We also note that the expansion in Theorem 11.3.5 is pointwise. Kubota [Kub 73] mentions such a formula and attributes it to Roelcke [Roe 66] in a footnote p. 62. But in these references, the formula is an L^2-formula, not pointwise. The difference between L^2 and pointwise convergence is blurred in Zagier, p. 316. Both smoothness conditions and decay conditions are needed to get the formula pointwise.

Concerning the "spectral decomposition formula for $L^2(\Gamma\backslash H)$" and other matters, Zagier states ([Zag 79] p. 312); "All of this material is standard and may be skipped by the expert reader. We will try to give at least a rough proof of all of the statements; for a more detailed exposition the reader is referred to Kubota's book [4]." However, in the above footnote, Kubota calls the formula one of "the deeper results" related to his own chapter, and *he does not give a proof.* Lang wrote him to ask whether he had notes from his course on which his book was based. He answered July 1, 2002; "The formula 5.3.12 is essentially contained in the works of Roelcke, and so I announced it in my course without proof. But, it must be very easy for you to make good use of Roelcke's work published some 35 years ago." But, as already pointed out at the end of the introduction to this chapter, we have not used Roelcke's method.

In Section 10.8 we described the extent to which the different eigenfunctions in the formula are orthogonal.

We shall extend the domain of validity of the formula to Gaussians instead of C_c^∞-functions. It will be seen in the rest of this book how the combination of the expansion formula and the Selberg trace formula leads to the desired theta inversion formula.

Finally, we note that the cuspidal subspace presents some questions. For instance, for arithmetic groups Γ, the cuspidal space $L^2_{\text{cusp}}(\Gamma\backslash G/K)$ is known to be infinite-dimensional, but work of Sarnak–Phillips led them to conjecture that it is frequently finite-dimensional in the nonarithmetic case, contrary to Kubota's assertion [Kub 73] p. 48: "It is easy to see that both Θ and H_0 are infinite–dimensional." See for instance Sarnak's lectures [Sar 03].

11.4 The Heat Kernel Eigenfunction Expansion

We shall apply the formalism of Chapters 2 and 3. To do that, we need the following:

Theorem 11.4.1. *For s not a pole* ($s \neq 0, 2$) *the function* ΛE_s *is in* LEG^∞. *The function* E_s *itself is in* LEG^∞ *for* $1 \leqq \text{Re}(s)$. *In terms of the Iwasawa coordinates* $(x, y) = (x_1,\ x_2,\ y)$ *let D be a monomial in the partial derivatives* $\partial/\partial x_1,\ \partial/\partial x_2,\ \partial/\partial y$.

There exist constants $c(D)$, $c'(D)$ such that for y outside a neighborhood of 1,

$$|DE_C(z)| \ll_D e^{c(D)|\log y||s|}|s|^{c'(D)}.$$

Proof. This is immediate from the Bessel expansion, Chapter 10, Theorem 10.5.3, and Chapter 3, Lemma 3.2.3

We may now specialize the theorems and lemmas of Chapters 2 and 3 to the Eisenstein functions.

In Theorem 11.3.1 we saw that the Eisenstein function E_s is an eigenfunction of Casimir, with eigenvalue

$$2s(s-2) = \mathrm{ev}(\omega, \chi_s).$$

Let $\mathbf{K}_t$ be the heat kernel as in Chapter 2, and $\mathbf{K}_t^\Gamma$ its periodization as in Chapter 4.

Theorem 11.4.2. *For $t > 0$, $\mathrm{Re}(s) = 1$, and the Γ-periodized heat kernel, we have*

$$(\mathbf{K}_t^\Gamma * E_s)(z) = e^{\mathrm{ev}(\omega,\chi_s)t} E_s(z).$$

The convolution on the left is convolution on $\Gamma\backslash G$. namely

$$(\mathbf{K}_t^\Gamma * E_s)(z) = \int_{\Gamma\backslash G} \mathbf{K}_t^\Gamma(z,w) E_s(w) d\mu(w)$$

with the same Haar measure used to define the heat kernel **Weierstrass–Dirac** *property.*

Proof. Special case of Chapter 3, Theorem 11.4.1, in light of Theorem 11.4.1.

Theorem 11.4.3. *For the Γ-periodized heat kernel with respect to the Iwasawa Haar measure $\mathbf{K}_t^\Gamma = \mathbf{K}_{t,\,\mathrm{Iw}}^\Gamma$, we have the expansion*

$$\mathbf{K}_t^\Gamma(z,w) = \mathbf{K}_{t,\mathrm{cusp}}^\Gamma(z,w) + c_0 + \frac{1}{16\pi}\int_{-\infty}^{\infty} e^{-2(1+r^2)t} E(1-\mathbf{i}r, Z) E(1+\mathbf{i}r, w)\, dr,$$

so

$$\mathbf{K}_t^\Gamma(z,z) = \mathbf{K}_{t,\mathrm{cusp}}^\Gamma(z,z) + c_0 + \frac{1}{16\pi}\int_{-\infty}^{\infty} e^{-2(1+r^2)t} |E(1+\mathbf{i}r, z)|^2 dr.$$

The constant term has the value

$$c_0 = \operatorname*{res}_{s=2}(\mathbf{K}_t^\Gamma * E_s) = \operatorname*{res}_{s=2} E_s = \frac{1}{\mathrm{Vol}(\Gamma\backslash G)},$$

with the Iwasawa measure volume.

Proof. Fix w and t and let $f(z) = \mathbf{K}_t^\Gamma(z,w)$. Let $\{f_n\}$ be a sequence in $C_c^\infty(\Gamma\backslash G/K)$, increasing with limit f. By Theorem 11.3.5 we have the expansion formula for each

f_n. The integral with f instead of f_n is absolutely convergent, and uniformly, because for z in a compact set, the series for $\mathbf{K}_t(z,w)$ is absolutely uniformly convergent. We can multiply $\mathbf{K}_t(z,w)$ by a cutoff function equal to 1 on increasing compact sets, and decreasing to 0 in some band beyond. We then get uniform approximation in the limit, and thus obtain the expansion formula for f itself. Then we invoke Theorem 11.4.2 to get the formula as stated with c_0 equal to the residue of $\mathbf{K}_t^\Gamma * E_s$ at $s=2$. By Theorem 11.3.1, we know the heat eigenvalue $e^{2s(s-2)} = 1$ at $s=2$. This shows that c_0 is also the residue of E_s at $s=2$. The value in terms of the volume comes from Chapter 10, Theorem 10.7.2. This concludes the proof.

Remark. In [Szm 83] Szmidt states the expansion formula precisely in the case of $\mathrm{SL}_2(\mathbf{C})$ and $\mathrm{SL}_2(\mathbf{Z}[\mathbf{i}])$, for functions "of sufficiently rapid decay," using Selberg's notation and convention for the spherical transform. See p. 400, line 7. Szmidt's formula amounts to ours taking into account different normalizations. His function on the left is the Γ-periodization of a point-pair invariant, but his Γ is the projective group $\mathrm{PSL}_2(\mathbf{Z}[\mathbf{i}])$; hence his periodization is 1/2 times ours. His Eisenstein series are 1/4 times ours, thus giving rise to a factor 1/16 in front of our Eisenstein integral. Multiplying both sides by 2 shows that our formula is compatible with his.

For the first term of the expansion, we connect with Chapter 8, Theorem 8.5.5. As in this theorem, we let $Q_t(z,w)$ be the kernel function in L^2 representing the Iwasawa measure convolution action of the heat kernel on $L^2_{\mathrm{cus}}(\Gamma\backslash G/K)$. Then Q_t is cuspidal in each variable. We called Q_t the **heat kernel's cuspidal component**.

Theorem 11.4.4. *The heat kernel's cuspidal component is equal to the cuspidal term in the eigenfunction expansion of the heat kernel. In other words, pointwise we have*

$$\mathbf{K}^\Gamma_{t,\mathrm{cus}} = Q_t,$$

and so

$$\mathbf{K}_t^\Gamma(z,w) = Q_t(z,w) + c_0 + \frac{1}{16\pi}\int_{-\infty}^{\infty} e^{-2(1+r^2)t} E(1-\mathbf{i}r,z)E(1+\mathbf{i}r,w)dr.$$

The function Q_t is continuous, bounded for one of the variables in a compact set.

Proof. Let $B_{\mathrm{cus}}(\Gamma\backslash G/K)$ denote the space of measurable bounded cuspidal functions. Let $f \in B_{\mathrm{cus}}(\Gamma\backslash G/K)$. Then $\mathbf{K}_t^\Gamma * f = Q_t * f$ by Chapter 8, Theorem 8.4.1 Convolution of a constant function with f gives 0 by Chapter 10, Theorem 10.8.2. The improper integral convolution of E with f is 0 by Chapter 10, Theorem 10.8.1. Hence for every $f \in B_{\mathrm{cus}}(\Gamma\backslash G/K)$ we have

$$\mathbf{K}^\Gamma_{t,\mathrm{cus}} *_{\Gamma\backslash G} f = Q_t *_{\Gamma\backslash G} f,$$

which implies the first stated equality. The rest of the theorem follows at once.

By Chapter 8, Theorem 8.5.5, we know the eigenexpansion of $Q_t\,(z,w)$, namely

$$Q_t(z,w) = \sum_{n=1}^{\infty} e^{-\lambda_n t} \psi_n(z)\overline{\psi_n(w)}, \tag{11.14}$$

$$Q_t(z,z) = \sum_{n=1}^{\infty} e^{-\lambda_n t} |\psi_n(z)|^2, \tag{11.15}$$

with an orthonormal basis of heat eigenfunctions $\{\psi_n\}$ on $\Gamma\backslash G/K$. These functions can be viewed as right K-invariant on $\Gamma\backslash G$ or left Γ-invariant on G/K.

Readers will appreciate that the continuous part of the expansion (the Eisenstein integral) has the same formal structure as the discrete Fourier-like series for Q_t, with the exponential term reflecting the eigencharacter, and the eigenfunctions themselves. In the discrete case, the eigenfunctions are in L^2_{cus}, and in the continuous case they are in a somewhat weaker space, namely LEG$^\infty$. From Theorem 11.4.3, we write down the expansion in full, on the diagonal.

Theorem 11.4.5. *With the Iwasawa Haar measure, for $z = w$ we have the formula*

$$\mathbf{K}_t^\Gamma(z,z) = \sum_{n=1}^{\infty} e^{-\lambda_n t} |\psi_n(z)|^2 + c_0 + \frac{1}{16\pi} \int_{-\infty}^{\infty} e^{-2(1+r^2)t} |E(1+\mathbf{i}r, z)|^2 dr.$$

The constant c_0 has the value stated in Theorem 11.4.3, namely $1/\mathrm{vol}\,(\Gamma\backslash G)$.

Note that all the terms in this formula are positive. Thus we obtain the following:

Corollary 11.4.6. *For all z,*

$$\sum_{n=1}^{\infty} e^{-\lambda_n t} |\psi_n(z)|^2 \leqq \mathbf{K}_t^\Gamma(z,z).$$

For each z, we have

$$\sum_{n=1}^{\infty} e^{-\lambda_n t} |\psi_n(z)|^2 = O(t^{-3/2}) \text{ for } t \to 0.$$

Proof. The first assertion comes directly from the positivity remark. The second comes from the formula for the heat kernel.

Theorem 11.4.4 also gives us a proof of Chapter 8, Theorem 8.5.1, as mentioned previously.

Theorem 11.4.7. *For all z, $w \in G$, and $\mathbf{K}_t^\Gamma = \mathbf{K}_{t,\,\mathrm{Iw}}^\Gamma$,* we have

$$\lim_{t\to\infty} \mathbf{K}_t^\Gamma(z,w) = c_0.$$

Proof. Immediate since all the terms go to 0 except the constant term c_0.

Part V
The Eisenstein–Cuspidal Affair

Chapter 12
The Eisenstein Y-Asymptotics

12.1 The Improper Integral of Eigenfunction Expansion over $\Gamma\backslash G$

Starting with the eigenfunction expansion of the Γ-periodized heat kernel at the end of the preceding chapter, we want to integrate each side over $\Gamma\backslash G$ to obtain the theta inversion relation. A convergence problem arises, which we solve by using the decomposition of the Γ-trace into the noncuspidal trace and the cuspidal trace. We abbreviate the notation, letting

$$\mathbf{K}_t^{\mathrm{NC}}(z,z) = \sum_{\gamma\in\mathrm{NCus}\Gamma} \mathbf{K}_t(\gamma z,z) \text{ and } \mathbf{K}_t^{\mathrm{Cus}}(z,z) = \sum_{\gamma\in\mathrm{Cus}\Gamma} \mathbf{K}_t(\gamma z,z).$$

The **eigenfunction expansion** of Chapter 11, Theorem 11.4.5, can then be written as a relation with five terms:

$$\begin{aligned}&\mathbf{K}_t^{\mathrm{NC}}(z,z)+\mathbf{K}_t^{\mathrm{Cus}}(z,z)\\ &= \sum_{n=1}^{\infty} e^{-\lambda_n t}|\psi_n(z)|^2 + c_0 + \frac{1}{16\pi}\int_{-\infty}^{\infty} e^{-2(1+r^2)t}|E(1+\mathbf{i}r,z)|^2 dr.\end{aligned}$$

We want to integrate this relation over $\Gamma\backslash G$ to obtain the theta inversion formula. *We use the Iwasawa heat kernel determined by the Iwasawa measure*

$$d\mu(z) = dx_1\, dx_2\, dy/y^3.$$

We determined explicitly the integral of the noncuspidal first term on the left in Chapter 5, Theorem 5.3.3, namely

$$\int_{\Gamma\backslash G} \mathbf{K}_t^{\mathrm{NC}}(z,z)\, d\mu(z) = e^{-2t}(4t)^{-1/2}\Theta^{\mathrm{NC}}(1/t), \tag{12.1}$$

the series on the right being called the **noncuspidal inverted theta series**.

12.1.1 $\mathbf{L^2}$-Cuspidal Trace

The series occurring as the first term on the right is trivially integrated since $\{\psi_n\}$ is an orthonormal basis, thus giving the L^2 **cuspidal theta series**

$$\theta_{\mathrm{cus}}(t) = \int_{\Gamma\backslash G} \sum_{n=1}^{\infty} e^{-\lambda_n t} |\psi_n(z)|^2 \, d\mu(z) = \sum_{n=1}^{\infty} e^{-\lambda_n t} \text{ with } \lambda_n > 0. \tag{12.2}$$

The constant c_0 is $1/\mathrm{vol}(\Gamma\backslash G)$, so its integral over $\Gamma\backslash G$ is 1.

This leaves the two other terms, which if integrated directly are divergent. We shall analyze each term separately. We use the fundamental domain as domain of integration, and view the integral as an improper integral:

$$\int_{\mathscr{F}} = \lim_{Y\to\infty} \int_{\mathscr{F}_{\leqq Y}} .$$

As in Section 10.8, the compact set $\mathscr{F}_{\leqq Y}$ with $Y > 1$ is the part of the fundamental domain lying between the unit sphere and $y = Y$. Also recall that $\mathscr{F}_U$ is the fundamental rectangle, the upper half of the fundamental square for $\Gamma_U\backslash U$. This fundamental square will be denoted by $\mathscr{F}_U^{(2)}$. Taking only half the square is due to the symmetry R sending $x \mapsto -x$.

In the next sections and the next chapter, we shall prove asymptotic properties of the integral over $\mathscr{F}_{\leqq Y}$ for the Eisenstein integral and the cuspidal trace respectively, using the Fourier series expansion. We shall see that asymptotically for $Y \to \infty$, the terms that cause divergence are the same for the cuspidal part as for the Eisenstein part. Thus they cancel, leaving the desired explicit formula.

Specifically, we let $\mathrm{Eis}_t(Y)$ be the Eisenstein integral obtained by integrating the Eisenstein term in the eigenfunction expansion of the heat kernel, that is,

$$\begin{aligned}\mathrm{Eis}_t(Y) &= \frac{1}{16\pi} \int_{\mathscr{F}_{\leqq Y}} \int_{-\infty}^{\infty} e^{-2(r^2+1)t} |E(1+\mathbf{i}r, z)|^2 \, dr \, d\mu(z) \\ &= \text{Eisenstein component of } \mathbf{K}_t^{\Gamma}(z,z) \text{ integrated over } \mathscr{F}_{\leqq \mathscr{Y}}.\end{aligned}$$

Let

$$\phi(s) = \frac{\Lambda\zeta_{\mathbf{Q}(\mathbf{i})}(2-s)}{\Lambda\zeta_{\mathbf{Q}(\mathbf{i})}(s)}$$

cf. Chapter 10, Theorem 10.5.3.

Theorem 12.1.1. *The Eisenstein integral has the asymptotic expansion*

$$\mathrm{Eis}_t(Y) = c_1(t)\log Y + c_{2,E}(t) + o(1) \; for \; Y \to \infty,$$

where

$$c_1(t) = \frac{1}{\pi}\int_{-\infty}^{\infty} e^{-2(r^2+1)t}\,dr = \frac{e^{-2t}}{(2\pi t)^{1/2}},$$

$$c_{2,E}(t) = -e^{-2t} - \frac{1}{\pi}\int_{-\infty}^{\infty} e^{-2(r^2+1)t}\phi'/\phi(1+\mathbf{i}r)dr.$$

As to the heat kernel cuspidal trace integrated over $\mathcal{F}_{\leqq Y}$, we use the notation

$$\mathrm{Cus}_t(Y) = \int_{\mathcal{F}_{\leqq Y}} \mathbf{K}_t^{\mathrm{Cus}}(z,z)\,d\mu(z).$$

Theorem 12.1.2. *The heat kernel cuspidal trace has the asymptotic expansion*

$$\mathrm{Cus}_t(Y) = c_1(t)\log Y + c_{2,\,\mathrm{Cus}}(t) + o(1)\ \textit{for}\ Y \to \infty,$$

with the same constant $c_1(t)$ defined in Theorem 12.1.1, and $c_{2,\mathrm{cus}}(t)$ is an inverted theta series which will be given explicitly in Chapter 13.

Theorem 12.1.1 concerning the Eisenstein term will be proved in Sections 12.2 and 12.3. Theorem 12.1.2 concerning the $(\Gamma\backslash G)$-integral of the cuspidal trace of the heat kernel will be proved in Chapter 13. Let

$$\theta_{\mathrm{Eis}}(t) = \frac{1}{2}\lim_{Y\to\infty}[\mathrm{Eis}_t(Y) - c_1\log Y] = \frac{1}{2}\,c_{2,E}(t).$$

The factor 1/2 comes from the fact that integrating over $\Gamma\backslash G$ gives 1/2 the integral over $\mathcal{F}$. The limit exists after we subtract $c_1(t)\log Y$. In Theorem 12.3.4, we shall prove that

$$2\theta_{\mathrm{Eis}}(t) = -e^{-2t} + \frac{1}{\pi}\int_{-\infty}^{\infty} e^{-2(r^2+1)t}\phi'/\phi(1+\mathbf{i}r)dr.$$

Similarly, let

$$\Theta^{\mathrm{Cus}}(1/t) = \frac{1}{2}\lim_{Y\to\infty}[\mathrm{Cus}_t(\mathrm{Y}) - c_1(t)\log\,Y] = \frac{1}{2}\,c_{2,\mathrm{cus}}(t).$$

Theorem 12.1.3. Theta Inversion Formula.

$$e^{-2t}(4t)^{-1/2}\Theta^{\mathrm{NC}}(1/t) + \Theta^{\mathrm{Cus}}(1/t) = \theta_{\mathrm{Cus}}(t) + 1 + \theta_{\mathrm{Eis}}(t).$$

In dealing with the Eisenstein integral, we use two arguments of Kubota [Kub 73], pp. 19 and 20, making use of Green's formula, and later for the asymptotics we use the argument on p. 107. Kubota uses Green's formula as a step toward what is known as the Maass–Selberg relations, but we take a different route, avoiding the added complications involved in these relations, which we bypass. We simply use the asymptotics.

Remark. The term e^{-2t} occurs as an explicit factor in all but the constant term $1/\text{vol}(\Gamma\backslash G)$. We see this in the above formulas, except possibly for the L^2 cuspidal series θ_{Cus}. However, it is a theorem already mentioned in Chapter 3, Proposition 3.2.7, that $\lambda_1 > \pi^2$ [Szm 83] following [Del 82] following Vigneras. In particular, $\lambda_1 > 2$. Thus we may define naturally $r_n = \lambda_n - 2 > 0$ (for $n \geqq 1$). Then we see that the L^2 cuspidal theta series is also divisible by e^{-2t}. What is the structural significance of the fact that all terms but c_0 are so divisible?

Readers may compare our explicit theta relation with Arthur's version of Selberg's trace formula [Art 74] in the context of representation theory and adelization, with test functions having compact support (Assumption 3.5). Also, Arthur is on adelized reductive groups, and we are on $\text{SL}_2(\mathbf{C})$.

The treatment of the trace formula in Elstrodt–Grunewald–Mennicke [EGM 98] is substantially more concrete and explicit than Arthur's, and does not assume compact support for the test functions. See their Chapter 6, Theorem 6.5.1 and Proposition 6.5.3. This proposition in particular gives asymptotic properties in a manner similar to the one we use. We carry out the explicit computations further because of our selection of the heat Gaussian. They work with an arbitrary cofinite discrete subgroup of $\text{PSL}_2(\mathbf{C})$ acting on hyperbolic 3-space, but with a holomorphic test function having some decay condition. With the heat Gaussian, we can push the explicit form of the theta inversion further, in preparation for taking the Gauss transform.

For the explicit trace formula on SL_2, there is an alternative path involving a different structure, due to Zagier [Zag 79] on $\text{SL}_2(\mathbf{R})$, and adapted to $\text{SL}_2(\mathbf{C})$ by Szmidt [Szm 83]. Zagier integrates the point-pair invariant minus its Eisenstein projection against the Eisenstein series, and takes the residue at $s = 2$. He then gets an interpretation of some of the terms occurring in the formula in terms of certain Dirichlet series associated with quadratic forms. Zagier writes that his proof "also gives more insight into the origin of the various terms in the trace formula; for instance, the class numbers occurring there now appear as residues of zeta-functions." Going further, we note that our version of the main term, the noncuspidal term, of the theta relation involves the conjugacy classes in Γ, instead of these residues. This situation is similar to situations as in class field theory, when objects on the base correspond to conjugacy classes in the Galois group. Here instead of a Galois group, we deal with an arithmetic group. Thus the whole Zagier approach is interesting for its own sake, and deserves a book exposition also to establish more precisely and structurally the correspondence between the terms coming from both approaches.

We note Zagier's additional statement that "[his] proof of the trace formula is more invariant and in some respects computationally simpler than the proofs involving truncation." To a large extent, the complications in the proof by what is called "truncation" are due to the mixing of more than one aspect of integrals over the fundamental domain by going through the Maass–Selberg relations, which are unnecessary for our purposes, as well as truncating functions rather than truncating only the fundamental domain.

Finally there is a second approach by Zagier [Zag 82]. Carrying out the proofs of the theta inversion formula following the Zagier methods, and also going on to the higher-dimensional SL_n, is a matter for other books.

12.2 Green's Theorem on $\mathscr{F}_{\leqq Y}$

Green's formula is an elementary formula from the basic theory of Riemannian manifolds. A complete treatment is given for instance in [Lan 02], Chapter 10, Theorem 12.3.4. The proof given there applies especially when the manifold has singularities on the boundary, by using the version of Stokes's theorem with singularities proved in Chapter 10, Theorem 10.3.3. Our manifold with boundary will be a truncated piece of the fundamental domain for $\Gamma\backslash G/K = \Gamma\backslash \mathbf{H}^3$.

The Riemannian metric is the one whose form for the length squared is

$$ds^2 = (dx_1^2 + dx_2^2 + dy^2)/y^2.$$

which we already used in Section 6.3, but we shall use it more deeply now. Under our normalizations, we have the following result:

Proposition 12.2.1. *The direct image of Casimir to $G/K = \mathbf{H}^3$ is given by*

$$\omega_{\mathbf{H}} = 2\boldsymbol{\Delta},$$

where $\boldsymbol{\Delta}$ is the differential operator given in coordinates by

$$\boldsymbol{\Delta} = y^2\left(\left(\frac{\partial}{\partial x_1}\right)^2 + \left(\frac{\partial}{\partial x_2}\right)^2 + \left(\frac{\partial}{\partial y}\right)^2\right) - y\frac{\partial}{\partial y}.$$

Proof. This is a routine computation from the formula for Casimir given in Section 2.2 (5). We start by computing the direct image of $\tilde{\mathrm{H}}_\alpha^2$.

First we claim that

$$\tilde{\mathrm{H}}_\alpha 2y\partial/\partial y. \tag{12.3}$$

For a function f on $\mathbf{h}_3$, we define the lift to G by

$$f_G\left(\begin{pmatrix}1 & x\\ 0 & 1\end{pmatrix}\begin{pmatrix}y^{1/2} & 0\\ 0 & y^{-1/2}\end{pmatrix}\right) = f(x,y).$$

Then for $\tilde{\mathrm{H}}_\alpha$ we have to find d/dt at $t=0$ of

$$f_G\left(\begin{pmatrix}1 & x\\ 0 & 1\end{pmatrix}\begin{pmatrix}y^{1/2} & \\ 0 & y^{-1/2}\end{pmatrix}\begin{pmatrix}1+t & 0\\ 0 & 1-t\end{pmatrix}\right) = f(x, y(1+t)^2)$$

because $\exp(tH_\alpha)$ is the furthest matrix on the right mod t^2. By the calculus chain rule, we obtain precisely (12.3).

Then $\tilde{H}_\alpha^2$ follows by freshman calculus,

$$\tilde{H}_\alpha^2 = 4y\left(\frac{\partial}{\partial y} + y\left(\frac{\partial}{\partial y}\right)^2\right), \tag{12.4}$$

so finally one gets

$$\frac{1}{2}\tilde{H}_\alpha^2 - 2\tilde{H}_\alpha = -2y\frac{\partial}{\partial y} + 2y^2\left(\frac{\partial}{\partial y}\right)^2, \tag{12.5}$$

The other derivatives are computed similarly. One may replace $\exp(tE_\alpha)$ by the matrix

$$\begin{pmatrix} 1 & t \\ 0 & 1 \end{pmatrix}.$$

Then one moves this matrix across the y-matrix, and conjugation by the y-matrix shows that to compute d/dt at $t = 0$, it suffices to do it for

$$f_G\left(\begin{pmatrix} 1 & x_1 + yt + \mathbf{i}x_2 \\ 0 & 1 \end{pmatrix}\begin{pmatrix} y^{1/2} & 0 \\ 0 & y^{1/2} \end{pmatrix}\right) = f(x_1 + yt + \mathbf{i}x_2, y).$$

Then the chain rule yields

$$\tilde{E}_\alpha^2 = y^2\left(\frac{\partial}{\partial y}\right)^2, \tag{12.6}$$

And similarly for the last term. This concludes the proof.

Remark. For those familiar with basic differential geometry, $\mathbf{\Delta}$ is the Laplacian for the above-mentioned metric.

The volume form gives rise to the Iwasawa measure

$$\Omega = d\mu = dx_1\, dx_2\, \frac{dy}{y^3}.$$

We shall deal with the compact truncated fundamental domain $\mathscr{F}_{\leqq Y}$ for $Y > 1$. There are singularities on the boundary, but they are of codimension $\geqq 2$, so Stokes's theorem with singularities holds as invoked above. The quotient space $\Gamma\backslash\mathbf{H}^3$ is a manifold without boundary but singularities on certain curves, which are of codimension 2. We can see this directly from the fundamental domain $\mathscr{F}$, defined by the inequalities

$$-1/2 \leqq x_1 \leqq 1/2,\ 0 \leqq x_2 \leqq 1/2,\ \text{and } x_1^2 + x_2^2 + y^2 \geqq 1.$$

The two vertical sides of the tube are identified by the translations T_1, $T_{\mathbf{i}}$. Half of the curved spherical part at the bottom is identified with the other half by the mapping $z \mapsto -1/z$.

The picture of the truncated fundamental domain on $\mathbf{H}^2$ looks like a 2-dimensional rectangular tube over part of a semicircle. The fundamental domain on $\mathbf{H}^3$ looks like a 3-dimensional tube over a portion of a sphere, lying at the bottom. At height $Y > 1$, the section is a rectangle Rec_Y lying above the fundamental rectangle $\mathscr{F}_U$. The truncated fundamental domain $\mathscr{F}_{\leqq Y}$ is therefore a manifold with boundary, singularities of codimension 2 on the boundary, and only one boundary component of codimension 1, namely the above rectangle above $\mathscr{F}_U$ after identifying sides (corresponding to representing $\Gamma\backslash\mathbf{H}^3$). Then:

The outward normal above this boundary rectangle at height y is $\partial_{\mathbf{n}} = y\partial/\partial_y$. The boundary volume form is $\Omega_y = dx\, dx_2/y^2$ on Rec_Y.

Applying [Lan 02], Chapter 10, Theorem 12.3.4, we get the following:

Theorem 12.2.2. Green's formula for $(\Gamma\backslash\mathbf{H}^3)(\mathbf{Y})$:

$$\int_{\mathscr{F}_{\leqq Y}} (\varphi\Delta\psi - \psi\Delta\varphi)\, d\mu = \int_{\mathrm{Rec}_Y} (\varphi\partial_{\mathbf{n}}\psi - \psi\partial_{\mathbf{n}}\varphi)\Omega_Y.$$

Let $\mathscr{F}^{(2)}_{\leqq Y}$ be the doubled domain $\mathscr{F}_{\leqq Y} \cup \mathrm{R}\mathscr{F}_{\leqq Y}$ lying above the fundamental square $\mathscr{F}^{(2)}_U$ for $\Gamma_U\backslash U$. The boundary at height Y is the fundamental square at height Y, which we denote by Sq_Y and which is a doubling of Rec_Y. Then Green's formula can be rewritten in terms of integration over this square:

$$\int_{\mathscr{F}^{(2)}_{\leqq Y}} (\varphi\Delta\psi - \psi\Delta\varphi) d\mu = \int_{\mathrm{Sq}_Y} (\varphi\partial_{\mathbf{n}}\psi - \psi\partial_{\mathbf{n}}\varphi)\Omega_Y.$$

We shall apply Green's formula under the following conditions.

Condition 1. *We suppose that φ and ψ are eigenfunctions of the Laplacian with eigenvalues λ_φ, λ_ψ.* Then Green's formula takes the form

$$(\lambda_\psi - \lambda_\phi) \int_{\mathscr{F}^{(2)}_{\leqq Y}} \varphi\psi d\mu = \int_{\mathrm{Sq}_Y} (\varphi\partial_{\mathbf{n}}\psi - \psi\partial_{\mathbf{n}}\varphi)\Omega_Y.$$

Rewriting this in terms of the coordinate y gives

$$(\lambda_\psi - \lambda_\varphi) \int_{\mathscr{F}^{(2)}_{\leqq Y}} \varphi\psi d\mu = \int_{\mathrm{Sq}_Y} [\varphi(z)\partial_y\psi(z) - \psi(z)\partial_y\varphi(z)]_{y(z)=Y} \frac{1}{Y} dx_1\, dx_2. \qquad (12.7)$$

Condition 2. *The functions φ, ψ admit Fourier series expansions*

$$\varphi(z) = \sum_{m\in o} a_m(y)e^{2\pi\mathbf{i}\langle m,x\rangle} \text{ and } \psi(z) = \sum_{m\in o} b_m(y)e^{2\pi\mathbf{i}\langle m,x\rangle}.$$

The first two derivatives of these series term by term are continuous and converge absolutely, uniformly for y in a compact set.

These conditions will be automatically satisfied in the applications to Eisenstein series in the next section. The arguments here will be purely formal.

Theorem 12.2.3. *Under the above conditions, we have the formula*

$$(\lambda_\psi - \lambda_\varphi) \int_{\mathscr{F}^{(2)}_{\leq Y}} \varphi(z)\, \psi(z) d\mu(z) = (a_0 b_0' - b_0 a_0')(Y)\frac{1}{Y} + \sum_{m\neq 0} (a_m b'_{-m} - a'_m b_{-m})(Y)\frac{1}{Y}.$$

Proof. The product of the Fourier series for φ and ψ is a double series, summing over $m,\, n \in \mathfrak{o}$. The integral over the square Sq_Y of a cross term with factor $e^{2\pi \mathbf{i}\langle m+n,\, x\rangle}$ is 0 if $m+n \neq 0$. Plugging the Fourier series in the last formula (12.7) proves the theorem.

We need somewhat more precise estimates in some cases as follows.

Condition 3. *For $y \to \infty$,*

$$\sum_{m\neq 0} |a_m b_{-m}(y)| = O(1),$$

$$\sum_{m\neq 0} \left|a'_m b_{-m}(y)\right| \text{ and } \sum_{m\neq 0} \left|a_m b'_{-m}(y)\right| = O(1/y).$$

Lemma 12.2.4. *Under all three conditions, for Y sufficiently large,*

$$\sum_{m\neq 0} -(a_m b'_{-m} - a'_m b_{-m})(Y)$$

$$= \sum_{m\neq 0} (\lambda_\psi - \lambda\varphi) \int_Y^\infty a_m b_{-m}(y) \frac{dy}{y^2} + \sum_{m\neq 0} \int_Y^\infty (a_m b'_{-m} - a'_m b_{-m})(y) \frac{dy}{y}.$$

Proof. The Fourier coefficients are given by

$$a_m(y) = \int_{C/\mathfrak{o}} \varphi(x,y) e^{-2\pi \mathbf{i}\langle m,x\rangle} dx.$$

The eigenfunction property of φ then implies a differential equation for the Fourier coefficients:

$$a''_m(y) = \left((2\pi|m|)^2 + \frac{\lambda_\varphi}{y^2}\right) a_m(y) + \frac{1}{y} a'_m(y), \tag{12.8}$$

$$b''_m(y) = \left((2\pi|m|)^2 + \frac{\lambda_\psi}{y^2}\right) b_{-m}(y) + \frac{1}{y} b'_{-m}(y).$$

We shall give the details of the proof below, but finish the proof of Lemma 12.2.4 assuming (12.8). We have

$$(a_m b'_{-m} - a'_m b_{-m})' = a_m b''_{-m} - a''_m b_{-m} \qquad \text{[trivially]}$$

$$= \frac{1}{y^2}(\lambda_\psi - \lambda_\varphi)(a_m b_{-m})(y) + \frac{1}{y}(a_m b'_{-m} - a'_m b_{-m})(y). \tag{12.9}$$

This second equality is immediate from the differential equation: multiply the first equation of (12.8) by b_{-m}, multiply the second equation by a_m, and subtract the first from the second. Finally, integrate back the derivative and note that the terms $a_m(y)$, $b_m(y)$ vanish at infinity. The relation of Lemma 12.2.4 follows.

We now show how (12.8) follows from the eigenfunction property. Write

$$\mathbf{\Delta} = D_1 + D_2,$$

where

$$D_1 = y^2\left((\partial/\partial x_1)^2 + (\partial/\partial x_2)^2\right) \text{ and } D_2 = y^2\left(\frac{\partial}{\partial y}\right)^2 - y\frac{\partial}{\partial y}.$$

We have identities

$$\begin{aligned}
&\int_{\mathbf{C}/o} \mathbf{\Delta}\varphi(x,y)e^{-2\pi\mathbf{i}\langle m,x\rangle}dx \\
&= \lambda_\varphi a_m(y) = \int_{\mathbf{C}/o} D_1\varphi(x,y)e^{-2\pi\mathbf{i}\langle m,x\rangle}dx + \int_{\mathbf{C}/o} D_2\varphi(x,y)e^{-2\pi\mathbf{i}\langle m,x\rangle}dx \\
&= \int_{\mathbf{C}/o} \varphi(x,y)D_1 e^{-2\pi\mathbf{i}\langle m,x\rangle}dx + D_2 a_m(y) \\
&= -y^2(2\pi|m|)^2 a_m(y) + y^2 a_m''(y) - y a_m'(y). \qquad (12.10)
\end{aligned}$$

This proves (12.8).

12.3 Application to Eisenstein Functions

We are now ready to carry out the asymptotic expansion of the $\mathcal{F}_{\leqq Y}$-integral of the Eisenstein term in the eigenfunction expansion of the heat kernel. We recall the precise result to be proved in (12.8) below. We shall apply the formulas of Section 12.2 to the Eisenstein series, making them explicit for the two functions

$$\varphi(z) = E_s(z) = E(s,z) \text{ and } \psi(z) = E_{s_1}(z) = E(s_1,z).$$

By Chapter 11, Theorem 11.3.1, we know that these functions are eigenfunctions of the Laplacian, with 1/2 the eigenvalue of Casimir by Proposition 12.2.1. Thus

$$\mathbf{\Delta}\varphi = \lambda_\varphi\varphi \text{ with } \lambda_\varphi = s(s-2) \text{ and } \mathbf{\Delta}\psi = \lambda_\psi\psi \text{ with } \lambda_\psi = s_1(s_1-2). \qquad (12.11)$$

We record the identity

$$\lambda_\psi - \lambda_\varphi = (s_1(s_1-2) - s(s-2)) = (s_1 - s)(s_1 + s - 2). \qquad (12.12)$$

For the Fourier coefficients of the Eisenstein series we write

$$a_m(y) = a_m(s,y) \text{ and } b_m(y) = a_m(s_1,y).$$

We write similarly as in Chapter 10, Theorem 10.5.3,

$$a_m(s,y) = A_m(s,y)/\Lambda\zeta_{\mathbf{F}}(s) \text{ with } \Lambda\zeta_{\mathbf{F}}(s) = \pi^{-s}\Gamma(s)\zeta_{\mathbf{F}}(s).$$

For $m \neq 0$,

$$A_m(s,y) = \sum_{cd=m} \left|\frac{d}{c}\right|^{s-1} yK_{s-1}(\pi y|m|).$$

Note that $A_{-m} = A_m$ because the pairs (c,d) with $cd = m$ correspond to pairs $(-c,d)$ with $-cd = -m$, and the expressions being summed are the same for m and $-m$. Then also $a_m = a_{-m}$. For the constant term, we recall that

$$a_0(s,y) = 4y^s + \phi(s)4y^{2-s} \text{ with } \phi(s) = \frac{\Lambda\zeta_{\mathbf{F}}(2-s)}{\Lambda\zeta_{\mathbf{F}}(s)}. \tag{12.13}$$

We get

$$\phi(s)\phi(2-s) = 1. \tag{12.14a}$$

In particular, for $s = 1 + \mathbf{i}r$, $2 - s = 1 - \mathbf{i}r$, we get

$$|\phi(1+\mathbf{i}r)| = 1. \tag{12.14b}$$

We now give estimates for the Fourier coefficients A_m and a_m with $m \neq 0$.

Lemma 12.3.1. *For $\sigma = Re(s)$ bounded, there exist constants c, c', c'' such that for all $m \neq 0$, and y sufficiently large, we have*

$$|A_m(s,y)| \text{ and } |A'_m(s,y)| \ll |m|^{c'} y^2 e^{-c''y|m|} \text{ or more simply } \ll e^{-cy|m|}.$$

Take $s = \sigma + \mathbf{i}r$, $s_1 = \sigma_1 + \mathbf{i}r_1$ with $1 \leqq \sigma, \sigma_1 \leqq 3/2$, say. Then

$$\sum_{m \neq 0} \left|a_m(s,y)a'_{-m}(s_1,y)\right| = O((rr_1)^{\varepsilon})O(e^{-cy}),$$

with some $c > 0$, and r, r_1, $y \to \infty$.

Proof. In the Bessel–Fourier expansion of Chapter 10, Theorem 10.5.3, the Bessel function has linear exponential decay by Section 10.3, **K7**. The same is verified similarly for its derivative. Hence the coefficients satisfy the first stated estimate for $|m| \neq 0$. Since $|m|$ and y are bounded away from 0, the constant c can be chosen such that the easier inequality is satisfied. One then takes the double sum, over lattice points in bands with $n \leqq |m| < n+1$ (there are $O(n)$ such points), and then a sum over the positive integers. Finally the geometric series gives the stated estimate. We get the estimate for the sum with a_m instead of A_m after dividing by the completed zeta function $\Lambda\zeta_{\mathbf{F}}(s)$ and using Chapter 10, Theorem 5.2 and Lemma 10.6.3.

Note that Lemma 12.3.1 gives even better estimates than those required for applying Theorem 12.2.3 or Lemma 12.2.4, which need only $O(1/y)$, whereas we have here $O(e^{-cy})$. We need the rough dependence only in an interval $1 \leqq \mathrm{Re}(s) \leqq 1 + \varepsilon$, say. In any case, we have the following result:

Theorem 12.3.2. *For all s, s_1 that are not poles of ϕ, Theorem 12.2.3 and Lemma 12.2.4 apply to the Eisenstein series $\varphi = E_s$ and $\psi = E_{s_1}$. Thus*

$$(s_1 - s)(s_1 + s - 2) \int\limits_{\mathscr{F}^{(2)}_{\leqq Y}} E(s,z)E(s_1,z)d\mu(z)$$

$$= [a_0(s,Y)a_0'(s_1,y) - a_0'(s,Y)a_0(s_1,Y)]\frac{1}{Y}$$

$$+ (s - s_1)(s_1 + s - 2) \int\limits_Y^\infty \sum_{m \neq 0} a_m(s,y)a_{-m}(s_1,y)\frac{dy}{y^2}$$

$$+ \sum_{m \neq 0} \int\limits_Y^\infty \left[a_m(s,y)a_{-m}'(s_1,y) - a_m'(s,y)a_{-m}(s_1,y)\right]\frac{dy}{y}.$$

This theorem will allow us to get the desired asymptotic expansion for the Eisenstein integral from an analysis of the integral of the constant term, which we shall make explicit. There will be a term of type $C \log Y$, so causing divergence as $Y \to \infty$, with an explicit constant C, plus another term independent of Y.

We divide the expressions in Theorem 12.3.2 by $(s_1 - s)(s_1 + s - 2)$. If not 0, this factor cancels in front of one integral on the right. By Lemma 12.3.1, this second integral is $O(1/Y)$. Now abbreviate, for $m = 0$ or $m \neq 0$,

$$F_m(s,s_1;y) = \frac{a_m(s,y)a_{-m}'(s_1,y) - a_m'(s,y)a_{-m}(s_1,y)}{(s_1 - s)(s_1 + s - 2)}$$

Note that this expression is symmetric in (s, s_1), because $a_{-m} = a_{-m}$, so

$$F_m(s,s_1;y) = F_m(s_1,s;y).$$

By Lemma 12.3.1, we may write Theorem 12.3.2 in a form simplified for our purposes.

Corollary 12.3.3. *Suppose* $(s_1 - s)(s_1 + s - 2) \neq 0$. *Then for* $Y \to \infty$,

$$\int\limits_{\mathscr{F}^{(2)}_{\leqq Y}} E(s,z)E(s_1,z)d\mu(z) = F_0(s,s_1;Y)\frac{1}{Y} + \int\limits_Y^\infty \sum_{m \neq 0} F_m(s,s_1;y)\frac{1}{y}dy + O(1/Y).$$

With $1 < \sigma \leqq 1 + \varepsilon$ we use the values

$$s = \sigma + \mathbf{i}r \text{ and } s_1 = \sigma - \mathbf{i}r = \bar{s},$$
$$\text{so } s_1 - s = -2\mathbf{i}r \text{ and } s_1 + s - 2 = 2\sigma - 2. \tag{12.15}$$

We want to investigate the asymptotics of the Eisenstein integral that arises from the eigenfunction expansion, namely

$$\mathrm{Eis}_t(Y) = \frac{1}{2}\frac{1}{16\pi} \int\limits_{-\infty}^{\infty} e^{-2(r^2+1)t} \left[\int\limits_{\mathscr{F}^{(2)}_{\leqq Y}} E(1 + \mathbf{i}r, z)E(1 + \mathbf{i}r, z)d\mu(z) \right] dr. \tag{12.16}$$

We have interchanged the order of integration from the Eisenstein component of $\mathbf{K}_t^{\Gamma}(z,z)$ integrated over $\mathscr{F}_{\leqq Y}$. We put the factor 1/2 because the inner integral is over the truncated double fundamental domain rather than the fundamental domain itself. This is more convenient when one is dealing with Fourier analysis. We shall give the first two terms of its asymptotic expansion for $Y \to \infty$ as stated in Theorem 12.1.1.

We use the decomposition of Corollary 12.3.3. We then consider for $1 < \sigma \leqq 1+\varepsilon$ the expression

$$2\mathrm{Eis}_t(\sigma,Y)_0 = \frac{1}{16\pi}\int_{-\infty}^{\infty} e^{-2(r^2+1)t} F_0(\sigma+\mathbf{i}r,\sigma-\mathbf{i}r;Y)\frac{1}{Y}\,dr, \tag{12.17}$$

$$2\mathrm{Eis}_t(\sigma,Y)_{\neq 0} = \frac{1}{16\pi}\int_{-\infty}^{\infty} e^{-2(r^2+1)t}\int_{Y}^{\infty}\sum_{m\neq 0} F_m(\sigma+\mathbf{i}r,\sigma-\mathbf{i}r;y)\frac{1}{y}\,dy\,dr \tag{12.18}$$

and let

$$2\mathrm{Eis}_t(\sigma,Y) = 2\mathrm{Eis}_t(\sigma,Y)_0 + 2\mathrm{Eis}_t(\sigma,Y)_{\neq 0}.$$

We let $\sigma \to 1$, and determine the Y-asymptotics of the two terms in the next two sections.

- In Section 12.4, we evaluate $\mathrm{Eis}_t(1,Y)_0$, giving the two main asymptotic terms.
- In Section 12.5, we estimate $\mathrm{Eis}_t(1,Y)_{\neq 0}$, and show that it is $o(1)$ for $Y \to \infty$. Thus

$$\mathrm{Eis}_t(Y) = \mathrm{Eis}_t(1,Y)_0 + o(1).$$

We repeat here the theorem stating our goal. Note that

$$\frac{1}{\pi}\int_{-\infty}^{\infty} e^{-2r^2t}\,dr = \frac{1}{\sqrt{2\pi t}}. \tag{12.19}$$

We let

$$c_1(t) = \frac{1}{\pi}\int_{-\infty}^{\infty} e^{-2(r^2+1)t}\,dr = \frac{e^{-2t}}{\sqrt{2\pi t}}, \tag{12.20}$$

$$c_{2,E}(t) = \frac{-1}{\pi}\int_{-\infty}^{\infty} e^{-2(r^2+1)t}\phi/\phi(1+\mathbf{i}r)\,dr - e^{-2t}. \tag{12.21}$$

The next sections will prove the following Theorem:

Theorem 12.3.4. *We have*

$$\mathrm{Eis}_t(Y) = c_1(t)\log Y + c_{2,E}(t) + o(1) \;\; for\; Y \to \infty.$$

12.4 The Constant-Term Integral Asymptotics

This section computes the limit as $\sigma \to 1$ of the first term in Section 12.3, (7), namely with $s = \sigma + \mathbf{i}r$, $1 < \sigma$,

$$2\mathrm{Eis}_t(1,Y)_0$$
$$= \lim_{\sigma\to 1} \frac{1}{16\pi} \int_{-\infty}^{\infty} e^{-2(r^2+1)t} [a_0(s,Y)a_0'(\bar{s},Y) - a_0'(s,Y)a_0(\bar{s},Y)] \frac{1}{Y} \frac{1}{2\sigma-2} \frac{-1}{2\mathbf{i}r} dr. \tag{12.22}$$

The expression in brackets is $2\mathbf{i}\ \mathrm{Im}(a_0\bar{a}_0') = a_0\bar{a}_0' - a_0'\bar{a}_0$. Recall Section 12.3, (12.24) and (12.25):

$$|\phi(1+\mathbf{i}r)| = 1 \text{ and } a_0(s,Y) = 4Y^s + 4\phi(s)Y^{2-s}.$$

Then each product $a_0\bar{a}_0'$ and $a_0'\bar{a}_0$ gives rise to four terms. Explicitly,

$$\begin{aligned}\frac{1}{16} a_0(s,Y)a_0'(\bar{s},Y)\frac{1}{Y} &= \bar{s}Y^{2\sigma-2} + (2-\bar{s})\,|\phi(s)|^2\,Y^{2-2\sigma} + \bar{s}\phi(s)Y^{-2\mathbf{i}r} \\ &\quad + (2-\bar{s})\phi(\bar{s})Y^{2\mathbf{i}r}.\end{aligned} \tag{12.23}$$

This yields the identity

$$\begin{aligned}\frac{1}{16}(a_0\bar{a}_0' - a_0'\bar{a}_0)\frac{1}{Y}\frac{1}{2\sigma-2}\frac{-1}{2\mathbf{i}r} &= \frac{Y^{2\sigma-2} - |\phi(s)|^2 Y^{-2\sigma+2}}{2\sigma-2} + \frac{\phi(\bar{s})Y^{2\mathbf{i}r} - \phi(s)Y^{-2\mathbf{i}r}}{2\mathbf{i}r} \\ &= Q_1(\sigma,r) + Q_2(\sigma,r), \text{ say.}\end{aligned} \tag{12.24}$$

Remark. The expression on the right corresponds to the expression in Kubota [Kub 73], bottom of p. 106 (mutatis mutandis, his 1 becomes our 2 because we are in the complex case). His expression on the left is given in terms of the scalar product of truncated Eisenstein series, which do not enter our considerations. We shall next deal with the two terms of (12.24) as in Kubota [Kub 73], p. 107.

We must compute the limit of each term as $\sigma \to 1$. We start with the term $Q_1(\sigma,r)$, especially the term having the factor $|\phi(s)|^2$. We need the start of the Taylor expansion of $|\phi(s)|^2$,

$$\begin{aligned}\phi(\sigma+\mathbf{i}r) &= \phi(1+\mathbf{i}r) + (\sigma-1)\phi'(1+\mathbf{i}r) + O(\sigma-1)^2, \\ \phi(\sigma+\mathbf{i}r) &= \phi(1+\mathbf{i}r) + (\sigma-1)\phi'(1+\mathbf{i}r) + O(\sigma-1)^2.\end{aligned}$$

So using $|\phi(1+\mathbf{i}r)| = 1$, we get

$$|\phi(\sigma+\mathbf{i}r)|^2 = 1 + (\sigma-1)\left[(\phi'(1+\mathbf{i}r)\phi(1-\mathbf{i}r) + \phi'(1-\mathbf{i}r)\phi(1+\mathbf{i}r)\right] + O(\sigma-1)^2. \tag{12.25}$$

Decomposing $|\phi(s)|^2$ as a sum

$$|\phi(s)|^2 = 1 + |\phi(s)|^2 - 1$$

leads to a decomposition

$$Q_1(\sigma, r) = \frac{Y^{2\sigma-2} - Y^{-2\sigma+2}}{2\sigma - 2} - \frac{(|\phi(s)|^2 - 1)Y^{-2\sigma+2}}{2\sigma - 2}$$

We determine the limit of each term on the right separately in (12.26) and (12.27). For the first term, we have

$$\lim_{\sigma \to 1} \frac{Y^{2\sigma-2} - Y^{2-2\sigma}}{2\sigma - 2} = 2\log Y. \tag{12.26}$$

Plugging (12.26) into (12.24), then (12.22), we see that we have just obtained the leading term with $\log Y$ in Theorem 12.3.4.

We have $\lim_{\sigma \to 1} Y^{-2\sigma+2} = 1$, so the second limit is

$$-\lim_{\sigma \to 1} \frac{|\phi(\sigma + \mathbf{i}r)|^2 - 1}{2\sigma - 2} = -(\phi'/\phi)(1 + \mathbf{i}r). \tag{12.27}$$

Proof. By the Taylor expansion,

$$\begin{aligned} \lim_{\sigma \to 1} \frac{|\phi(\sigma + \mathbf{i}r)|^2 - 1}{\sigma - 1} &= \phi'(1 + \mathbf{i}r)\phi(1 - \mathbf{i}r) + \phi'(1 - \mathbf{i}r)\phi(1 + \mathbf{i}r) \\ &= (\phi'/\phi)(1 + \mathbf{i}r) + (\phi'/\phi)(1 - \mathbf{i}r) \quad [\text{by } \phi(s)\phi(2 - s) = 1] \\ &= 2(\phi'/\phi)(1 + \mathbf{i}r) \; [\text{by } \phi'/\phi(s) - \phi'/\phi(2 - s) = 0]. \end{aligned}$$

This proves (12.27).

Plugging (12.27) into (12.24), then into (12.22), we see that we have just obtained the constant $c_{2,E}(t)$ in Theorem 12.3.4.

There remains the term involving $2\mathbf{i}r$ in (12.24). Abbreviate

$$h(r) = e^{-2(r^2+1)t}.$$

Lemma 12.4.1. *We have*

$$\lim_{Y \to \infty} \frac{1}{\pi} \int_{-\infty}^{\infty} h(r) \frac{\phi(1 - \mathbf{i}r)Y^{2\mathbf{i}r} - \phi(1 + \mathbf{i}r)Y^{-2\mathbf{i}r}}{2\mathbf{i}r} dr = -h(0) = -e^{-2t}.$$

Proof. Let $I(Y)$ be the integral on the left. Writing down real and imaginary parts, and using the fact that the integral is real, we get

$$I(Y) = \frac{1}{\pi}\int_{-\infty}^{\infty} h(r)\mathrm{Re}\phi(1-\mathbf{i}r)\frac{\sin(2r\cdot\log Y)}{r}dr$$
$$+\frac{1}{\pi}\int_{-\infty}^{\infty} h(r)\frac{\mathrm{Im}\phi(1-\mathbf{i}r)}{r}\cos(2r\cdot\log Y)dr.$$

Note that $\phi(1) = -1$, and in particular ϕ (12.22) is real, so $\mathrm{Im}\phi(1-\mathbf{i}r)$ goes to 0 as $r \to 0$. In the second integral, we put the r under this imaginary part, so that the quotient is regular at $r = 0$. It is otherwise smooth and rapidly decreasing, so by the Riemann–Lebesgue lemma, the second integral (with the cosine) goes to 0 as $Y \to \infty$. As to the first integral, we know that for a smooth rapidly decreasing function,

$$f(0) = \lim_{A\to\infty}\frac{1}{\pi}\int_{-\infty}^{\infty} f(x)\frac{\sin Ax}{x}dx.$$

Here the function is $f(r) = h(r)\mathrm{Re}\phi(1-\mathbf{i}r)$, so $f(0) = -e^{-2t}$. This proves the lemma. It also concludes the proof of the two main asymptotic terms in Theorem 12.3.4.

12.4.1 Appendix

For the convenience of the reader we recall the proof of the Fourier identity. Define

$$f_A(\mathrm{O}) = \frac{1}{\sqrt{2\pi}}\int_{-A}^{A} f^{\wedge}(u)du.$$

For smooth rapidly decreasing functions f, Fourier inversion means that the limit of $f_A(\Omega)$ is $f(0)$. Write down the Fourier integral for $f^{\wedge}(u)$ and use Fubini. Then we get precisely

$$f_A(0) = \frac{1}{\pi}\int_{-\infty}^{\infty} f(x)\frac{\sin Ax}{x}dx,$$

proving the desired identity.

The Fourier integral is the one normalized by

$$f^{\wedge}(u) = \frac{1}{\sqrt{2\pi}}\int_{-\infty}^{\infty} f(x)e^{-\mathbf{i}xu}dx.$$

There remains to prove that the remaining term coming from $2\mathrm{Eis}_t\ (\sigma, Y)_0$ at $\sigma = 1$ is $o(1)$. We do this in the next and final section of this chapter.

12.5 The Nonconstant-Term Error Estimate

This section completes the proof of Theorem 12.3.4 by dealing with the term with $m = 0$ in Theorem 12.3.2, and showing that it gives rise to part of the error term $o(1)$. We shall prove

$$\lim_{\sigma\to 1} \mathrm{Eis}_t(\sigma, Y)_{\neq 0} = \mathrm{Eis}_t(1, Y)_{\neq 0} = o(1) \text{ for } Y \to \infty.$$

We start by analyzing the term with $m = 0$ in Theorem 12.3.2 in light of the formulas obtained in Section 12.4.

Proposition 12.5.1. *Let $s = \sigma + \mathbf{i}r$ with $\sigma > 1$, and set*

$$G(\sigma, r, Y) = \int_Y^\infty \sum_{m\neq 0} [a_m(s,y)a'_m(\bar{s},y) - a'_m(s,y)a_m(\bar{s},y)]\frac{dy}{y}.$$

Then for fixed r and Y, we have that

$$\lim_{\sigma\to 1} \frac{G(\sigma, r, Y)}{\sigma - 1}$$

exists and equals $\partial_1 G(1, r, Y)$.

Proof. We take $s = \sigma + \mathbf{i}r$ and $s_1 = \bar{s}$, with $\sigma > 1$. With this, we combine the identity from Theorem 12.3.2 and equation (12.24) from Section 12.4 to arrive at the identity

$$\begin{aligned} G(\sigma, r, Y) &= 32\mathbf{i}r(Y^{2\sigma-2} - |\phi(s)|^2 Y^{-2\sigma+2}) + 32(\sigma - 1)\left(\phi(\bar{s})Y^{2\mathbf{i}r} - \phi(s)Y^{-2\mathbf{i}r}\right) \\ &\quad + 4\mathbf{i}r(\sigma - 1)\int_Y^\infty \sum_{m\neq 0} [a_m(s,y)a_m(\bar{s},y)]\frac{dy}{y^2} \\ &\quad - 4\mathbf{i}r(\sigma - 1)\int_{\mathscr{F}^{(2)}_{\leq Y}} E(s,z)E(\bar{s},z)d\mu(z). \end{aligned} \tag{12.28}$$

Obviously, the last three terms on the right-hand side of (12.22) are divisible by $(\sigma - 1)$. Regarding the first term, recall that $|\phi(1 + \mathbf{i}r)| = 1$ for all r, so

$$\lim_{\sigma\to 1} (Y^{2\sigma-2} - |\phi(s)|^2 Y^{-2\sigma+2}) = 0 \text{ for all } r.$$

Clearly, this term is real-analytic in σ near $\sigma = 1$, so

$$Y^{2\sigma-2} - |\phi(s)|^2 Y^{-2\sigma+2} = (\sigma - 1)H(\sigma, r, Y) \text{ near } \sigma = 1$$

for some function H. Finally, since all functions in (12.22) are holomorphic at $s = 1$ (the Eisenstein series and the coefficients a_m), the first assertion follows.

To conclude, observe that we have also shown that $G(1, r, Y) = 0$. Therefore, the limit in question is simply the definition of the derivative of G with respect to the first variable, evaluated at 1.

Lemma 12.5.2. *With notation as above, we have, for any σ near 1, that*

$$\lim_{r\to 0} \frac{G(\sigma,r,Y)}{r}$$

exists and equals $\partial_2 G(\sigma,0,Y)$. In addition, the limit

$$\lim_{r\to 0} \frac{\partial_1 G(1,r,Y)}{r}$$

exists and equals $\partial_2\partial_1 G(1,0,Y)$.

Proof. The proof is identical to that of Proposition 12.5.1, in this instance using that

$$\lim_{r\to 0}(\phi(1-\mathbf{i}r)Y^{2\mathbf{i}r} - \phi(1+\mathbf{i}r)Y^{-2\mathbf{i}r}) = 0.$$

Theorem 12.5.3. *We have*

$$\lim_{\sigma\to 1} \mathrm{Eis}_t(\sigma,Y)_{\neq 0} = \mathrm{Eis}_t(1,Y)_{\neq 0} = O(1) \text{ as } Y \to \infty.$$

Proof. From (8), §3, we need to show that

$$\lim_{\sigma\to 1} \int_{-\infty}^{\infty} e^{-2(r^2+1)t} \frac{G(\sigma,r,Y)}{\sigma-1}\frac{dr}{r} = o(1) \text{ as } Y \to \infty.$$

Since

$$\frac{\partial}{\partial\sigma} = \frac{\partial}{\partial s} + \frac{\partial}{\partial \bar{s}},$$

we have that

$$\begin{aligned}
\lim_{\sigma\to 1} \frac{G(\sigma,r,Y)}{\sigma-1} &= \left(\frac{\partial}{\partial s} + \frac{\partial}{\partial \bar{s}}\right) \int_Y^{\infty} \sum_{m\neq 0} \left[a_m(s,y)a'_m(\bar{s},y) - a'_m(s,y)a_m(\bar{s},y)\right] \frac{dy}{y}\Bigg|_{\sigma=1} \\
&= \int_Y^{\infty} \sum_{m\neq 0} \left[a_m(s,y)a''_m(\bar{s},y) - a''_m(s,y)a_m(\bar{s},y)\right] \frac{dy}{y}\Bigg|_{\sigma=1} . \qquad (12.29)
\end{aligned}$$

The proof of Lemma 12.3.1 easily extends to prove the necessary (12.23) in order to complete the theorem.

Chapter 13
The Cuspidal Trace Y-Asymptotics

This chapter gives the proof for Chapter 12, Theorem 12.1.2, the asymptotic expansion of the cuspidal trace of the heat kernel. Even more, we determine the cuspidal theta series explicitly. From Section 5.4, we have an explicit tabulation of the Γ-conjugacy classes of Γ_∞, that is, the cuspidal Γ-conjugacy classes. These split into the nonregular ones and the regular ones, so the cuspidal trace will decompose into a sum of the nonregular and regular classes respectively, thus splitting this chapter in two parts.

The cuspidal trace integral over $\Gamma\backslash G$ does not converge, so we integrate over the truncated fundamental domain $\mathscr{F}_{\leqq Y}$, and determine the asymptotic expansion up to $o(1)$ for $Y \to \infty$. To do this, we reduce the integral to the tube $\mathscr{T}_{\leqq Y}$, which is more symmetric, using the tiling of Chapter 9 Theorems 9.8.2 and 9.8.5. The procedure starts with Proposition 13.1.1. We deal with the nonregular and regular cases separately. Explicit evaluations of integrals with the heat kernel will be made in Chapter 14. These evaluations give the values for the constant coefficients of the asymptotic expansion. The coefficients involve specific integrals of the form

$$\int_0^\infty f_\varphi(r) r h(r) dr,$$

where $h(r) = 1$, or $\log\, r$, or $\log(r^2+4)$. See Section 14.5.

We shall need a piece of analytic number theory, giving a version of a classical Littlewood theorem concerning the partial sums of the zeta function approximating the zeta function itself, and its analogue with the Hurwitz constant in addition to the Euler constant. We give complete proofs in the next and final chapter over $\mathbf{Z}$ and $\mathbf{Z}[\mathbf{i}]$, which are the relevant cases for us. As far as we know, the corresponding results for number fields are not in the literature.

Although we apply this chapter to the heat Gaussian $\mathbf{g}_t$ and heat kernel $\mathbf{K}_t$, a lot is just abstract nonsense of general integration formulas, so we start with more general functions.

13.1 The Nonregular Cuspidal Integral over $\mathcal{F}_{\leqq Y}$

Let φ be a continuous, K-bi-invariant function with quadratic exponential decay. Recall that we want to analyze asymptotically the cuspidal integral

$$\mathrm{Cus}_{\varphi}(y) = \int_{\mathcal{F}_{\leqq Y}} \mathbf{K}_{\varphi}^{\mathrm{Cus}}(z,z)d\mu(z).$$

The measure is the Iwasawa measure, and

$$\mathbf{K}_{\varphi}(z,w) = \varphi(z^{-1}w).$$

Then by definition,

$$\mathbf{K}_{\varphi}^{\mathrm{Cus}}(z,z) = \sum_{\gamma\in\mathrm{Cus}(\Gamma)} \mathbf{K}_{\varphi}(z,\gamma z) = \sum_{\gamma\in\mathrm{Cus}(\Gamma)} \varphi(z^{-1}\gamma z).$$

We recall that $\mathrm{Cus}(\Gamma) = \mathbf{c}(\Gamma)\Gamma_{\infty}$ is $\mathbf{c}(\Gamma)$-invariant. We decompose it into the regular and nonregular $\mathbf{c}(\Gamma)$-invariant subsets. Thus

$$\begin{aligned}\mathbf{K}_{\varphi}^{\mathrm{Cus}}(z,z) &= \mathbf{K}_{\varphi}^{\mathrm{RegCus}}(z,z) + \mathbf{K}_{\varphi}^{\mathrm{NRCus}}(z,z)\\ &= \sum_{\gamma\in\mathrm{RegCus}(\Gamma)} \varphi(z^{-1}\gamma z) + \sum_{\gamma\in\mathrm{NRCus}(\Gamma)} \varphi(z^{-1}\gamma z). \qquad (13.1)\end{aligned}$$

The nonregular cuspidal set itself has the $\pm$ decomposition

$$\mathrm{NRCus}(\Gamma) = \pm\mathrm{c}(\Gamma)\Gamma_{\mathrm{U}} = \mathbf{c}(\Gamma)\Gamma_U \cup \mathbf{c}(\Gamma)(-\Gamma_U) \text{ (disjoint union)}, \qquad (13.2)$$

whose conjugacy classes are represented by elements as shown in Section 5.4:

$u(0) = I$ (so $u(m)$ with $m = 0$);
$u(m)$ with m ranging over $\mathfrak{o}^{\bullet}/\pm 1$;
minus the above elements representing the classes of $\mathbf{c}(\Gamma)(-\Gamma_U)$.

We let

$\mathrm{CC}_{\mathrm{Cus,U}}$ = set of Γ – conjugacy classes in $\mathbf{c}(\Gamma)\Gamma_U$.
$\mathrm{CC}_{\mathrm{Cus,-U}}$ = set of Γ – conjugacy classes in $\mathbf{c}(\Gamma)(-\Gamma_U)$.
$\mathrm{CC}_{\mathrm{NRCus}}$ = union of the previous two, i.e., conjugacy classes of $\pm\mathbf{c}(\Gamma)\Gamma_U$.

Note that for all $u \in U$ (so for all $m \in \mathfrak{o}$), we have for $u = u(m)$,

$$\varphi(-u) = \varphi(u).$$

Hence the classes with a minus sign give the same values of φ as the classes without the minus sign. The **nonregular cuspidal trace** $\mathbf{K}_{\varphi}^{\mathrm{NRCus}}(z,z)$ can thus be rewritten

$$\sum_{\gamma\in \mathrm{NRCus}(\Gamma)} \varphi(z^{-1}\gamma z) = 2 \sum_{\mathfrak{c}\in \mathrm{CC}_{\mathrm{NRCus}}} \sum_{\gamma\in\mathfrak{c}} \varphi(z^{-1}\gamma z)$$

$$= 2\varphi(I) + \frac{1}{2} \sum_{0\neq m\in\mathfrak{o}} \sum_{\gamma\in\Gamma_U\backslash\Gamma} \varphi(z^{-1}\gamma^{-1}u(m)\gamma z) \tag{13.3}$$

The factor 2 disappears in the sum over m on the right because of Chapter 7, Proposition 7.4.2, whereby m ranging over $\mathfrak{o}^\bullet/\pm 1$ is 1/2 of the full sum over $m\neq 0$. The 1/2 appears because the isotropy group of $u(m)$ is $\pm\Gamma_U$, so the sum over $\gamma\in\Gamma_U\backslash\Gamma$ is twice what it is supposed to be.

We want the asymptotics of the nonregular cuspidal trace in (13.3) integrated over $\mathscr{F}_{\leqq Y}$ with $Y > 1$, $Y\to\infty$. Let

$$I_1(y) = \int_{\mathscr{F}_{\leqq Y}} \sum_{0\neq m\in\mathfrak{o}} \sum_{\gamma\in\Gamma_U\backslash\Gamma} \varphi(z^{-1}\gamma^{-1}u(m)\gamma z)d\mu(z).$$

With the Iwasawa measure $d\mu(z) = dx\,dy/y^3$, we get

$$\int_{\mathscr{F}_{\leqq Y}} \mathbf{K}_\varphi^{\mathrm{NRCus}}(z,z)d\mu(z) = 2\varphi(I)\mu(\mathscr{F}_{\leqq Y}) + \frac{1}{2}I_1(Y), \tag{13.4}$$

Note that here we are integrating over $\mathscr{F}_{\leqq Y}$. We remind readers of Section 6.4, (13.2), which gives the difference between an integral over $\Gamma\backslash G$ and an integral over $\mathscr{F}$. They differ by a factor of 2.

Given coset representatives of $\Gamma_U\backslash\Gamma$, we can sum over γ ranging over these representatives, and then interchange the integral and the sums. Representatives can be selected also to take into account some symmetry, using Γ_∞ instead of Γ_U.

As usual, we let $H = \{\pm I, \pm R\}$ so that

$$\Gamma_\infty = \Gamma_U H = H\Gamma_U.$$

We use the subgroups filtration

$$\Gamma_U \subset \Gamma_\infty \subset \Gamma,$$

and let $\mathrm{Rep}(\Gamma_\infty\backslash\Gamma)$ be a set of representatives for $\Gamma_\infty\backslash\Gamma$. Then a set of representatives for $\Gamma_U\backslash\Gamma$ is given by

$$\mathrm{Rep}(\Gamma_U\backslash\Gamma) = \{\eta\gamma | \eta\in H \text{ and } \gamma\in\mathrm{Rep}(\Gamma_\infty\backslash\Gamma)\}.$$

As in Section 9.6 and 9.7, we let $\mathscr{S}$ be the part of the tube over $\mathscr{F}_U$ lying below the unit sphere. Then the tube $\mathscr{T}$ is the union

$$\mathscr{T} = \mathscr{F}\cup\mathscr{S}.$$

The set of representatives $\mathrm{Rep}(\Gamma_\infty\backslash\Gamma)$ will then be taken as the set

$$\mathrm{Rep}_{\mathscr{T}} = I\cup\mathrm{Rep}_{\mathscr{S}},$$

where $\text{Rep}_{\mathscr{S}}$ was defined in Chapter 9, Theorem 9.8.5, so the images $\gamma(\mathscr{F})$, $\gamma \in \text{Rep}_{\mathscr{S}}$, give a tiling of $\mathscr{S}$. In particular, $\text{Rep}_{\mathscr{S}}\mathscr{F} = \mathscr{S}$.

The **truncated tube** at Y is defined as in Section 9.9 by

$$\mathscr{T}_{\leqq Y} = \mathscr{F}_{\leqq Y} \cup \mathscr{S} = \{z = x + \mathbf{j}y \text{ such that } x \in \mathscr{F}_U \text{ and } y \leqq Y\} = \mathscr{F}_U \times (0, Y].$$

As in that chapter, we let

$\mathscr{F}_{>Y}$ =subset of the fundamental domain consisting of points (x,y) with $x \in \mathscr{F}_U$ and $y > Y$.

We record the useful identity

$$\text{Rep}_{\mathscr{T}}\mathscr{F}_{\leqq \mathscr{Y}} = \mathscr{T}_{\leqq Y} - \text{Rep}_{\mathscr{S}}\mathscr{F}_{>Y}. \tag{13.5}$$

This comes from $\text{Rep}_{\mathscr{T}}\mathscr{F}_{\leqq Y} = \mathscr{F}_{\leqq Y} + \text{Rep}_{\mathscr{S}}\mathscr{F}_{\leqq Y} = \mathscr{F}_{\leqq Y} + \text{Rep}_{\mathscr{S}}\mathscr{F} - \text{Rep}_{\mathscr{S}}\mathscr{F}_{>Y}$.

Proposition 13.1.1. *We have*

$$\mathrm{I}_1(\mathrm{Y}) = 4 \sum_{0 \neq \mathrm{m} \in \mathfrak{o}} \int_{\mathscr{T}_{\leqq Y}} \varphi(\mathrm{z}^{-1}\,\mathrm{u}(\mathrm{m})\mathrm{z})d\mu(\mathrm{z}) + \mathrm{o}(1) \textit{ for } \mathrm{Y} \to \infty.$$

Proof. In the definition of $\mathrm{I}_1(\mathrm{Y})$ we use the representatives $\eta\gamma$ as described above for $\Gamma_{\mathrm{U}} \backslash \Gamma$. Then

$$\mathrm{I}_1(\mathrm{Y}) = \sum_{\eta \in \mathrm{H}} \int_{\mathscr{F}_{\leqq Y}} \sum_{0 \neq \mathrm{m}} \sum_{\gamma \in \text{Rep}_{\mathscr{T}}} \varphi(\mathrm{z}^{-1}\gamma^{-1}\eta^{-1}\mathrm{u}(\mathrm{m})\eta\gamma z)d\mu(z).$$

Conjugation by a given η permutes the family $\{\mathrm{u}(\mathrm{m})\}_{\mathrm{m} \neq 0}$, so

$$\begin{aligned}
\mathrm{I}_1(\mathrm{Y}) &= 4 \int_{\mathscr{F}_{\leqq \mathrm{Y}}} \sum_{\mathrm{m} \neq 0} \sum_{\gamma \in \text{Rep}_{\mathscr{T}}} \varphi((\gamma z)^{-1}\mathrm{u}(\mathrm{m})\gamma z)d\mu(z) \\
&= 4 \sum_{\mathrm{m} \neq 0} \int_{\text{Rep}_{\mathscr{T}}\mathscr{F}_{\leqq Y}} \varphi(z^{-1}\mathrm{u}(\mathrm{m})z)d\mu(z) \\
&= 4 \sum_{\mathrm{m} \neq 0} \int_{\mathscr{T}_{\leqq Y}} \varphi(z^{-1}\mathrm{u}(\mathrm{m})z)d\mu(z) - 4\mathrm{Err}(\mathrm{Y}) \text{ by (13.5)},
\end{aligned} \tag{13.6}$$

where Err(Y) is the error term, defined by

$$\mathrm{Err}(\mathrm{Y}) = \int_{\text{Rep}_{\mathscr{S}}\mathscr{F}_{>\mathrm{Y}}} \sum_{\mathrm{m} \neq 0} \varphi(z^{-1}\mathrm{u}(\mathrm{m})z)d\mu(z).$$

We have to show that

$$\mathrm{Err}(\mathrm{Y}) = \mathrm{o}(1) \text{ for } \mathrm{Y} \to \infty.$$

This will take two lemmas. The first lemma gives a general criterion for an integral over $\mathrm{Rep}_{\mathscr{S}}\mathscr{F}_{>Y}$ to be o(1). The second lemma proves that the criterion is satisfied for functions with quadratic exponential decay.

Lemma 13.1.2. *Let $\psi \in L^1(\mathscr{S})$. Then*

$$\int_{\mathrm{Rep}_{\mathscr{S}}\mathscr{F}_{>Y}} \psi(z)d\mu(z) = o(1) \text{ for } Y \to \infty.$$

Proof. Let χ_Y denote the characteristic function of the set $\mathrm{Rep}_{\mathscr{S}}\mathscr{F}_{>Y}$, and write

$$\int_{\mathrm{Rep}_{\mathscr{S}}\mathscr{F}>Y} \psi(z)d\mu(z) = \int_{\mathscr{S}} \chi_Y(z)\psi(z)d\mu(z).$$

Since the family $\{\gamma\mathscr{F}\}$ with $\gamma \in \mathrm{Rep}_{\mathscr{S}}$ is locally finite, for all $z \in \mathscr{S}$ there exists Y sufficiently large such that $z \notin \mathrm{Rep}_{\mathscr{S}}\mathscr{F}_{>Y}$. In other words,

$$\lim_{Y\to\infty} \chi_Y = 0 \text{ pointwise on } \mathscr{S}.$$

Since $\chi_Y\psi \in L^1(\mathscr{S})$ because it is bounded by $|\psi| \in L^1(\mathscr{S})$, applying the dominated convergence theorem concludes the proof of the lemma.

Lemma 13.1.3. *Let φ be K-bi-invariant with quadratic exponential decay. Then the function*

$$z \mapsto \sum_{m\neq 0} \varphi(z^{-1}u(m)z)$$

is in $L^1(\mathscr{S})$.

Proof. Using the Iwasawa decomposition of $z = u(x)a_y k$, we get

$$\begin{aligned} z^{-1}u(m)z &= k^{-1}a_y^{-1}u(-x)u(m)u(x)a_y k \\ &= k^{-1}u(m/y)k. \end{aligned}$$

Then

$$\sum_{m\neq 0} \varphi(z^{-1}u(m)z) = \sum_{m\neq 0} \varphi(u(m/y)).$$

By quadratic exponential decay (cf. Chapter 4, Proposition 4.1.2), we get the estimate

$$\int_{\mathscr{S}} \sum_{m\neq 0} \varphi(z^{-1}u(m)z)d\mu(z) \ll \sum_{m\neq 0} \int_0^1 \mathrm{e}^{-(\log(|m|/y))^2 c} \frac{dy}{y^3}. \tag{13.7}$$

It will suffice to prove that the integral on the right is estimated by

$$\int_0^1 e^{-(\log(|m|/y))^2 c} \frac{dy}{y^3} = 0\left(\frac{1}{|m|^3}\right)$$

for $|m| \to \infty$. This is simple calculus. For instance, because of the $\log^2$, the integral is unchanged if we replace $|m|/y$ by $y/|m|$. Then change variables, putting $v = y/|m|$, $dv = dy/|m|$. The integral becomes

$$\frac{1}{|m|^2} \int_0^{1/|m|} e^{-(\log v)^2} c \frac{dv}{v^3}.$$

The function under the integral sign is bounded on [0,1], so the whole term is $O(1/|m|^3)$ as desired. This proves Lemma 13.1.3, which together with Lemma 13.1.2 concludes the proof of Proposition 13.1.1. (Of course, the quadratic exponential decay condition can be considerably weakened.)

By Proposition 13.1.1, we give a coordinatized form to the integral, namely writing $z = u(x)a_y k$,

$$I_1(Y) = 4 \sum_{m \neq 0} \int_{x \in \mathscr{F}_U} \int_0^Y \varphi(a_y^{-1} u(x)^{-1} u(m) u(x) a_y) dx \frac{dy}{y^3} + o(1).$$

Since $u(x)$ and $u(m)$ commute, we get the following result:

Theorem 13.1.4.

$$I_1(Y) = 4 \sum_{0 \neq m \in \mathfrak{o}} \frac{1}{2} \int_0^Y \varphi(u(m/y)) \frac{dy}{y^3} + o(1).$$

The new factor 1/2 is the measure of $\mathscr{F}_U$ (half the measure of the standard square).

Now let

$$I_2(Y) = \sum_{m \neq 0} \int_0^Y \varphi(u(m/Y)) \frac{dy}{y^3}.$$

Combining Theorem 13.1.4 with (13.4), we get the following:

Corollary 13.1.5.

$$\int_{\mathscr{F}_{\leqq Y}} \mathbf{K}_\varphi^{\mathrm{NRCus}}(z,z) d\mu(z) = 2\varphi(I)\mu\mathscr{F}_{\leqq Y}) + I_2(Y) + o(1).$$

Theorem 13.2.5 in the next section will provide an asymptotic expansion up to $o(1)$ for $I_2(Y)$. However, we state here the part dealing with the coefficient of the divergent asymptotic term.

Theorem 13.1.6. *Writing*

$$\int_{\mathscr{F}_{\leqq Y}} \mathbf{K}_\varphi^{\mathrm{NRCus}}(z,z) d\mu(z) = \mathrm{c}_{1,\mathrm{NRC}}(\varphi) \log Y + O(1),$$

the coefficient $c_{1,\mathrm{NRC}}(\mathbf{g}_t)$ *(so for the heat Gaussian) is given by*

$$c_{1,\mathrm{NRC}}(\mathbf{g}_t) = 2\pi \int_0^\infty rf(r)dr = \frac{1}{2}\frac{e^{-2t}}{\sqrt{2\pi t}}.$$

This will come by putting together Theorem 13.2.5 below and the evaluation of the integral in Chapter 14, Proposition 14.3.1.

13.2 Asymptotic Expansion of the Nonregular Cuspidal Trace

In this section we determine the Y-asymptotics of the term that appeared in Corollary 13.1.5, namely

$$I_2(Y) = \sum_{0 \neq m \in \mathfrak{o}} \int_0^Y \varphi(u(m/y))\frac{dy}{y^3}. \tag{13.8}$$

The result is given in Theorem 13.2.5. We start with a change of variables taking into account the symmetry properties of φ.

Lemma 13.2.1. *There exists a* C^∞ *function* $f = f_\varphi$ *on* $\mathbf{R}^+$ *such that for all* $x \in \mathbf{C} \neq 0$

$$\varphi(u(x)) = f(|x|).$$

Proof. Matrix multiplication gives

$$u(x)u(x)^* = \begin{pmatrix} 1+|x|^2 & x \\ \bar{x} & 1 \end{pmatrix},$$

so $\mathrm{tr}(u(x)u(x)^*) = 2+|x|^2$. Let $u(x) = k_1bk_2$ be the polar decomposition with $b_1 \geqq 1$ $(b = \mathrm{diag}(b_1, b_1^{-1})$, so

$$u(x)u(x)^* = k_1b^2k_1^{-1}.$$

Taking the trace yields

$$b_1^2 + b_1^{-2} = 2+|x|^2.$$

Solving the quadratic equation yields

$$b_1^2 = 1+\frac{1}{2}|x|^2+\frac{1}{2}\sqrt{4|x|^2+|x|^4}.$$

Since φ is a C^∞ function of b_1, the lemma follows.

Remarks. From the quadratic exponential decay of φ, we conclude that $f(r)$ has quadratic exponential decay for $r \to \infty$. Also, f extends continuously at 0, and $f(0) = \varphi(I)$. In particular, the functions $rf(r)$ and $r(\log r)f(r)$ are in $L^1(0,\infty)$.

The function $f = f_\varphi$ will be computed explicitly in Section 14.3 when $\varphi = \mathbf{g}_t$ is the heat Gaussian, for some specific evaluations of certain integrals.

In terms of the function $f = f_\varphi$, we can then rewrite

$$
\begin{aligned}
I_2(Y) &= \sum_{m \neq 0} \int_0^Y f(|m|/y)\frac{dy}{y^3} \\
&= \sum_{m \neq 0} \frac{1}{|m|^2} \int_{|m|/Y}^{\infty} rf(r)dr \quad [\text{letting } r = |m|/y]. \qquad (13.9)
\end{aligned}
$$

We split the sum over $m \neq 0$ into $|m| \geqq Y$ and $|m| < Y$, and describe the asymptotics in each of these two cases.

Throughout, we shall use the **lattice-point counting theorem** on disks, due to Sierpiński–Van der Corput–Vinogradov–Hua; cf. [Lan 97], second or third edition, Section 14.3, p. 366, worked out exercise, namely:

Lattice-Point Counting Theorem. *In the disk and the annulus respectively,*

$$
\begin{aligned}
&\#\{\mathrm{m} \in \mathbf{Z}[\mathbf{i}] \mid |m| < R\} = \pi R^2 + O(R^{2/3}) && \textit{for } R \to \infty, \\
&\#\{\mathrm{m} \in \mathbf{Z}[\mathbf{i}] \mid R \leqq |m| < R+1\} = 2\pi R + O(R^{2/3}) && \textit{for } R \to \infty.
\end{aligned}
$$

The statement for the annulus is an immediate consequence of the statement for the disk, by subtraction (taking the derivative...). One may select R to be an integer or not, the difference being absorbed by the error term. A classical conjecture says that the error term is $O(R^{1/2})$, but the exponent 2/3 suffices amply for our purposes.

The next two lemmas will encounter a piece of analytic number theory, involving a result of Hardy–Littlewood that we extend to $\mathbf{Z}[\mathbf{i}]$. We give a self-contained treatment of what we need in Sections 14.1 and 14.3. This opens up the possibility of a new area of "analytic number theory" in a geometric context involving special functions.

Case 1. $|m| \geqq Y$.

Lemma 13.2.2. *For $Y \to \infty$,*

$$
\sum_{|m| \geqq Y} \frac{1}{|m|^2} \int_{|m|/Y}^{\infty} rf(r)dr = 2\pi \int_1^{\infty} \frac{1}{v} \int_v^{\infty} rf(r)\,dr\,dv + o(1).
$$

Proof. Let I be the sum on the left, over $|m| \geqq Y$. We want to give its asymptotic behavior in Y as stated. The error $o(1)$ suffices, but the power $Y^{-1/3}$ comes up in the proof. The lattice-point counting theorem gives (see the remark below)

$$\begin{aligned}
I &= 2\pi \sum_{\substack{n \geqq Y \\ n\in\mathbf{Z}}} \frac{1}{n^2} n \int_{n/Y}^{\infty} rf(r)dr + O(1) \sum_{n\geqq Y} \frac{1}{n^2} n^{2/3} \int_{n/Y}^{\infty} rf(r)dr \\
&= 2\pi \sum_{n\geqq Y} \frac{1}{n} \int_{n\backslash Y}^{\infty} rf(r)dr + O(Y^{-1/3}).
\end{aligned} \tag{13.10}$$

To see this, note first that the integral on the right is bounded independently of n; and second,

$$\sum_{n\geqq Y} n^{-4/3} = O(Y^{-1/3}),$$

thus proving (13.10).

Remarks. When we applied the lattice-point counting theorem, we replaced $|m|$ by n when $n \leqq |m| \leqq n+1$. This is harmless, because the difference $1/n^2 - 1/(n+1)^2$ is $O(1/n^3)$, so after multiplying by n, it is $O(1/n^2)$. The error coming from this replacement has the order of magnitude

$$\ll \sum_{n\geqq Y} \frac{1}{n^2} \int_{n/Y}^{\infty} rf(r)dr,$$

which is smaller than the other error term.

There remains to replace the sum in (13.10) by the integral as stated in the lemma. To do this, let

$$h_1(v) = \int_{v}^{\infty} rf(r)dr,$$

and let $v_n = v_n(Y) = n/Y$ for $n \geqq Y$. Then $v_{n+1} - v_n = 1/Y$, and

$$2\pi \sum_{n\geqq Y} \frac{1}{n} \int_{n/Y}^{\infty} rf(r)dr = 2\pi \sum_{n\geqq Y} \frac{1}{v_n}(v_{n+1} - v_n)h_i(v_n).$$

We view this expression as a Riemann sum for the integral

$$2\pi \int_{1}^{\infty} \frac{1}{v} h_1(v)dv$$

over $[1,\infty)$. The integral is absolutely convergent (with room to spare). Indeed, the quadratic exponential decay of f implies that given a positive integer N, we have $f(r) = O(1/r^N)$ for $r \to \infty$, so

$$h_1(v) \ll \int_v^\infty r^{1-N} dr = O(v^{2-N}) \text{ for } v \to \infty.$$

So the absolute convergence follows. Then the Riemann sum converges to the integral on $[1,\infty)$ as in the finite case, which proves Lemma 13.2.2.

Case 2. $|m| < Y$

Next we make similar arguments to settle the case of $|m| < Y$. We meet the **Euler constant** for **Q(i)**, which for present purposes is defined by the relation

$$\sum_{0<|m|<Y} \frac{1}{|m|^2} = 2\pi \log Y + 4\gamma_{\mathbf{Q}(i)} + o(1) \text{ for } Y \to \infty.$$

See Chapter 14, Theorem 14.3.3.

Lemma 13.2.3.

$$\sum_{0<|m|<Y} \frac{1}{|m|^2} \int_{|m|/Y}^\infty rf(r)dr$$

$$= (2\pi \log Y + 4\gamma_{\mathbf{Q(i)}}) \int_0^\infty rf(r)dr - 2\pi \int_0^1 \frac{1}{v} \int_0^v rf(r)drdv + o(1).$$

Proof. We write

$$\int_{|m|/Y}^\infty rf(r)dr = \int_0^\infty rf(r)dr - \int_0^{|m|/Y} rf(r)dr.$$

The first integral on the right is a constant, which occurs as a factor of the partial sum of the Dedekind zeta function for **Q(i)**. By the definition of $\gamma_{\mathbf{Q(i)}}$, we get the first term in the expansion of the lemma. There remains to show that

$$\sum_{0<|m|<Y} \frac{1}{|m|^2} \int_0^{|m|/Y} rf(r)dr = 2\pi \int_0^1 \frac{1}{v} \int_0^v rf(r)drdv + o(1).$$

As in Case 1, we use the lattice-point-counting error term to get

$$\sum_{0<|m|<Y} \frac{1}{|m|^2} \int_0^{|m|/Y} rf(r)dr = \sum_{0<n<Y} \frac{1}{n^2}(2\pi n + O(n^{2/3})) \int_0^{n/Y} rf(r)dr$$

$$(*) \qquad = 2\pi \sum_{0<n<Y} \frac{1}{n} \int_0^{n/Y} rf(r)dr + O(1) \sum_{0<n<Y} \frac{1}{n^{4/3}} \int_0^{n/Y} rf(r)dr.$$

The legitimacy of replacing $|m|$ by n is justified in the same manner as in the preceding case.

In the present Case 2, we let

$$h_2(v) = \int_0^v rf(r)dr.$$

As before, $v_n = n/Y$, but now $0 < n < Y$. Then $v_n < 1$, and $v_{n+1} - v_n = 1/Y$ as before. The first sum in $(*)$ is then

$$2\pi \sum_{0<n<Y} \frac{1}{v_n}(v_{n+1} - v_n)h_2(v_n),$$

which is a Riemann sum for the integral

$$2\pi \int_0^1 \frac{1}{v} h_2(v)dv.$$

This integral is absolutely convergent (near 0, $h_2(v) = O(v)$), so the Riemann sums converge to the integral, thus giving the double integral of the lemma, with a minus sign coming from the first step of the proof.

The second sum in $(*)$ can be estimated as follows. For $n < Y$, the integrand of the second term in $(*)$ is bounded on [0,1], so the second sum on the right of $(*)$ is bounded by

$$0(1) \sum_{0<n<Y} \frac{1}{n^{4/3}} \frac{n}{Y} \ll \frac{1}{Y} \sum_{n<Y} \frac{1}{n^{1/3}} \ll Y^{-1/3}.$$

This concludes the proof of Lemma 13.2.3.

Lemmas 13.2.2 and 13.2.3 contain double integrals. We change the order of integration in order to transform them into single integrals, which will combine. For Lemma 13.2.2, we have a double integral over an infinite sector

$$v \leqq r < \infty \text{ and } 1 \leqq v < \infty.$$

We get the interchange

$$\int_1^\infty \int_v^\infty (\ldots)dr\,dv = \int_1^\infty \int_1^r (\ldots)dv\,dr.$$

Hence

$$\int_1^\infty \frac{1}{v} \int_v^\infty rf(r)dr\,dv = \int_1^\infty r(\log r)f(r)dr. \tag{13.11}$$

For Lemma 13.2.3 we change the order of integration over the finite triangle defined by the inequalities

$$0 \leqq r \leqq v \text{ and } 0 \leqq v \leqq 1,$$

so that

$$\int_0^1 \int_0^v (\ldots) dr\, dv = \int_0^1 \int_r^1 (\ldots) dv\, dr.$$

Then for the double integral we get

$$-\int_0^1 \frac{1}{v} \int_0^v r f(r) dr\, dv = \int_0^1 r(\log r) f(r) dr. \tag{13.12}$$

Adding (13.11) and (13.12) yields the following:

Lemma 13.2.4.

$$2\pi \int_1^\infty \frac{1}{v} \int_v^\infty r f(r)\, dr\, dv - 2\pi \int_0^1 \frac{1}{v} \int_0^v r f(r)\, dr\, dv = 2\pi \int_0^\infty r(\log r) f(r) dr.$$

Putting all the lemmas together, we obtain the asymptotics for Corollary 13.1.5.

Theorem 13.2.5. *For $Y \to \infty$, and $f = f_\varphi$ (Lemma 13.2.1), we have*

$$\begin{aligned} I_2(Y) := & \sum_{m \neq 0} \int_0^Y \varphi(u(m/Y)) \frac{dy}{y^3} \\ = & (2\pi \log Y + 4\gamma_{\mathbf{Q}(\mathbf{i})}) \int_0^\infty r f(r) dr + 2\pi \int_0^\infty r(\log r) f(r) dr + o(1). \end{aligned}$$

13.3 The Regular Cuspidal Integral over $\mathscr{F}_{\leqq Y}$

We carry out the analogue of Section 13.1 and 13.2 for regular cuspidal conjugacy classes. In Chapter 5, Theorem 5.4.6, we determined that there are precisely four such classes, $\mathfrak{c}_j$ $(j = 1, \ldots, 4)$, with representatives γ_j taken as

$$\gamma_1 = Ru(1),\ \gamma_2 = Ru(\mathbf{i}),\ \gamma_3 = Ru(1 + \mathbf{i}),\ \gamma_4 = R.$$

The isotropy group of γ_j in Γ under conjugation is the group H_j, cyclic of order 4, generated by γ_j (Chapter 5, Proposition 5.4.3). We have a bijection

$$H_j \backslash \Gamma \to \mathfrak{c}_j \text{ by } \gamma \mapsto \gamma^{-1} \gamma_j \gamma. \tag{13.13}$$

Let φ be a K-bi-invariant continuous function on G, with exponential quadratic decay. In the application, φ is the heat Gaussian $\mathbf{g}_t$, but for the moment we just want to tabulate the algebraic formalism of the trace in the regular cuspidal case. In light of (13.13), we have the formula

$$\sum_{\gamma \in \mathrm{RefCus}(\Gamma)} \varphi(z^{-1}\gamma z) = \sum_{j=1}^{4} \sum_{\gamma \in H_j \backslash \Gamma} \varphi(z^{-1}\gamma^{-1}\gamma_j \gamma z). \tag{13.14}$$

The product map

$$H_j \times \Gamma_u \longrightarrow \Gamma_\infty = H_j \Gamma_U$$

is a bijection, so Γ_U can be taken as a set of coset representatives for $H_j \backslash \Gamma_\infty$. Thus (13.14) can be rewritten

$$\sum_{\gamma \in \mathrm{RegCus}(\Gamma)} \varphi(z^{-1}\gamma z) = \sum_{j=1}^{4} \sum_{\gamma \in \Gamma_\infty \backslash \Gamma} \sum_{m \in \mathfrak{o}} \varphi(z^{-1}\gamma^{-1}u(m)^{-1}\gamma_j u(m)\gamma z). \tag{13.15}$$

The determination of this expression will roughly follow Proposition 13.1.1 and its two lemmas, except that we shall have to deal separately with a finite number of cases $2x - m = 0$, $x \in \mathscr{F}_U$, instead of just the single case $m = 0$.

We recall from Section 9.7 that the images $\gamma\mathscr{F}$ ($\gamma \in \mathrm{Rep}_{\mathscr{T}}$) tile the tube. As in Sections 13.1 and 13.2 we use the notation

$$\sum_{\gamma \in \mathrm{Rep}_{\mathscr{T}}} \int_{\gamma\mathscr{F}_{\leqq Y}} = \int_{\mathrm{Rep}_{\mathscr{T}}\mathscr{F}_{\leqq Y}}.$$

Lemma 13.3.1. *Write $z = u(x)a_y k$ in Iwasawa coordinates. Then*

$$\int_{\mathscr{F}_{\leqq Y}} \sum_{\gamma \in \mathrm{RegCus}(\Gamma)} \varphi(z^{-1}\gamma z) d\mu(z) = \int_{\mathrm{Rep}_{\mathscr{T}}\mathscr{F}_{\leqq Y}} \sum_{m \in \mathfrak{o}} \varphi\left(u\left(\frac{2x - m}{y}\right)\right) \frac{dx\,dy}{y^3}.$$

(As usual, $dx = dx_1\, dx_2$.)

Proof. The expression (13.15) for the integrand is a function of γ_z. We move γ from z to $\mathscr{F}_{\leqq Y}$ to get the sum over $\gamma \in \mathrm{Rep}_{\mathscr{T}}$ of the integral over $\gamma\mathscr{F}_{\leqq Y}$. We replace z by its Iwasawa coordinates, and use again $\gamma_j = Ru(m_j)$ ($j = 1, \ldots, 4$). The only difference with the analogous step leading to Theorem 13.1.4 is the presence of R. We move R to the left across $u(-m)$ and $u(-x)$, each move having the effect of changing the minus sign to a plus sign. The K-bi-invariance of φ finally gets rid of R on both ends. This time, we observe that we get $u(m)$ twice, which is $u(2m)$, together with the $u(m_j)$. The product of u's in the middle is then

$$u(2x + 2m + m_j).$$

As m ranges over $\mathfrak{o}$, $2m+m_j$ ranges over $\mathfrak{o}$ because the m_j were selected as representatives of $\mathfrak{o}/2\mathfrak{o}$. So we are back to summing over $\mathfrak{o}$. Taking the conjugation with a_y^{-1} introduces the denominator y as usual. This proves the lemma. *Note*: it does not matter whether we write $+m$ or $-m$, and we prefer $-m$ for later purposes, dealing with special values $x = m/2$.

In Lemma 13.3.1, we have a sum over $m \in \mathfrak{o}$. It turns out that there are some special values of m that are going to be treated separately, when $2x - m = 0$ for some $x \in \mathscr{F}_U$. This is precisely the set $\mathfrak{o}_2^+$ defined by

$$\mathfrak{o}_2^+ = \text{subset of } m \in \mathfrak{o} \text{ such that } |m| < 2,\ \mathrm{Im}(m) \geqq 0.$$

There are 6 elements in $\mathfrak{o}_2^+$, of which $1, -1, \mathbf{i}$ are relatively prime to 2, and $0, \pm 1 + \mathbf{i}$ are not. We call $\mathfrak{o}_2^+$ the **special subset** of $\mathfrak{o}$.

We now split the sum over $m \in \mathfrak{o}$ in Lemma 13.3.1 into two sums,

$$\sum_{m \in \mathfrak{o}} = \sum_{m \in \mathfrak{o}_2^+} + \sum_{m \notin \mathfrak{o}_2^+}.$$

The term with the first sum over $\mathfrak{o}_2^+$ will be handled separately in Section 13.6. It is the analogue of the term with $m = 0$ in the simpler case of Section 13.1. The next step shows how the second sum gives rise to a term similar to the term found in Proposition 13.1.1. We start with the expression on the right of Lemma 13.3.1.

Proposition 13.3.2. *For $Y \to \infty$,*

$$\int_{\mathrm{Rep}_{\mathscr{T}}\mathscr{F}_{\leqq Y}} \sum_{m \in \mathfrak{o}} \varphi\left(u\left(\frac{2x-m}{y}\right)\right) \frac{dx\,dy}{y^3}$$

$$= \sum_{m \in \mathfrak{o}_2^+} \int_{\mathrm{Rep}_{\mathscr{T}}\mathscr{F}_{\leqq Y}} \varphi\left(u\left(\frac{2x-m}{y}\right)\right) \frac{dx\,dy}{y^3} + \sum_{m \notin \mathfrak{o}_2^+} \int_{\mathscr{T}_{\leqq Y}} \varphi\left(u\left(\frac{2x-m}{y}\right)\right) \frac{dx\,dy}{y^3} + o(1).$$

Proof. The first sum occurs as part of the sum in the lemma. As to the second, we note that for $x \in \mathscr{F}_U$ and $m \in \mathfrak{o}$, $m \notin \mathfrak{o}_2^+$, we have

$$|2x - m| \geqq |m|/\sqrt{5}.$$

The worst case is when $m = \pm 2 + \mathbf{i}$, $|m| = \sqrt{5}$. Since $\varphi(u(x)) = f(|x|)$, we get with some constant $c > 0$,

$$\sum_{m \notin \mathfrak{o}_2^+} \varphi\left(u\left(\frac{2x-m}{y}\right)\right) \ll \sum_{m \notin \mathfrak{o}_2^+} e^{-\log(|m|/2y))^2 c}.$$

The arguments from Proposition 13.1.1 and its two lemmas apply, to give

$$\int_{\mathrm{Rep}_{\mathscr{T}}\mathscr{F}_{\leqq Y}} \sum_{m\notin \mathfrak{o}_2^+} \varphi\left(u\left(\frac{2x-m}{y}\right)\right)\frac{dx\,dy}{y^3}$$

$$= \int_{\mathscr{T}_{\leqq Y}} \sum_{m\notin \mathfrak{o}_2^+} \varphi\left(u\left(\frac{2x-m}{y}\right)\right)\frac{dx\,dy}{y^3} + o(1) \text{ for } Y \to \infty,$$

which proves the proposition.

We will give the asymptotics of the nonspecial term in Proposition 13.4.3. The special term with the sum over $\mathfrak{o}_2^+$ will be handled separately in Section 13.6. It is the analogue of the term with $m = 0$ in the simpler case of Section 13.1.

13.4 Nonspecial Regular Cuspidal Asymptotics

We go on with the case that $m \in \mathfrak{o}$, $m \notin \mathfrak{o}_2^+$, i.e., m is **nonspecial**. We let I_{NS} be the second integral on the right in Proposition 13.3.2, that is,

$$I_{\mathrm{NS}} = \sum_{m\notin \mathfrak{o}_2^+} \int_{\mathscr{T}_{\leqq Y}} \varphi\left(u\left(\frac{2x-m}{y}\right)\right)\frac{dy\,dx}{y^3} = \sum_{m\notin \mathfrak{o}_2^+} \int_{\mathscr{F}_U} \int_0^Y \varphi\left(u\left(\frac{2x-m}{y}\right)\right)\frac{dy\,dx}{y^3}. \tag{13.16}$$

The arguments are similar to those in Section 13.2 when $m \neq 0$. As usual, we write by Lemma 13.2.1,

$$\varphi\left(u\left(\frac{2x-m}{y}\right)\right) = f\left(\frac{|2x-m|}{y}\right).$$

We change variables, $r = |2x-m|/y$ to get

$$\mathrm{I}_{\mathrm{NS}} = \sum_{m\notin \mathfrak{o}_2^+} \int_{\mathscr{F}_U} \frac{1}{|2x-m|^2} \int_{|2x-m|/Y}^{\infty} f(r)r\,dr\,dx$$

$$= \int_{\mathscr{F}_U} \left(\sum_{m\notin \mathfrak{o}_2^+} \frac{1}{|2x-m|^2} \int_{|2x-m|/Y}^{\infty} rf(r)dr \right) dx. \tag{13.17}$$

We shall determine the asymptotics of the inside integral in Y, then integrate dx. We have Case 1 and Case 2 as in Section 13.2. It is crucial that we have the lower bound

$$|2x-m| \geqq \frac{1}{2}|m| \geqq 1 > 0 \text{ for all } x \in \mathscr{F}_U \text{ and } m \notin \mathfrak{o}_2^+.$$

Lemma 13.4.1. *For $Y \to \infty$,*

$$\sum_{|m| \geqq Y} \frac{1}{|2x-m|^2} \int_{|2x-m|/Y}^{\infty} rf(r)dr = 2\pi \int_1^{\infty} \frac{1}{v} \int_v^{\infty} rf(r)dr + o(1)$$

$$= 2\pi \int_1^{\infty} r(\log r)f(r)dr + o(1).$$

The error term is uniform for $x \in \mathcal{F}_U$.

Proof. Simply note that

$$|m| - \sqrt{2} \leqq |m - 2x| \leqq |m| + \sqrt{2} \text{ for all } x \in \mathcal{F}_U,$$

and apply the proof of Lemma 13.2.2. The second evaluation comes from (13.4) of Section 13.2.

We define the **perturbation of the Hurwitz constant** $\gamma_{\mathrm{Hur},\mathbf{z}[\mathbf{i}]}(2x,2)$ to be the constant term in the asymptotic relation (see Chapter 14, Proposition 14.4.2)

$$\sum_{\substack{|m| < Y \\ m \notin \mathfrak{o}_2^+}} \frac{1}{|2x-m|^2} = 2\pi \log Y + \gamma_{\mathrm{Hur},\mathbf{Z}[\mathbf{i}]}(2x,2) + o(1) \text{ for } Y \to \infty.$$

Lemma 13.4.2. *For $x \in \mathcal{F}_U$ and $Y \to \infty$,*

$$\sum_{\substack{|m| < Y \\ m \notin \mathfrak{o}_2^+}} \frac{1}{|2x-m|^2} \int_{|2x-m|/Y}^{\infty} rf(r)dr$$

$$= (2\pi \log Y + \gamma_{\mathrm{Hur},\mathbf{Z}[\mathbf{i}]}(2x,2)) \int_0^{\infty} rf(r)dr$$

$$- 2\pi \int_0^1 \frac{1}{v} \int_0^v f(r)dr dv + o(1)$$

$$= (2\pi \log Y + \delta_{\mathrm{Hur},\mathbf{Z}[\mathbf{i}]}(2x,2)) \int_0^{\infty} rf(r)dr$$

$$+ 2\pi \int_0^1 r(\log r)f(r)dr + o(1).$$

The error term is uniform for $x \in \mathcal{F}_U$.

Proof. The proof of Lemma 13.2.3 applies to get the first expression. Then we use (5) of Section 13.2 for the second.

Next comes the **average of the Hurwitz constant** in the present nonspecial case. We let

$$\gamma^{\mathrm{Av},2}_{\mathrm{Hur},\mathbf{Z}[\mathbf{i}]} = \int_{\mathscr{F}_U} \gamma_{\mathrm{Hur},\mathbf{Z}[\mathbf{i}]}(2x,2)dx.$$

Proposition 13.4.3.

$$\sum_{m\notin \mathfrak{o}_2^+} \int_{\mathscr{F}_U} \int_0^Y \varphi\left(u\left(\frac{2x-m}{y}\right)\right)\frac{dy\,dx}{y^3} = I_{\mathrm{NS}}$$

$$= \left(\pi \log Y + \gamma^{\mathrm{Av},2}_{\mathrm{Hur},\mathbf{Z}[\mathbf{i}]}\right)\int_0^\infty rf(r)dr + \pi\int_0^\infty r(\log r)f(r)dr + o(1).$$

Proof. Immediate by combining Lemmas 13.4.1 and 13.4.2, and using the uniformity of the error term and the fact that $\mathscr{F}_U$ has measure 1/2.

13.5 Action of the Special Subset

We next come to the **special subset** $\mathfrak{o}_2^+$ of $\mathfrak{o} = \mathbf{Z}[\mathbf{i}]$, consisting of the six elements in $2\mathscr{F}_U$ having absolute value < 2, that is $0, \mathbf{i}, \pm 1, \pm 1 + \mathbf{i}$. For each $m \in \mathfrak{o}_2^+$, we define a **special element** γ_m as follows:

$$\gamma_0 = \begin{pmatrix} 0 & -1 \\ 1 & 0 \end{pmatrix} \quad \text{if } m = 0,$$

$$\gamma_m = \begin{pmatrix} 1 & 0 \\ 2/m & 1 \end{pmatrix} \quad \text{if } |m| \neq 0, \text{ so } |m| = 1 \text{ or } \sqrt{2}.$$

We abbreviate c_{γ_m} by c_m. So $c_0 = 1$ and $c_m = 2/m$ for $m \neq 0$.

Similarly, shortening the notation of Section 9.8, for $Y > 1$ (in practice Y is large positive), we define

$$\delta_{m,Y} = \frac{1}{2}|c_m|^2 Y.$$

so

$$\delta_{m,Y} = \begin{cases} \frac{1}{2}Y & \text{if } |m| = 0 \\ \frac{1}{8}Y & \text{if } |m| = 1 \\ \frac{1}{4}Y & \text{if } |m| = \sqrt{2}. \end{cases}$$

Writing $\mathbf{B}_m$ instead of $\mathbf{B}_{\gamma_m}$ as in Section 9.8, we let

$\mathbf{B}_{m,Y}$ = open Euclidean ball with center $(m/2,\ \delta_{m,Y})$ and radius $\delta_{m,Y}$, in $\mathbf{H}^3$.

It is the ball tangent to **C** at $m/2$, of radius $\delta_{m,Y}$, defined by the equation

$$|x - \frac{m}{2}|^2 + y^2 < \delta_{m,Y}^2.$$

We give the special case of Chapter 9, Proposition 9.9 in the present context.

Proposition 13.5.1. *The special element γ_m maps the plane $\mathbf{C}_{\mathbf{j}Y}$ to the sphere of radius $\delta_{m,Y}$ tangent to the point $m/2$ (except for this point, which is the image of $\mathbf{j}\infty$), and maps the region above the plane to the ball $\mathbf{B}_{m,Y}$ (the interior of the above sphere).*

As in Section 9.8, we shall consider the tube

$$\mathscr{T} = \mathscr{F}_U \times \mathbf{R}^+,$$

and for $Y > 1$, similarly to $\mathscr{F}_{\leqq Y}$ we let

$$\mathscr{T}_{\leqq Y} = \mathscr{F}_U \times \mathbf{R}^+_{\leqq Y}.$$

For $m \in \mathfrak{o}_2^+$ only a portion of the ball $\mathbf{B}_{m,Y}$ lies in $\mathscr{T}$. To determine this portion, we are led to define the interval I_m of the interior angles at $m/2$ within $\mathscr{F}_U$, with $m \in \mathfrak{o}_2^+$. Thus we let

$$I_m = \begin{cases} (0,\pi) & \text{for } m = 0, \\ (\frac{\pi}{2},\pi) & \text{for } m = 1, \\ (0,\frac{\pi}{2}) & \text{for } m = -1, \\ (\pi,2\pi) & \text{for } m = \mathbf{i}, \\ (\pi,\frac{3\pi}{2}) & \text{for } m = 1+\mathbf{i}, \\ (\frac{3\pi}{2},2\pi) & \text{for } m = -1+\mathbf{i}. \end{cases}$$

Let θ_m denote the length of the interval. Then

$$\theta_m = \begin{cases} \pi & \text{if } m = 0 \text{ or } \mathbf{i}, \\ \pi/2 & \text{otherwise.} \end{cases}$$

In the subsequent calculations, only the length will be relevant. The diagram illustrating the interior angles is as follows. The rectangle represents $\mathscr{F}_U$.

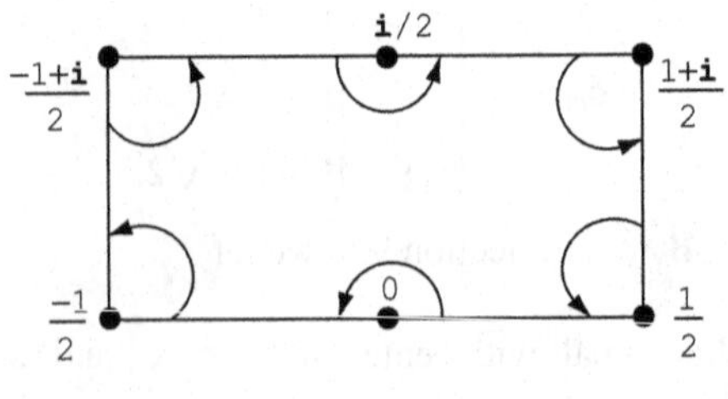

For the ultimate application to the asymptotic expansion, we note that

$$\sum_{m \in \mathfrak{o}_2^+} -\theta_m \log \delta_{m,Y} = 4\pi \log Y + 9\pi \log 2. \tag{13.18}$$

For each $m \in \mathfrak{o}_2^+$ we introduce cylindrical coordinates (ρ, θ, y), where

$$x - \frac{m}{2} = \rho e^{i\theta}.$$

The center of the open ball $\mathbf{B}_{m,Y}$ can be viewed as lying above the point $m/2$ with $m \in \mathfrak{o}_2^+$. "Above" means in the perpendicular direction above this sheet of paper, A cross section with one great circle is shown in the next figure.

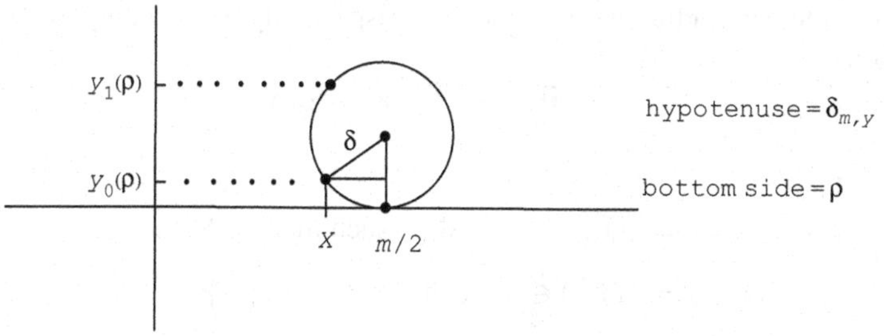

Proposition 13.5.2. *Let $Y > 1$. Then γ_m induces a bijection of $\Gamma_\infty \mathscr{F} > Y$ with $\mathbf{B}_{m,Y}$. The intersection*

$$\mathbf{B}_{m,Y} \cap \mathscr{T}$$

is the set of points (ρ, θ, y) such that

$$0 < \rho < \delta_{m,Y}, \ \theta \in I_m, \ y_0(\rho) < y < y_1(\rho),$$

where

$$y_0(\rho) = \delta_{m,Y} - \sqrt{\delta_{m,Y}^2 - \rho^2} \text{ and } y_1(\rho) = \delta_{m,Y} + \sqrt{\delta_{m,Y}^2 - \rho^2}.$$

In brief, the fraction of the ball that lies in $\mathscr{T}$ is equal to $\theta_m/2\pi$.

Proof. The first assertion is a repetition of Chapter 9, Proposition 9.9.2, in our special case. The assertion about the intersection is clear from the figure and Pythagoras. In terms of a formula,

$$\frac{\text{vol}\,(\mathbf{B}_{m,Y} \cap \mathscr{T})}{\text{vol}\,(\mathbf{B}_{m,Y})} = \theta_m/2\pi.$$

13.6 Special Regular Cuspidal Asymptotics

We apply the geometry of Section 13.5 to determine the asymptotics of the second sum in Proposition 13.3.2. We first get formulas for a fixed $m \in \mathfrak{o}_2^+$, and then take the finite sum over $m \in \mathfrak{o}_2^+$. The truncated trace integral will be shown in Theorem 13.6.3 to be an integral over $\mathcal{T} - \mathbf{B}_{m,Y} + o(1)$. The set $\mathcal{T} - \mathbf{B}_{m,Y}$ is decomposed as follows.

Define the **sector** $\mathrm{Sec}_{m,Y}$ and its complement in $\mathcal{F}_U$:

$$\mathrm{Sec}_{m,Y} = \left\{ x \in \mathcal{F}_U \text{ such that } \left| x - \frac{m}{2} \right|^2 < \delta_{m,Y}^2 \right\}$$

$$\mathrm{Sec}'_{m,Y} = \left\{ x \in \mathcal{F}_U \text{ such that } \left| x - \frac{m}{2} \right|^2 \geqq \delta_{m,Y}^2 \right\}.$$

Thus the center and radius are $m/2$ and $\delta_{m,Y}$ respectively. We decompose $\mathcal{T} - \mathbf{B}_{m,Y}$ into subsets,

$$\mathcal{T} - \mathbf{B}_{m,Y} = \mathcal{T}_{1,m,Y} \cup \mathcal{T}_{2,m,Y}, \tag{13.19}$$

where

$$\mathcal{T}_{1,m,Y} = \{(x,y) \in \mathcal{T} - \mathbf{B}_{m,Y} \text{ such that } x \in \mathrm{Sec}_{m,Y}\}$$

$$\mathcal{T}_{2,m,Y} = \{(x,y) \in \mathcal{T} \text{ such that } x \in \mathrm{Sec}'_{m,Y}\}$$

Note that the condition on x for $\mathcal{T}_{2,m,Y}$ automatically puts (x,y) outside $\mathbf{B}_{m,Y}$, and allows all $y > 0$; in other words, $\mathcal{T}_{2,m,Y}$ has the product structure over its base in the $\mathbf{C}$-plane,

$$\mathcal{T}_{2,m,Y} = \mathrm{Sec}'_{m,Y} \times \mathbf{R}^+.$$

On the other hand, for $\mathcal{T}_{1,m,Y}$ the values of y are restricted as indicated in the second figure of Section 13.5, i.e., outside the interval between $y_0(\rho)$ and $y_1(\rho)$.

We continue the story with the function φ as in the introduction to this chapter, so φ is continuous, K-bi-invariant, with quadratic exponential decay. Furthermore, f is the associated function as in Lemma 13.2.1, $f(r) = \varphi(u(r))$.

Lemma 13.6.1. *Let $m \in \mathfrak{o}_2^+$. Then*

$$\int_{\mathcal{T}_{1,m,Y}} \varphi\left(u\left(\frac{2x-m}{y}\right)\right) \frac{dxdy}{y^3} = \frac{1}{4}\theta_m \int_0^\infty r\log\left(\frac{r^2+4}{4r}\right) f(r)dr.$$

Proof. The integral over $\mathcal{T}_{1,m,Y}$ on the left in cylindrical coordinates of Section 13.5 is

$$= \int_{I_m} \int_0^{\delta_{m,Y}} \left(\int_0^{y_0(\rho)} + \int_{y_1(\rho)}^\infty \right) f\left(\frac{2\rho}{y}\right) \frac{dy\, \rho\, d\rho\, d\theta}{y^3} \quad \text{because } 2x - m = 2\rho e^{i\theta}$$

$$= \theta_m \int_0^{\delta_{m,Y}} \left(\int_0^{y_0(\rho)} + \int_{y_1(\rho)}^{\infty} \right) f\left(\frac{2\rho}{y}\right) \frac{\rho\, dy\, d\rho}{y^3}$$

$$= \frac{1}{4}\theta_m \int_0^{\delta_{m,Y}} \frac{1}{\rho} \left(\int_0^{2\rho/y_1(\rho)} + \int_{2\rho/y_0(\rho)}^{\infty} \right) r f(r)\, dr\, d\rho \text{ by } r = 2\rho/y.$$

We change variables once more, putting $v = \delta_m/\rho$. The limits of integration come out of the expressions for y_0 and y_1 in Proposition 13.5.2. Then the last integral is

$$= \frac{1}{4}\theta_m \int_1^{\infty} \frac{1}{v} \left(\int_0^{2/(v+\sqrt{v^2-1})} + \int_{2/(v+\sqrt{v^2+1})} \right) r f(r)\, dr\, dv$$

$$= \frac{1}{4}\theta_m \int_1^{\infty} \frac{1}{v} \left(\int_0^{2(v-\sqrt{v^2-1})} + \int_{2(v+\sqrt{v^2+1})}^{\infty} \right) r f(r)\, dr\, dv.$$

We interchange the limits of integration in each one of the two inner integrals.

First integral.

$$\int_1^{\infty} \int_0^{2(v-\sqrt{v^2-1})} \cdots dr\, dv = \int_0^{2} \int_1^{(r^2+4)/4r} \cdots dv\, dr.$$

This comes from the inverse functions

$$r = 2(v - \sqrt{v^2-1}) \text{ and } v = (r^2+4)/4r,$$

defined for $v \geqq 1,\ r \leqq 2$. Then the first integral is trivially evaluated,

$$\int_1^{\infty} \frac{1}{v} \int_0^{2(v-\sqrt{v^2-1})} r f(r)\, dr\, dv = \int_0^{2} r \log\left(\frac{r^2+4}{4r}\right) f(r)\, dr.$$

Second integral.

$$\int_1^{\infty} \int_{2(v+\sqrt{v^2-1})}^{\infty} \ldots dr\, dv = \int_2^{\infty} \int_1^{(r^2+4)/4r} \ldots dv\, dr.$$

The inverse functions are now

$$r = 2(v + \sqrt{v^2-1}) \text{ and } v = (r^2+4)/4r,$$

defined for $v \geqq 1,\ r \geqq 2$. The second integral is trivially evaluated, with the same integrand, but the limits of integration are $[2, \infty)$.

Taking the sum of the first and second integrals yields precisely the value stated in Lemma 13.6.1, thus concluding the proof.

There remains to evaluate the integral corresponding to the one of Lemma 13.6.1, but over the other set $\mathcal{T}_{2,m,Y}$. We do part of the job next, and postpone the rest to Chapter 14, because it requires another technique.

Lemma 13.6.2. *Fix $1 < Y_0 < Y$. Then*

$$\int_{\mathcal{T}_{2,m,Y}} \varphi\left(u\left(\frac{2x-m}{y}\right)\right)\frac{dx\,dy}{y^3} = C_{m,Y}\int_0^\infty rf(r)dr,$$

where

$$C_{m,Y} = \frac{1}{4}\int_{\mathrm{Sec}'_{\mathrm{m,Y_0}}} \frac{1}{|x-m/2|^2}\,dx + \frac{1}{4}\theta_m(\log\,\delta_{m,Y_0} - \log\,\delta_{m,Y}).$$

Proof. We have the decomposition

$$\mathcal{T}_{2,m,Y} = (\mathrm{Sec}'_{m,Y_0} \times \mathbf{R}^+) \cup (A_{m,Y_0,Y} \times \mathbf{R}^+),$$

where $A_{m,Y_0,Y}$ is the set of elements $x \in \mathcal{F}_U$ such that

$$\delta_{m,Y} < |x - \frac{m}{2}| < \delta_{m,Y_0}.$$

Both sets in this decomposition are product spaces. The integral is with respect to $dx\,dy/y^3$, and we may take the double integral in any order. Let the inner integral be over dy/y^3, and change variable. For fixed x, let $r = |2x-m|/y$, so

$$\frac{dr}{y} = -2|x-m/2|\frac{dy}{y^3}.$$

Then using Lemma 13.2.1,

$$\int_0^\infty \varphi\left(u\left(\frac{2x-m}{y}\right)\right)\frac{dy}{y^3} = \int_0^\infty f(r)\frac{1}{2|x-m/2|y}dr$$
$$= \frac{1}{4}\frac{1}{|x-m/2|^2}\int_0^\infty rf(r)dr.$$

Integrating with respect to $x \in \mathrm{Sec}'_{m,Y_0}$, we obtain the first part of the constant $C_{m,Y}$ as stated. For the second part, we can perform the integration with respect to dx over the annulus sector $A_{m,Y_0,Y}$, with polar coordinates

$$dx_1\,dx_2 = \rho\,d\rho\,d\theta,$$

so

$$\frac{1}{4}\int\limits_{A_m \cdot Y_0, Y} \frac{1}{|x-m/2|^2}dx = \frac{1}{4}\int\limits_0^{\theta_m}\int\limits_{\delta_{m,Y}}^{\delta_{m,Y_0}} \frac{1}{\rho^2}\rho\, d\rho\, d\theta,$$

which yields the second part as stated, This concludes the proof of the lemma.

Remark. The integral remaining in the constant C_m (the first part) will combine with the Hurwitz constant from Section 13.4.

Lemma 13.6.2 shows in particular that the function being integrated is in $L^1(\mathcal{T}_{2,m,Y})$. The next theorem will use only this property, not the specific evaluation of the integral. So we formulate this theorem more generally for any function $\psi = \psi(x,y)$ in $L^1(\mathcal{T}_{2,m,Y_0})$ for the measure

$$d\mu(x,y) = \frac{dx\,dy}{y^3}.$$

The application will be to the function

$$\psi(x,y) = \varphi\left(u\left(\frac{2x-m}{y}\right)\right) \quad \text{with } m \in \mathfrak{o}_2^+.$$

Theorem 13.6.3. *Let* $\psi \in L^1(\mathcal{T}_{2,m,Y_0})$. *Then*

$$\int\limits_{Rep_{\mathcal{T}}\mathcal{F}_{\leqq Y}} \psi\, d\mu = \int\limits_{\mathcal{T}-\mathbf{B}_{m,Y}} \psi\, d\mu + o(1) \quad \textit{for}\ Y \to \infty.$$

Proof. We use the set defined in Chapter 9, Corollary 9.8.3, in the present context, so we let

$$S_m = \{\gamma \in \mathrm{Rep}_{\mathcal{S}} \text{ such that } \gamma(\mathbf{j}\infty) = m/2\}.$$

We have the disjoint union of S_m and its complement S'_m in $\mathrm{Rep}_{\mathcal{T}}$,

$$\mathrm{Rep}_{\mathcal{T}} = S_m \cup S'_m.$$

Then

$$\begin{aligned}\mathcal{T} = \mathrm{Rep}_{\mathcal{T}}\mathcal{F}_{\leqq Y} \cup \mathrm{Rep}_{\mathcal{T}}\mathcal{F}_{>Y} &= \mathrm{Rep}_{\mathcal{T}}\mathcal{F}_{\leqq Y} \cup S_m\mathcal{F}_{>Y} \cup S'_m\mathcal{F}_{>Y} \\ &= \mathrm{Rep}_{\mathcal{T}}\mathcal{F}_{\leqq Y} \cup (\mathbf{B}_{m,Y} \cap \mathcal{T}) \cup S'_m\mathcal{F}_{>Y} \qquad (13.20a)\end{aligned}$$

by Chapter 9, Theorems 9.9.3 and 9.9.4. The union is disjoint up to boundary points, which have measure 0. Hence

$$\mathcal{T} - \mathbf{B}_{m,Y} = \mathrm{Rep}_{\mathcal{T}}\mathcal{F}_{\leqq Y} \cup S'_m\mathcal{F}_{>Y}. \qquad (13.20b)$$

Integrating over these sets yields the desired formula by the following lemma:

Lemma 13.6.4.

$$\int_{S'^{\mathscr{F}}_{m>Y}} \psi\, d\mu = o(1) \ for\ Y \to \infty.$$

Proof. We use the same argument as for Lemma 13.1.2. Directly from the definitions, we have

$$S'_m \mathscr{F}_{>Y} \subset \mathscr{T}_{2,m,Y_0} \text{ for } Y \geqq Y_0$$

For every $z \in \mathscr{T} - \mathbf{B}_{m,Y}$ there exists $\gamma \in S'_m$ and $w \in \mathscr{F}$ such that $z = \gamma w$. Let $\chi_{m,Y}$ be the characteristic function of $S'_m \mathscr{F}_{>Y}$. Then pointwise $\chi_{m,Y} \to 0$ as $Y \to \infty$, so by the dominated convergence theorem,

$$\lim_{Y\to\infty} \int_{S'_m F_{>Y}} \psi\, d\mu = \lim_{Y\to\infty} \int_{\mathscr{T}_{2,m,Y_0}} \chi_{m,Y} \psi\, d\mu = 0,$$

which proves the lemma, and concludes the proof of the theorem.

Using the specific integral expressions of Lemmas 13.6.1 and 13.6.2, we get the following theorem:

Theorem 13.6.5. *Let $m \in \mathfrak{o}_2^+$, For $Y \to \infty$,*

$$\int_{\mathrm{Rep}_{\mathscr{T}} \mathscr{F}_{\leqq Y}} \varphi\left(u\left(\frac{2x-m}{y}\right)\right) \frac{dx\,dy}{y^3}$$

$$= \frac{1}{4}\theta_m \int_0^\infty r \log\left(\frac{r^2+4}{4r}\right) f(r) dr + C_{m,Y} \int_0^\infty r f(r) dr + o(1),$$

where $C_{m,Y}$ is defined in Lemma 13.6.2.

We may now take the sum over $m \in \mathfrak{o}_2^+$. From Section 13.5 (1) we see that 4 cancels in the last sum, namely

$$\sum_{m\in\omega_2^+} -\frac{1}{4}\theta_m \log \delta_{m,Y} = \pi \log Y + \frac{9\pi}{4} \log 2.$$

This gives the asymptotic term $\pi \log Y$ that we want. A priori, the dependence of the penultimate term on Y_0 is canceled by the dependence coming from the first part of the constant C_m.

We shall now put the results of this chapter together. In Lemma 13.2.1, the function $f = f_\varphi$ was defined, and at the beginning of Section 13.1 we also defined

$$\mathrm{Cus}_\varphi(Y) = \int_{\mathscr{F}_{\leqq Y}} \mathbf{K}_\varphi^{\mathrm{Cus}}(z,z)\, d\mu(z)$$

with a test function φ that is continuous, K-bi-invariant, with quadratic exponential decay.

Theorem 13.6.6. *We have the asymptotic relation for $Y \to \infty$,*

$$\mathrm{Cus}_{\varphi}(Y) = 2\varphi(Y)\mu(\mathscr{F}) + a_1 \int_0^{\infty} rf(r)dr + 3\pi \int_0^{\infty} r\log(r)f(r)dr$$

$$+\pi \int_0^{\infty} r\log\left(\frac{r^2+4}{4r}\right) f(r)dr + o(1),$$

where

$$a_1 = 4\pi \log Y + 4\gamma_{Q(i)} + \gamma_{\mathrm{Hur},Z(i)}^{\mathrm{Av},2} - \pi \log Y_0 C'_{Y_0}$$

with

$$C'_{Y_0} = \frac{1}{4} \sum_{m \in \mathfrak{o}_2^+} \int_{Sec'_{m,Y_0}} \frac{1}{|x - m/2|^2} dx.$$

Proof. The above formula is a summary of the computations given in Theorem 13.2.5, Proposition 13.4.3, and Theorem 13.6.5. Important preliminary results are given in Corollary 13.1.5 and Proposition 13.3.2.

Of course, Theorem 13.6.6 will be applied when $\varphi = \mathbf{g}_t$ is the heat Gaussian. In this case, $f_{\mathbf{g}_t}$ is also written $\mathbf{f}_t$, and will be evaluated explicitly in Section 14.5, (14.2). Using this evaluation, we get the following result:

Corollary 13.6.7. *For $\varphi = \mathbf{g}_t$, $\mathbf{K}_{\varphi} = \mathbf{K}_t$ being the heat kernel, we have*

$$\int_{\mathscr{F}_{\leq Y}} \mathbf{K}_t^{\mathrm{Cus}}(z,z)d\mu(z) = 4\pi \log Y \int_0^{\infty} \mathbf{rf}_t(r)dr + \Theta^{\mathrm{Cus}}(1/t) + o(1) \text{ as } Y \to \infty.$$

In Chapter 14, Corollary 14.5.2, we will show that

$$\int_0^{\infty} \mathbf{rf}_t(r)dr = \frac{e^{-2t}}{2\pi(8\pi t)^{1/2}},$$

so the coefficient of $\log Y$ from Theorem 13.6.6 equals the coefficient of $\log Y$ in Chapter 12, Theorem 12.3.4. Writing

$$\log\left(\frac{r^2+4}{4r}\right) = \log(r^2+4) - \log(r) - 2\log(2),$$

we obtain the following:

Corollary 13.6.8.

$$\Theta^{\mathrm{Cus}}(1/t) = 2\mathbf{g}_t(I)\mu(\mathscr{F}) + C_1 \int_0^\infty r\mathbf{f}_t(r)dr$$

$$+ 2\pi \int_0^\infty r\log(r)\mathbf{f}_t(r)dr + \pi \int_0^\infty r\log(r^2+4)\mathbf{f}_t(r)dr,$$

where

$$C_1 = 4\gamma_{\mathbf{Q}(\mathbf{i})} + \gamma^{\mathrm{Av},2}_{\mathrm{Hur},\mathbf{Z}[\mathbf{i}]} - 2\pi\log(2) - \pi\log Y_0 + C'_{Y_0}.$$

Directly from the definition of $\mathbf{g}_t$, we have the value

$$\mathbf{g}_t(I) = \frac{e^{-2t}}{(8\pi t)^{3/2}}.$$

The volume of the fundamental domain $\mu(\mathscr{F})$ is given in Section 10.7 expressed in terms of $\zeta_{\mathbf{Q}(\mathbf{i})}(2)$, specifically

$$\mu(\mathscr{F}) = \mathrm{Vol}(\Gamma\backslash\mathbf{H}^3) = \frac{2}{\pi^2}\zeta_{\mathbf{Q}(\mathbf{i})}(2).$$

The computations in Chapter 14 are aimed at further evaluating the constant C_1, as well as the three integrals that appear in $\Theta^{\mathrm{Cus}}(1/t)$.

In Chapter 14, Corollary 14.6.5, we shall see how the seeming dependence of C_1 on Y_0 disappears.

Chapter 14
Analytic Evaluations

This chapter is essentially an appendix that puts together some explicit evaluations used in Chapter 13. The main results of Sections 14.1 and 14.2 over **Q** and **Q**(**i**) respectively are needed in situations such as Chapter 13, Lemma 13.2.2 and Lemma 13.2.3, which in a sense climb the ladder from the arithmetic level to the geometric level.

The first three sections are concerned with the asymptotic behavior of the partial sums of the Dedekind zeta function for **Z**, **Z**[**i**] and other asymptotic properties. The first result, Theorem 14.1.1, is due to Hardy–Littlewood [HaL 21]. Hardy–Littlewood climb the ladder of complexification in Section 14.3, and the geometric ladder in Chapter 13.

It turns out that a "constant term" in two apparently distinct expansions are actually equal. One of these is the constant in the analytic expansion of the zeta function at 1, and the other is the constant in an asymptotic expansion of the partial sums of the zeta function. We recall these constants in the cases of interest to us, namely over **Q** and over **Q**(**i**). We give complete proofs for the convenience of the reader.

The Euler constant first comes up in the Dedekind zeta function context. It comes up in Section 14.4 in the more complicated Hurwitz zeta function context, as a function on **C**/**Z**[**i**], or rather the representative square centered at the origin, of side 1.

14.1 Partial Sums Asymptotics for $\zeta_{\mathbf{Q}}$ and the Euler Constant

Let $\zeta_{\mathbf{Q}}$ be the Riemann zeta function. For Theorem 14.1.1, we use a method different from those of both Hardy–Littlewood and Titchmarsh [Tit 51], Theorem 4.11, with the theta function and Mellin transform, applicable to other situations involving this kind of structure.

Theorem 14.1.1. *Let* $\sigma_1 > 1$. *For* $0 < \sigma \leqq \sigma_1$, $\sigma \neq 1$, *we have*

$$\zeta_{\mathbf{Q}}(\sigma) = \sum_{n<x} \frac{1}{n^\sigma} + \frac{x^{1-\sigma}}{\sigma - 1} + O(x^{-\sigma}).$$

The implied constant in the O-term depends only on σ_1.

Proof. Without loss of generality, we may assume that x is an integer. For $\sigma > 1$, we write the Mellin integral as a sum of integrals over (0,1] and $[1,\infty)$,

$$\zeta_{\mathbf{Q}}(\sigma) - \sum_{n<x} \frac{1}{n^\sigma} = \frac{1}{\Gamma(\sigma)} \int_0^1 \sum_{n \geq x} e^{-nt} t^\sigma \frac{dt}{t} + \frac{1}{\Gamma(\sigma)} \int_1^\infty \sum_{n \geq x} e^{-nt} t^\sigma \frac{dt}{t}. \tag{14.1}$$

With $0 < \mathrm{t} \leqq 1$ (for the first integral), we have

$$\sum_{n \geq x} e^{-nt} = e^{-xt} \sum_{n=0}^\infty e^{-nt} = \frac{e^{-xt}}{1 - e^{-t}} = e^{-xt} \left(\frac{1}{t} + h(t) \right), \tag{14.2}$$

with $h(t)$ bounded. Then we write the first integral of (14.1) in the form

$$\frac{1}{\Gamma(\sigma)} \int_0^1 \sum_{n \geqq x} e^{-nt} t^\sigma \frac{dt}{t} = \frac{1}{\Gamma(\sigma)} \int_0^1 e^{-xt} t^{\sigma - t} \frac{dt}{t} + \frac{1}{\Gamma(\sigma)} \int_0^1 h(t) e^{-xt} t^\sigma \frac{dt}{t}. \tag{14.3}$$

The first integral on the right is reduced to an integral over $[1,\infty)$ via the identity

$$\frac{1}{\Gamma(\sigma)} \int_0^1 e^{-xt} t^{\sigma-1} \frac{dt}{t} = \frac{x^{1-\sigma}}{\sigma - 1} - \frac{1}{\Gamma(\sigma)} \int_1^\infty e^{-xt} t^{\sigma-1} \frac{dt}{t},$$

using $\Gamma(\sigma) = (\sigma - 1)\Gamma(\sigma - 1)$. Then (14.1) becomes

$$\zeta_{\mathbf{Q}}(\sigma) - \sum_{n<x} \frac{1}{n^\sigma} = \frac{x^{1-\sigma}}{\sigma - 1} - \frac{1}{\Gamma(\sigma)} \int_1^\infty e^{-xt} t^{\sigma-1} \frac{dt}{t} + \frac{1}{\Gamma(\sigma)} \int_0^1 \mathrm{h(t)e^{-xt}} t^\sigma \frac{dt}{t}$$
$$+ \frac{1}{\Gamma(\sigma)} \int_1^\infty \sum_{n \geq x} e^{-nt} t^\sigma \frac{dt}{t}. \tag{14.4}$$

All three integrals converge for $\sigma > 0$. Replacing σ by its complexification s, we see that (14.4) becomes a relation between meromorphic functions of s, holding for $\sigma = \mathrm{Re}(s) > 0$. We estimate the three integrals.

We reduce an estimate of the middle integral over [0,1] to an estimate of the integral over $[1,\infty)$ by means of the identity

$$\frac{1}{\Gamma(\sigma)} \int_0^1 e^{-xt} t^\sigma \frac{dt}{t} = x^{-\sigma} - \frac{1}{\Gamma(\sigma)} \int_1^\infty e^{-xt} t^\sigma \frac{dt}{t}.$$

Thus we are reduced to estimate all three integrals over $[1,\infty)$, in which case $t \geqq 1$. To estimate the first integral, we can replace $t^{\sigma-1}$ by t^{σ}. For the geometric series of the third integral, we have

$$\sum_{n\geq x} e^{-nt} = O(e^{-xt}),$$

so uniformly for $\sigma > 0$ we have inequalities (using $\ll$ instead of the O notation)

$$\begin{aligned} \frac{1}{\Gamma(\sigma)}\int_1^\infty \sum_{n\geq x} e^{-nt} t^{\sigma}\frac{dr}{t} &\ll \frac{1}{\Gamma(\sigma)}\int_1^\infty e^{-xt}t^{\sigma}\frac{dt}{t} \\ &\ll \frac{1}{\Gamma(\sigma)}\int_1^\infty e^{-(x-1)t}e^{-t}t^{\sigma}\frac{dt}{t} \\ &\ll \frac{e^{-x}}{\Gamma(\sigma)}\int_1^\infty e^{-t}t^{\sigma}\frac{dt}{t} \leqq e^{-x}. \end{aligned} \tag{14.5}$$

These estimates show that all three integrals are $O(e^{-x})$, so finally

$$\zeta_{\mathbf{Q}}(\sigma) - \sum_{n<x}\frac{1}{n^{\sigma}} = \frac{x^{1-\sigma}}{\sigma-1} + O(x^{-\sigma}) + O(e^{-x}).$$

Theorem 14.1.1 follows since $0 < \sigma \leqq \sigma_1$.

Remarks. By further expanding (14.2), one can obtain a further development in the asymptotic series of Theorem 14.1.1.

We allow $\sigma > 0$ for the uniformity, but Titchmarsh's method needs σ away from 0. On the other hand, we strive toward Theorems 14.3.1 and 14.3.4 Here we don't care about large values of σ.

Define the **Euler constant** $\gamma_{\mathbf{Q}}$ to be the constant term of the Laurent expansion of $\zeta_{\mathbf{Q}}(s)$ at $s = 1$, that is,

$$\zeta_{\mathbf{Q}}(s) = \frac{1}{s-1} + \gamma_{\mathbf{Q}} + O(s-1) \text{ as } s \to 1. \tag{14.6}$$

Theorem 14.1.2.

$$\sum_{n<x}\frac{1}{n} = \log x + \gamma_{\mathbf{Q}} + O(1/x) \textit{ as } x \to \infty,$$

Proof. We use the expansion of the zeta function (14.6) at $s = 1$ with $s = \sigma$, and the expansion

$$\frac{x^{1-\sigma}}{\sigma-1} = \frac{1}{\sigma-1} - \log x + O_x(\sigma-1) \text{ as } \sigma \to 1.$$

We substitute these in Theorem 14.1.1, cancel $1/(\sigma-1)$, and take the limit as $\sigma \to 1$. Then Theorem 14.1.2 drops out.

14.2 Estimates Using Lattice-Point Counting

This section gives some estimates to be used in the next main section. These estimates involve both some asymptotic expansions and "big oh" estimates of what amounts to error terms. We reduce some of these to the case considered in Section 14.1. We use the annulus version of the **lattice-points counting theorem** of Section 13.2, namely

$$\#\{a|\mathbf{N}a \in [x^2,(x+1)^2]\} = \frac{\pi}{2}x + O(x^{2/3}) \text{ as } x \to \infty.$$

Lemma 14.2.1. *For* $0 < t \leqq 1$, *we have*

$$\sum_{\mathbf{N}a \geqq x^2} e^{-\mathbf{N}at} = \frac{\pi}{2}\sum_{n \geqq x} ne^{-n^2t} + O\left(\sum_{n \geqq x} n^{2/3}e^{-n^2t}\right) + O(xe^{-x^2t/2}).$$

Proof. Directly from the lattice-point counting, we have the following upper and lower bounds:

$$\textit{upper bound}: \quad \sum_{\mathbf{N}a \geq x^2} e^{-\mathbf{N}at} \leq \frac{\pi}{2}\sum_{n \geq x} ne^{-n^2t} + O\left(\sum_{n \geq x} n^{2/3}e^{-n^2t}\right);$$

$$\textit{lower bound}: \quad \sum_{\mathbf{N}a \geq x^2} e^{-\mathbf{N}at} \geq \frac{\pi}{2}\sum_{n \geq x} ne^{-(n+1)^2t} + O\left(\sum_{n \geq x} n^{2/3}e^{-n^2t}\right).$$

Observe that

$$\sum_{n \geq x} ne^{-(n+1)^2t} = \sum_{n \geq x}(n+1)e^{-(n+1)^2t} - \sum_{n \geq x} e^{-(n+1)^2t}.$$

Now write

$$\sum_{n \geq x}(n+1)e^{-(n+1)^2t} = \sum_{n \geq x} ne^{-n^2t} + O\left(xe^{-x^2t/2}\right),$$

where the O term accounts for the first term in the series on the left-hand side. Thus, the lower bound can be rewritten as

$$\sum_{\mathbf{N}a \geq x^2} e^{-\mathbf{N}at} \geq \sum_{n \geq x} ne^{-n^2t} + O\left(xe^{-x^2t/2}\right) - \sum_{n \geq x} e^{-(n+1)^2t} + O\left(\sum_{n \geq x} n^{2/3}e^{-n^2t}\right).$$

Trivially,

$$\sum_{n \geq x} e^{-(n+1)^2t} = O\left(\sum_{n \geq x} n^{2/3}e^{-n^2t}\right),$$

so

$$\sum_{\mathbf{N}a \geq x^2} e^{-\mathbf{N}at} \geq \sum_{n \geq x} ne^{-n^2t} + O\left(xe^{-x^2t/2}\right) + O\left(\sum_{n \geq x} n^{2/3}e^{-n^2t}\right),$$

which, when combined with the upper bound, proves Lemma 14.2.1.

Next we deal with the case $t \geqq 1$, which is easier. Directly from the annulus lattice point counting theorem, we get the estimate

$$\sum_{\mathbf{N}a \geqq x^2} e^{-\mathbf{N}at} = O\left(\sum_{n \geqq x} ne^{-n^2 t}\right) = O(xe^{-x^2}) \text{ for } x \to \infty. \tag{14.7}$$

We can then estimate integrals over $[1, \infty)$ for $\sigma > 0$, starting with

$$\frac{1}{\Gamma(\sigma)} \int_1^\infty \sum_{\mathbf{N}a \geqq x^2} e^{-\mathbf{N}at} t^\sigma \frac{dt}{t} \ll \frac{1}{\Gamma(\sigma)} \int_1^\infty \sum_{n \geqq x} ne^{-n^2 t} t^\sigma \frac{dt}{t} = O(xe^{-x^2}) \tag{14.8}$$

for $x \to \infty$. This is immediate, for instance by writing

$$e^{-n^2 t} = e^{-(n^2-1)t} e^{-t} \leqq e^{-n^2} ee^{-t}.$$

Note that the integrals in (14.8) are holomorphic in s for $\mathrm{Re}(s) = \sigma > 0$, because the integrals are absolutely convergent.

In the next section, we meet variations of the integrals in (14.8), namely

$$\frac{1}{\Gamma(\sigma)} \int_1^\infty \sum_{n \geqq x} n^{2/3} e^{-n^2 t} t^\sigma \frac{dt}{t} \text{ and } \frac{1}{\Gamma(\sigma)} \int_1^\infty xe^{-x^2 t/2} t^\sigma \frac{dt}{t}.$$

Both of these integrals are holomorphic in s for $\mathrm{Re}(s) = \sigma > 0$, again because of the absolute convergence. We tabulate two more estimates with $\sigma > 0$.

$$\frac{1}{\Gamma(\sigma)} \int_1^\infty \sum_{n \geqq x} n^{2/3} e^{-n^2 t} t^\sigma \frac{dt}{t} = O(x^{2/3} e^{-x^2}). \tag{14.9}$$

Proof. First we put in the factor $e^t e^{-t}$ as in (14.7) above. Then it suffices to estimate the series

$$\sum_{n \geqq x} n^{2/3} e^{-(n^2-1)t} \leqq \sum_{n \geqq x} n^{2/3} e^{-(n^2-1)} = \sum_{n \geqq x} n^{2/3} e^{-n^2} e.$$

We use an integral comparison, starting with the integral

$$\int_x^\infty e^{-u^2 u^{-4/3}} du.$$

Integration by parts gets it to the integral we want, with $u^{2/3}$. This gives a lower estimate than stated in (14.9), but we have to add the first term of the Riemann sum, which brings it to $O(x^{2/3} e^{-x^2})$. The remaining factor is $\leqq 1$. This takes care of (14.9).

$$\frac{1}{\Gamma(\sigma)} \int_1^\infty xe^{-x^2 t/2} t^\sigma \frac{dt}{t} \leqq 2^\sigma xe^{-(x^2-1)/2}. \tag{14.10}$$

Proof. Immediate, as for (14.8).

14.3 Partial-Sums Asymptotics for $\zeta_{\mathbf{Q(i)}}$ and the Euler Constant

We carry out the analogous results of Section 14.1 for the cases of **Q(i)** and **Z[i]**. Let $\zeta_{\mathbf{Q(i)}}$ be the Dedekind zeta function associated with **Q(i)**.

Theorem 14.3.1. *Let $\sigma_1 > 1$. For $5/6 < \sigma \leqq \sigma_1$, $\sigma \neq 1$, we have*

$$\zeta_{\mathbf{Q(i)}}(\sigma) = \sum_{\mathbf{N}a<x^2} \frac{1}{\mathbf{N}a^\sigma} + \frac{\pi}{2}\frac{x^{2-2\sigma}}{\sigma-1} + O(x^{1-2\sigma}) \text{ as } x \to \infty.$$

The constant implicit in the O-term is independent of x and σ.

Proof. Our first main step for the proof of Theorem 14.3.1 is like the one of Theorem 14.1.1, namely for $\sigma > 1$,

$$\zeta_{\mathbf{Q(i)}}(\sigma) - \sum_{\mathbf{N}a<x^2} \frac{1}{\mathbf{N}a^\sigma} = \frac{1}{\Gamma(\sigma)} \int_0^\infty \sum_{\mathbf{N}a \geqq x^2} e^{-\mathbf{N}at} t^\sigma \frac{dt}{t}$$

$$= \frac{1}{\Gamma(\sigma)} \int_0^1 \sum_{\mathbf{N}a \geqq x^2} e^{-\mathbf{N}at} t^\sigma \frac{dt}{t} + \frac{1}{\Gamma(\sigma)} \int_1^\infty \sum_{\mathbf{N}a \geqq x^2} e^{-\mathbf{N}at} t^\sigma \frac{dt}{t}. \tag{14.11}$$

Integrals over $[1,\infty)$ like the one on the right will contribute to error terms. As noted in the last section, it is holomorphic in s with $\mathrm{Re}(s) = \sigma > 0$.

More seriously, for $\sigma > 1$ we get for the (0,1] integral,

$$\frac{1}{\Gamma(\sigma)} \int_0^1 \sum_{\mathbf{N}(a) \geq x^2} e^{-\mathbf{N}(a)t} t^\sigma \frac{dt}{t} = \frac{\pi}{2} \frac{1}{\Gamma(\sigma)} \int_0^1 \sum_{n \geq x} n e^{-n^2 t} t^\sigma \frac{dt}{t} + \mathrm{Err}(x,\sigma), \tag{14.12}$$

where the error term is

$$\mathrm{Err}(x,\sigma) = \frac{1}{\Gamma(\sigma)} \int_0^1 O\left(\sum_{n \geq x} n^{2/3} e^{-n^2 t}\right) t^\sigma \frac{dt}{t} + \frac{1}{\Gamma(\sigma)} \int_0^1 O(x e^{-x^2 t/2}) t^\sigma \frac{dt}{t}.$$

Note that

$$\frac{\pi}{2} \frac{1}{\Gamma(\sigma)} \int_0^1 \sum_{n \geq x} n e^{-n^2 t} t^\sigma \frac{dt}{t} = \frac{\pi}{2} \zeta_{\mathbf{Q}}(2\sigma - 1) - \frac{\pi}{2} \frac{1}{\Gamma(\sigma)} \int_1^\infty \sum_{n \geq x} n e^{-n^2 t} t^\sigma \frac{dt}{t}. \tag{14.13}$$

Similarly,

$$\frac{1}{\Gamma(\sigma)} \int_0^1 \sum_{n \geq x} n^{2/3} e^{-n^2 t} t^\sigma \frac{dt}{t} = \zeta_{\mathbf{Q}}(2\sigma - 2/3) - \frac{1}{\Gamma(\sigma)} \int_1^\infty \sum_{n \geq x} n^{2/3} e^{-n^2 t} t^\sigma \frac{dt}{t} \tag{14.14}$$

and

$$\frac{1}{\Gamma(\sigma)}\int_0^1 xe^{-x^2t/2}t^{\sigma}\frac{dt}{t} = \frac{2^{\sigma}}{x^{2\sigma-1}} - \frac{1}{\Gamma(\sigma)}\int_1^{\infty} xe^{-x^2t/2}t^{\sigma}\frac{dt}{t}. \tag{14.15}$$

By (14.12–14.15) we get a meromorphic identity for

$$\zeta_{\mathbf{Q(i)}}(\sigma) - \sum_{\mathbf{N}a<x^2} \mathbf{N}a^{-\sigma}$$

involving the terms on the right side of (14.12), valid in the range

$$2\sigma - 2/3 > 1 \text{ or equivalently } \sigma > 5/6.$$

We also get a *holomorphic* expression for

$$\zeta_{\mathbf{Q(i)}}(\sigma) - \frac{\pi}{2}\zeta_{\mathbf{Q}}(2\sigma-1)$$

down to $\sigma > 5/6$.

For the main integral over (0,1] in (14.11), we get

$$\frac{1}{\Gamma(\sigma)}\int_0^1 \sum_{\mathbf{N}a \geqq x^2} e^{-\mathbf{N}at}t^{\sigma}\frac{dt}{t} = \frac{\pi}{2}\cdot\frac{1}{\Gamma(\sigma)}\int_0^1 \sum_{n\geqq x} ne^{-n^2t}t^{\sigma}\frac{dt}{t} + \mathrm{Err}(x,\sigma) \tag{14.16}$$

with the error term

$$\mathrm{Err}(x,\sigma) = O\left(\frac{1}{\Gamma(\sigma)}\int_0^1 \sum_{n\geqq x} n^{2/3}e^{-n^2t}t^{\sigma}\frac{dt}{t}\right) + O\left(\frac{1}{\Gamma(\sigma)}\int_0^1 xe^{-x^2t/2}t^{\sigma}\frac{dt}{t}\right). \tag{14.17}$$

For the main integral on the right of (14.16) we have the equalities

$$\frac{\pi}{2}\cdot\frac{1}{\Gamma(\sigma)}\int_0^1 \sum_{n\geqq x} ne^{-n^2t}t^{\sigma}\frac{dt}{t}$$

$$= \frac{\pi}{2}\cdot\frac{1}{\Gamma(\sigma)}\int_0^{\infty} \sum_{n\geqq x} ne^{-n^2t}t^{\sigma}\frac{dt}{t} - \frac{\pi}{2}\cdot\frac{1}{\Gamma(\sigma)}\int_1^{\infty} \sum_{n\geqq x} ne^{-n^2t}t^{\sigma}\frac{dt}{t} \tag{14.18}$$

$$= \frac{\pi}{2}\left(\zeta_{\mathbf{Q}}(2\sigma-1) - \sum_{n<x}\frac{1}{n^{2\sigma-1}}\right) + O(xe^{-x^2}) \quad \text{[by (14.7) of Section 14.2]} \tag{14.19}$$

$$= \frac{\pi}{2}\frac{x^{1-(2\sigma-1)}}{2\sigma-2} + O\left(x^{-(2\sigma-1)}\right) + O\left(xe^{-x^2}\right) \quad \text{[by Theorem 14.1.1]} \tag{14.20}$$

$$= \frac{\pi}{2}\frac{x^{1-(2\sigma-1)}}{2\sigma-2} + O\left(x^{-(2\sigma-1)}\right). \tag{14.21}$$

We have now proved the main term in Theorem 14.3.1.

For the first integral in the error term (14.17), we can argue in the same way as (14.18)–(14.21) to get the estimate

$$\frac{1}{\Gamma(\sigma)}\int_0^1 \sum_{n\geq x} n^{2/3} e^{-n^2 t} t^{\sigma} \frac{dt}{t} = O\left(x^{1/3-2\sigma}\right). \tag{14.22}$$

For the second integral of the error term (14.17), we use the identity

$$\frac{1}{\Gamma(\sigma)}\int_0^1 x e^{-x^2 t/2} t^{\sigma} \frac{dt}{t} = x^{1-2\sigma} 2^{\sigma} - \frac{1}{\Gamma(\sigma)}\int_1^{\infty} x e^{-x^2 t/2} t^{\sigma} \frac{dt}{t}.$$

Using the bound (14.14) of Section 14.2, we get

$$\frac{1}{\Gamma(\sigma)}\int_0^1 x e^{-x^2 t/2} t^{\sigma} \frac{dt}{t} = O\left(x^{1-2\sigma}\right).$$

Thus (14.16) above becomes

$$\frac{1}{\Gamma(\sigma)}\int_0^1 \sum_{\mathbf{N}a\geq x^2} e^{-\mathbf{N}at} t^{\sigma} \frac{dt}{t}$$

$$= \frac{\pi}{2}\frac{1}{\Gamma(\sigma)}\int_0^1 \sum_{n\geq x} n e^{-n^2 t} t^{\sigma} \frac{dt}{t} + O\left(x^{1/3-2\sigma}\right) + O\left(x^{1-2\sigma}\right). \tag{14.23}$$

Putting together (14.12), (14.16), (14.21), and the above error term estimate (14.23) concludes the proof of Theorem 14.3.1.

The asymptotics coming after (14.16) may also be summarized in the following form:

Theorem 14.3.2. *For $5/6 < \sigma < \sigma_1$ and $x \to \infty$,*

$$\zeta_{\mathbf{Q}(\mathbf{i})}(\sigma) = \sum_{\mathbf{N}(a)<x^2} \frac{1}{\mathbf{N}(a)^{\sigma}}$$

$$+ \frac{1}{\Gamma(\sigma)}\int_1^{\infty} \sum_{\mathbf{N}(a)\geq x^2} e^{-\mathbf{N}(a)t} t^{\sigma} \frac{dt}{t} + \frac{\pi}{2}\zeta_{\mathbf{Q}}(2\sigma-1) - \frac{\pi}{2}\frac{1}{\Gamma(\sigma)}\int_1^{\infty} \sum_{n\geq x} n e^{-n^2 t} t^{\sigma} \frac{dt}{t}$$

$$+ O\left(\zeta_{\mathbf{Q}}(2\sigma - 2/3)\right) - \frac{1}{\Gamma(\sigma)}\int_1^{\infty} O\left(\sum_{n\geq x} n^{2/3} e^{-n^2 t}\right) t^{\sigma} \frac{dt}{t}$$

$$+ O\left(\frac{2^{\sigma}}{x^{2\sigma-1}}\right) - \frac{1}{\Gamma(\sigma)}\int_1^{\infty} O\left(x e^{-x^2 t/2}\right) t^{\sigma} \frac{dt}{t}.$$

Remarks. In the case of the upper half-plane, the Kronecker limit formula computes the constant term in the Laurent expansion at $s = 1$ for the Eisenstein series. The possibility now opens up to do the the same type of expansion as Theorem 14.1.1 and 14.3.1 for the Eisenstein series, for $\mathrm{SL}_2(\mathbf{R})$, $\mathrm{SL}_2(\mathbf{C})$ and up the geometric SL_n ladder, and beyond (generalizing the Lie group and the discrete subgroup), as well as for zetas associated with thetas or generalized thetas such as those that arise in this book.

Define the **Euler constant** $\gamma_{\mathbf{Q(i)}}$ to be the constant term in the expansion of $\zeta_{\mathbf{Q(i)}}(s)$ at $s = 1$, that is,

$$\zeta_{\mathbf{Q(i)}}(s) = \frac{\pi/4}{s-1} + \gamma_{\mathbf{Q(i)}} + O(s-1) \text{ as } s \to 1.$$

Theorem 14.3.3.

$$\sum_{\mathrm{N}a<x^2} \frac{1}{\mathrm{N}a} = \frac{\pi}{4} \log x^2 + \gamma_{\mathbf{Q(i)}} + O(1/x) \textit{ as } x \to \infty.$$

Proof. Similar to the proof of Theorem 14.1.2. This time we substitute

$$\frac{\pi}{2} \frac{x^{1-(2\sigma-1)}}{2\sigma-2} = \frac{\pi}{4} \cdot \frac{e^{-(\sigma-1)\log x^2}}{\sigma-1} = \frac{\pi}{4} \cdot \frac{1}{\sigma-1} - \frac{\pi}{4} \log x^2 + O_x(\sigma-1) \text{ as } \sigma \to 1,$$

in Theorem 14.3.1, cancel $(\pi/4)/(\sigma-1)$, and let $\sigma \to 1$ to conclude the proof.

In the applications, we deal with a context where it is not so much the ideals that matter as lattice points. Thus we reformulate the theorem in terms of lattice points.

Theorem 14.3.4. *Let $m = m_1 + \mathbf{i}m_2$ with $m_1, m_2 \in \mathbf{Z}$ denote the elements of the lattice $\mathbf{Z[i]}$ in $\mathbf{C}$. Then*

$$\sum_{0<|m|<R} \frac{1}{|m|^2} = 2\pi \log R + 4\gamma_{\mathbf{Q(i)}} + O(1/R) \textit{ as } R \to \infty.$$

Proof. Simply observe that

$$\sum_{0<|m|<R} \frac{1}{|m|^2} = 4 \sum_{\mathrm{N}a<x^2} \frac{1}{\mathrm{N}a}$$

upon taking $R = x$.

Remarks. The convention about $\gamma_{\mathbf{F}}$ for a number field $\mathbf{F}$, here equal to $\mathbf{Q(i)}$, doesn't seem to be universal. For instance, [Szm 83] uses the extra factor 4. Also, he says he is taking the partial sums over all integral ideals, but actually he's taking the sum over the lattice points (4 times as many), p. 404.

We don't know whether the Euler constant has been treated in the literature other than for imaginary quadratic fields; even more so for the Hurwitz constant obtained by making a translation, which we treat in the next section for **Z[i]**. Such a treatment would be a nice new chapter in analytic number theory.

14.4 The Hurwitz Constant

We consider the Hurwitz constant over **Z** and over **Z[i]** successively. For $0 < \alpha < 1$, we define the **Hurwitz constant** $\gamma^{+}_{\mathrm{Hur},\mathbf{Z}}(\alpha)$ to be the constant term in the asymptotic relation

$$\sum_{1 \leqq n \leqq N} \frac{1}{n+\alpha} = \log N + \gamma^{+}_{\mathrm{Hur},\mathbf{Z}}(\alpha) + o(1) \text{ for } N \to \infty.$$

If $\alpha = 0$, then $\gamma^{+}_{\mathrm{Hur},\mathbf{Z}}(0)$ is the Euler constant $\gamma_{\mathbf{Z}}$. If $\alpha = 1$, then

$$\gamma^{+}_{\mathrm{Hur},\mathbf{Z}}(1) = \gamma_{\mathbf{Z}} - 1.$$

A proof of the existence of such a constant will be given in the slightly more complicated case of **Z[i]** below (Proposition 14.4.2).

Proposition 14.4.1.

$$\int_0^1 \gamma^{+}_{\mathrm{Hur},\mathbf{Z}}(\alpha) d\alpha = 0.$$

Proof. We write

$$\begin{aligned}
\int_0^1 \sum_{1 \leqq n \leqq N} \frac{1}{n+\alpha} d\alpha &= \int_0^1 \sum_{1 \leqq n \leqq N} \int_0^\infty e^{-(n+\alpha)t}\, dt\, d\alpha \\
&= \int_0^\infty \sum_{1 \leqq n \leqq N} \int_0^1 e^{-(n+\alpha)t}\, d\alpha\, dt \\
&= \int_0^\infty \int_1^{N+1} e^{-\alpha t}\, d\alpha\, dt \\
&= \int_0^\infty \frac{e^{-t} - e^{-(N+1)t}}{t} dt.
\end{aligned}$$

However, for $x > 0$,

$$\log x = \int_0^\infty \frac{e^{-t} - e^{-xt}}{t} dt. \tag{14.24}$$

This is verified by showing that the derivatives are equal, and the values at $x = 1$ agree. Thus

$$\int_0^1 \sum_{1 \leqq n \leqq N} \frac{1}{n+\alpha} d\alpha = \log(N+1).$$

This gives

$$\log N + \int_0^1 \gamma^+_{\mathrm{Hur},\mathbf{Z}}(\alpha) d\alpha + o(1) = \log(N+1),$$

or

$$\int_0^1 \gamma^+_{\mathrm{Hur},\mathbf{Z}}(\alpha) d\alpha = \log \frac{N+1}{N} + o(1),$$

which proves the proposition.

14.4.1 The Complex Case, with Z[i]

Let $x \in \mathbf{C}$, $x \notin o = \mathbf{Z}[\mathbf{i}]$, except possibly $x = 0$. We let m denote elements of $\mathbf{Z}[\mathbf{i}]$. We start with the alternative definition of the **Hurwitz constant** $\gamma_{\mathrm{Hur},\mathbf{Z}[\mathbf{i}]}(x)$ via the asymptotic relation for $R \to \infty$,

$$\sum_{1 \leqq |m| < R} \frac{1}{|m+x|^2} = 2\pi \log R + \gamma_{\mathrm{Hur},\mathbf{Z}[\mathbf{i}]}(x) + o(1). \tag{14.25}$$

The next proposition proves the existence of such a constant, and relates it to the Euler constant.

Proposition 14.4.2. *Let $x \in \mathbf{C}$, $x \notin \mathbf{Z}[\mathbf{i}]$ except possibly $x = 0$. There is a constant $\gamma_{Hur,\mathbf{Z}[\mathbf{i}]}(x)$ such that the asymptotic relation* (14.25) *is satisfied. This constant is given by the series*

$$\gamma_{Hur,\mathbf{Z}[\mathbf{i}]}(x) = 4\gamma_{\mathbf{Q}(\mathbf{i})} + \sum_{|m| \neq 0} \left(\frac{1}{|m+x|^2} - \frac{1}{|m|^2} \right).$$

Proof. The above series converges absolutely, uniformly for x in a compact set not containing an element of $\mathbf{Z}[\mathbf{i}]$ except possibly 0. In fact, we have trivially for x in the compact set,

$$\left| \frac{1}{|m+x|^2} - \frac{1}{|m|^2} \right| = O\left(\frac{1}{|m|^3} \right) \text{ for } |m| \to \infty.$$

By an integral test or lattice-point counting, we have

$$\sum_{|m|\geqq R}\left(\frac{1}{|m+x|^2}-\frac{1}{|m|^2}\right)=O(1/R) \text{ for } R\to\infty.$$

Then

$$\begin{aligned}\sum_{|m|\neq 0}\left(\frac{1}{|m+x|^2}-\frac{1}{|m|^2}\right) &= \sum_{0<|m|<R}\left(\frac{1}{|m+x|^2}-\frac{1}{|m|^2}\right)+O(1/R)\\ &= \sum_{0<|m|<R}\frac{1}{|m+x|^2}-\sum_{0<|m|<R}\frac{1}{|m|^2}+O(1/R)\\ \text{[by Theorem 14.3.4]} &= \sum_{0<|m|<R}\frac{1}{|m+x|^2}-\left(2\pi\log R+4\gamma_{\mathbf{Q(i)}}\right)+O(1/R),\end{aligned}$$

which proves the proposition.

Proposition 14.4.2 suffices for our purposes. For a complete tabulation, one would have still to show that the Hurwitz constant with the parameter x is the constant term of the Hurwitz zeta function at $s=1$.

14.4.2 Average of the Hurwitz Constant

We represent $\mathbf{C}/\mathbf{Z}[\mathbf{i}]$ by the fundamental square Sq centered at the origin, of sides 1. If $\|x\|_\infty$ denotes the sup norm of x (max of $|x_1|$, $|x_2|$), then Sq is defined by

$$\mathrm{Sq}=\{x\in\mathbf{C},\ \|x\|_\infty\leqq 1/2\}.$$

Since $x\in\mathrm{Sq}$ and $m\neq 0$, we have $m+x\neq 0$. We now determine the average of the Hurwitz constant. We consider the complement of the disk of radius 1/2 in the square, so let

$$S=\mathrm{Sq}-D_{1/2}(0).$$

We let

$$I(S)=\int_0^\infty\int_S e^{-|x|^2 t}dx\,dt.$$

Proposition 14.4.3.

$$\int_{\mathrm{Sq}}\gamma_{\mathrm{Hur},\mathbf{Z}[\mathbf{i}]}(x)dx=2\pi\log 2-I(S).$$

Proof. The area of the square is 1. We integrate (14.25) over the square, and get

$$\begin{aligned}
2\pi \log N + \int_{\mathrm{Sq}} \gamma_{\mathrm{Hur},\mathbf{Z}[\mathbf{i}]}(x)dx + o(1) &= \int_{\mathrm{Sq}} \sum_{0<|m|\leqq N} \frac{1}{|m+x|^2} dx \\
&= \int_{\mathrm{Sq}} \sum_{0<|m|\leqq N} \int_0^\infty e^{-|m+x|^2 t} dt\, dx \\
&= \int_0^\infty \left[\sum_{0<|m|\leqq N} \int_{\mathrm{Sq}} e^{-|m+x|^2 t} dx \right] dt \qquad (14.26) \\
&= I_N, \text{ say.}
\end{aligned}$$

Let

$$I'_N = \int_0^\infty \int_{1/2<|x|\leqq N} e^{-|x|^2 t} dx\, dt.$$

The images of the square Sq by translation with $|m| \leqq N$ lie in the disk of radius $N+1$ and contain the disk of radius $N-1$. Thus we obtain the inequalities

$$I_{N-1} - I(S) \leqq I_N \leqq I'_{N+1} - I(S). \qquad (14.27)$$

We shall now give asymptotics for I_N. Note that the above inequalities take into account the difference between regions defined by sup norm inequalities and regions defined by Euclidean norm inequalities. The latter are fitted to use polar coordinates, whereas the former have the usual advantages of Riemann sums.

Lemma 14.4.4. $I'_N = 2\pi \log 2N$.

Proof. Changing to polar coordinates, we get

$$I'_N = \int_0^\infty \int_0^{2\pi} \int_{1/2}^N e^{-r^2 t} r\, dr\, d\theta\, dt.$$

The integral is exact and gives the stated answer using (14.24).

From the lemma and (14.27) we get

$$I_N = I'_N - I(S) + o(1) \quad \text{for } N \to \infty. \qquad (14.28)$$

We now put (14.25), (14.26), (14.28) together, and use Lemma 14.4.4. Then $2\pi \log N$ cancels because it occurs on both sides of the equality sign in (14.26). The rest gives precisely the expression of Proposition 14.4.2, thus concluding the proof of Proposition 14.4.3.

The integral $I(S)$ can be given in somewhat different form, which will not be needed because it canceled in the application. We give this form anyhow, just in case:

Lemma 14.4.5. $I(S) = 8\int_0^{\pi/4} \log(1/\cos\theta)\, d\theta.$

Proof. We interchange the order of integration in the definition of $I(S)$, getting

$$I(S) = \int_S \frac{1}{|x|^2} dx.$$

The set S consists of four pieces between the circle of radius 1/2 and the square, in the four quadrants. Each piece has symmetry across the diagonal. We let S_1 be the piece in the first quadrant lying below the diagonal and between the circle and the vertical side (see the Figure). In polar coordinates,

$$S_1 = \left\{ r,\theta \,\middle|\, 0 \leqq \theta \leqq \pi/4 \text{ and } 1/2 \leqq r \leqq \frac{1/2}{\cos\theta} \right\}.$$

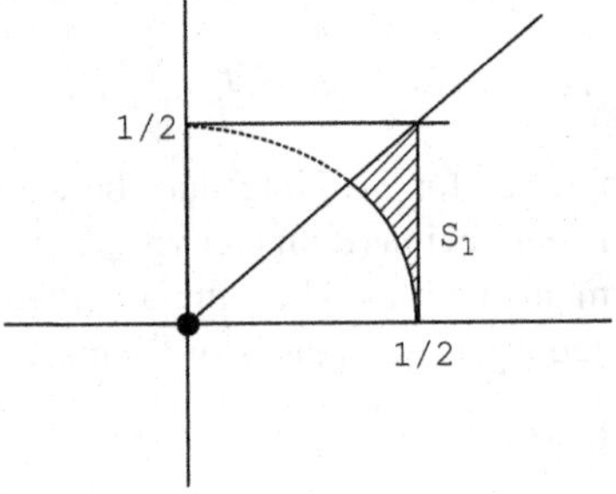

Then

$$I(S) = 8\int_{S_1} \frac{1}{|x|^2} dx = 8\int_0^{\pi/4} \int_{1/2}^{1/2\cos\theta} \frac{1}{r^2} r\, dr\, d\theta,$$

which gives the value stated in the lemma.

Remarks. Recall that in our evaluation of $\Theta^{\mathrm{Cus}}(1/t)$ we had the constant $\gamma^{\mathrm{Av},2}_{\mathrm{Hur},\mathbf{Z}(\mathbf{i})}$, which was defined in Section 13.4 as

$$\gamma^{\mathrm{Av},2}_{\mathrm{Hur},\mathbf{Z}(\mathbf{i})} = \int_{\mathscr{F}U} \gamma_{\mathrm{Hur},\mathbf{Z}(\mathbf{i})}(2x,x)dx, \tag{14.29}$$

where

$$\sum_{m \notin o_2^+, |m|<Y} \frac{1}{|2x-m|^2} = 2\pi \log Y + \gamma_{\mathrm{Hur},\mathbf{Z}(\mathbf{i})}(2x,x) + o(1) \text{ as } Y \to \infty.$$

As an exercise, one could use the computations and method of proof from this section in order to express the constant $\gamma^{\mathrm{Av},2}_{\mathrm{Hur},\mathbf{Z}(\mathbf{i})}$ in terms of the Hurwitz constant $\gamma_{\mathbf{Q}(\mathbf{i})}$ and various integrals, such as $I(S)$ from Lemma 14.4.5. However, the expansion will be quite lengthy, not nearly as precise as Proposition 14.4.3. Indeed, by unfolding the integral in (14.29), one sees that some aspects of the complex plane parametrized by x are covered twice, some once, and some not at all. Also, these regions are not symmetric about the x_1-axis since the omitted set in the summation, namely o_2^+ itself, is not symmetric about the x_1-axis. Finally, the Hurwitz constant $\gamma_{\mathbf{Q}(\mathbf{i})}$ will appear along with the nonzero terms in the zeta function sum that are omitted by excluding o_2^+. All of these considerations make for a lengthy computation, which, although it follows the pattern set in this section, will result in an evaluation of $\gamma^{\mathrm{Av},2}_{\mathrm{Hur},\mathbf{Z}(\mathbf{i})}$ that does not add to the precision of the resulting theta inversion formula.

14.5 $\int_0^\infty f_\varphi(r)rh(r)dr$ when $\varphi = \mathbf{g}_t$

We have met three integrals occurring as constant factors in asymptotic relations. We determine explicitly the first one in a brief section here.

First, we have an explicit value for the function f_φ when $\varphi = \mathbf{g}_t$. Indeed, as in Lemma 14.2.1 of Chapter 13, using the trace of $u(r)u(r)^*$, writing $\mathbf{f}_{\mathbf{g}_t} = \mathbf{f}_t$, $u(r) = k_1bk_2$, and using $y = y_b$, we have

$$y + y^{-1} = 2 + r^2 \tag{14.30}$$

$$\mathbf{f}_t(r) = \frac{e^{-2t}}{(8\pi t)^{3/2}} e^{-(\log y)^2/8t} \frac{\log y}{\sinh(\log y)}. \tag{14.31}$$

Differentiating (14.30) gives

$$\frac{dy}{dr} - y^{-2}\frac{dy}{dr} = 2r,$$

whence

$$\frac{dy}{dr} = \frac{2ry}{y - y^{-1}}.$$

Thus

$$rdr = \sinh(\log y)\frac{dy}{y}. \tag{14.32}$$

The term $\sinh(\log\ y)$ cancels in $rf(r)dr$, and we obtain

$$\mathbf{f}_t(r)rdr = \frac{e^{-2t}}{(8\pi t)^{3/2}} e^{-(\log y)^2/8t} (\log y)\frac{dy}{y}. \tag{14.33}$$

We shall need the integral of the next lemma when the function $h(r)$ is one of the functions

$$h(r)=1,\ h(r)=\log r,\ h(r)=\log(r^2+4)\quad \text{[Section 13.6]}.$$

The argument is formal and needs only absolute convergence.

Proposition 14.5.1.

$$\int_0^\infty \mathbf{f}_t(r)h(r)r\,dr=\frac{4e^{-2t}}{(8\pi t)^{3/2}}\int_0^\infty xe^{-x^2/2t}h(2\sinh x)dx.$$

Proof. We have

$$\int_0^\infty \mathbf{f}_t(r)h(r)r\,dr=\frac{e^{-2t}}{(8\pi t)^{3/2}}\int_1^\infty(\log y)e^{-(\log y)^2/8t}h(2\sinh(\log y^{1/2}))\frac{dy}{y}.$$

We set $x=\log y^{1/2}$, so $2dx=dy/y$, to get the stated result.

Corollary 14.5.2.

$$\int_0^\infty \mathbf{f}_t(r)r\,dr=\frac{e^{-2t}}{2\pi(8\pi t)^{1/2}}.$$

Proof. The result follows from taking $h=1$ and the trivial computation

$$\int_0^\infty xe^{-x^2/2t}dx=t.$$

Corollary 14.5.3.

$$\int_0^\infty \mathbf{f}_t(r)(\log r)r\,dr=\frac{e^{-2t}\log(2)}{(8\pi t)^{1/2}}+\frac{e^{-2t}}{2\pi(8\pi t)^{1/2}}\int_0^\infty e^{-x^2/2t}\coth(x)dx.$$

Proof. If $h(r)=\log\ r$, then $h(2\sinh x)=\log 2+\log(\sinh x)$, so using Corollary 14.5.2,

$$\int_0^\infty \mathbf{f}_t(r)(\log r)r\,dr=\frac{e^{-2t}\log(2)}{(8\pi t)^{1/2}}+\frac{4e^{-2t}}{(8\pi t)^{3/2}}\int_0^\infty xe^{-x^2/2t}\log(\sinh x)dx.$$

Now, we integrate by parts to get

$$\int_0^\infty xe^{-x^2/2t}\log(\sinh x)dx=t\int_0^\infty e^{-x^2/2t}\coth(x)dx,$$

from which the result follows.

Corollary 14.5.4.

$$\int_0^\infty \mathbf{f}_t(r)\log(r^2+4)r\,dr = \frac{2e^{-2t}\log(2)}{(8\pi t)^{1/2}} + \frac{2e^{-2t}}{2\pi(8\pi t)^{1/2}}\int_0^\infty e^{-x^2/2t}\tanh(x)dx.$$

Proof. If $h(r) = \log(r^2+4)$, then $h(2\sinh(x)) = 2\log 2 + 2\log(\cosh(x))$, so using Corollary 14.5.2 we have

$$\int_0^\infty \mathbf{f}_t(r)\log(r^2+4)r\,dr = \frac{2e^{-2t}\log(2)}{(8\pi t)^{1/2}} + \frac{8e^{-2t}}{(8\pi t)^{3/2}}\int_0^\infty xe^{-x^2/2t}\log(\cosh x)dx.$$

Now we integrate by parts to get

$$\int_0^\infty xe^{-x^2/2t}\log(\cosh x)dx = t\int_0^\infty e^{-x^2/2t}\tanh(x)dx,$$

from which the result follows.

Remarks. At this point, we choose to end our computations of the integrals in Corollary 14.5.3 and Corollary 14.5.4 in the present form. Indeed, looking forward in our work, the computation of the Gauss transform of the integrals in Corollary 14.5.3 and Corollary 14.5.4 are easily derived, which is what we look toward in the development of our theta inversion formula given in the present work. Note that although the integrals in Corollary 14.5.3 and 14.5.4 are not exact, their Gauss transforms will be exact integrals.

14.6 Evaluation of C'_{Y_0} and C_1

In this section, we combine computations from the previous sections in order to evaluate the constant

$$C'_{Y_0} = \frac{1}{4}\sum_{m\in O_2^+}\int_{\mathrm{Sec}'_{m,Y_0}} \frac{1}{|x-m/2|^2}dx,$$

which appeared in Chapter 13, Theorem 6.6. We recall that O_2^+ consists of six elements: 0, $\mathbf{i}$, -1, $-1+\mathbf{i}$, 1, $1+\mathbf{i}$, that is, the corners of $2\mathscr{F}_U$ and the central points 0, $\mathbf{i}$ of the top and bottom boundaries. We deal with the three pairs successively. We recall the values of $\delta_{m,Y}$ from Section 13.5:

$$\delta_{\mathrm{m,Y}} = \begin{cases} \frac{1}{2}Y & \text{if } m = 0 \\ \frac{1}{8}Y & \text{if } |m| = 1 \\ \frac{1}{4}Y & \text{if } |m| = \sqrt{2}. \end{cases}$$

For the first pair, we have

$$\mathrm{Sec}'_{0,Y_0} = A_0 \cup S_0,$$

where

$A_0 =$ the part in $\mathscr{F}_U$ of the annulus of x with $\delta_{0,Y_0} \leqq |x| \leqq 1/2$,
$S_0 =$ the part of $\mathscr{F}_U$ outside the circle of radius $1/2$ centered at 0.

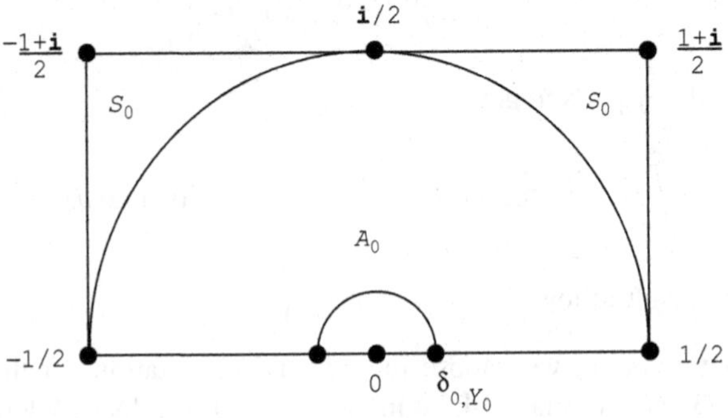

The rectangle in the figure is the fundamental rectangle $\mathscr{F}_U$.

Lemma 14.6.1.

$$\frac{1}{4}\int_{\mathrm{Sec}'_{0,Y_0}} \frac{1}{|x|^2}dx = \frac{\pi}{4}\log Y_0 + \int_0^{\pi/4} \log(1/\cos(\theta)d\theta) \tag{14.34}$$

and

$$\frac{1}{4}\int_{\mathrm{Sec}'_{\mathbf{i}Y_0}} \frac{1}{|x-\mathbf{i}/2|^2}dx = \frac{\pi}{4}\log(4Y_0) + \int_0^{\pi/4} \log(1/\cos(\theta))d\theta. \tag{14.35}$$

Proof. For (14.30) we write

$$\int_{\mathrm{Sec}'_{0,Y_0}} = \int_{A_0} + \int_{S_0}.$$

Then

$$\int_{A_0} \frac{dx}{|x|^2} = \int_0^{\pi}\int_{\delta_{0,Y_0}}^{1/2} \frac{r\,dr\,d\theta}{r^2} = \pi \log Y_0$$

by freshman calculus and the definition of δ_{0,Y_0}.

For the S_0 integral, note that when $x_1 = 1/2$ along the right boundary of $\mathscr{F}_U$, then $1/2 = r\cos\theta$, that is, $r = 1\sqrt{2\cos\theta}$. Thus

$$\int_{S_0} \frac{dx}{|x|^2} = 4 \int_0^{\pi/4} \int_{1/2}^{1/2\cos\theta} \frac{r\,dr\,d\theta}{r^2} = 4 \int_0^{\pi/2} \log(1/\cos\theta)d\theta.$$

This proves (14.30).

For (14.31), we use the decomposition

$$\mathrm{Sec}'_{\mathbf{i},Y_0} = A_{\mathbf{i}} \cup S_{\mathbf{i}},$$

where $A_{\mathbf{i}}$, $S_{\mathbf{i}}$ are as shown in the figure.

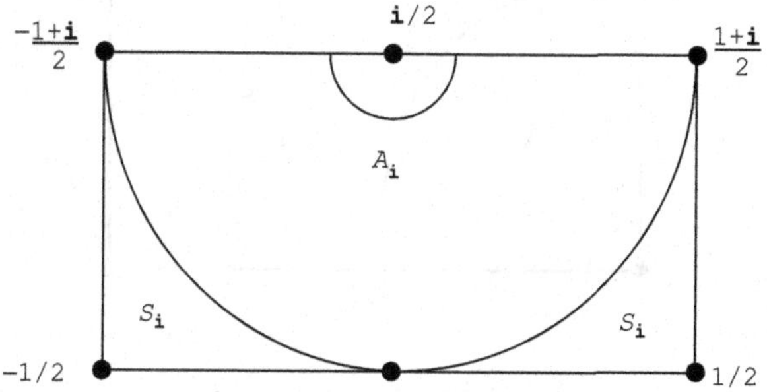

The radius of the small circle centered at $\mathbf{i}/2$ is $\delta_{\mathbf{i},Y_0}$. We write

$$\int_{\mathrm{Sec}'_{\mathbf{i},Y_0}} \frac{dx}{|x-\mathbf{i}/2|^2} = \int_{A_{\mathbf{i}}} + \int_{S_{\mathbf{i}}}.$$

Then

$$\int_{A_{\mathbf{i}}} \frac{dx}{|x-\mathbf{i}/2|^2} = \int_{\pi}^{2\pi} \int_{\delta_{\mathbf{i}},Y_0}^{1/2} \frac{r\,dr\,d\theta}{r^2}$$
$$= \pi \log(1/2\delta_{\mathbf{i},Y_0}) = \pi \log(4Y_0).$$

For the $S_{\mathbf{i}}$ integral, since $x_1 = -1/2$ for $\pi \leqq \theta \leqq 5\pi/4$, we get

$$\int_{S_{\mathbf{i}}} \frac{dx}{|x-\mathbf{i}/2|^2} = 4 \int_{\pi}^{5\pi/4} \int_{1/2}^{-1/2\cos\theta} \frac{r\,dr\,d\theta}{r^2} = 4 \int_0^{\pi/4} \log(1/\cos\theta)d\theta.$$

This proves (14.35) and concludes the proof of Lemma 14.6.1.

The next two lemmas will deal with the other two pairs $(-1,-1+\mathbf{i})$ and $(1,1+\mathbf{i})$. For these, it is convenient to use a slightly different notation when decomposing Sec'_{m,Y_0}, as follows. We do explicitly one case, and get the others as corollaries by a change of variables.

For δ small > 0, we let

$$\mathrm{Sec}'_\delta = \{x \in \mathscr{F}_U \text{ such that } |x+1/2| \geqq \delta\}.$$

We then have

$$\mathrm{Sec}'_\delta = A_\delta \cup S$$

where A_δ, S are as shown in the figure.

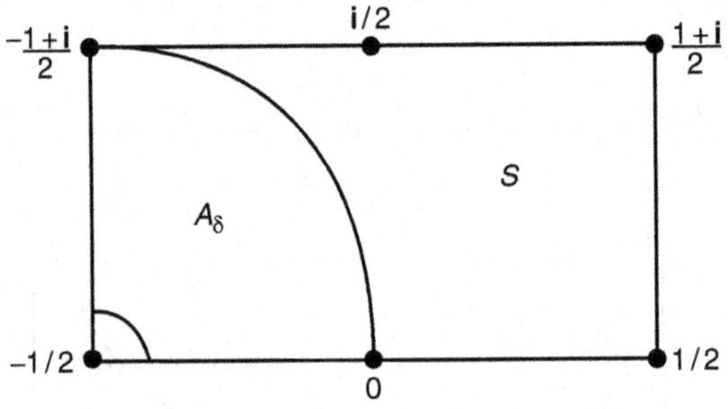

The small quarter-circle has radius δ, and A_δ is the quarter-annulus as shown. The set S is the part of $\mathscr{F}_U$ outside the circle of radius 1/2, centered at $-1/2$. Note that S does not depend on δ, but A_δ does, and $S = S_{-1}$ in the previous notation.

Lemma 14.6.2. *Let $0 < \theta_0 < \pi/2$ be such that $\tan\theta_0 = 1/2$ (θ_0 is the angle of the diagonal). Then*

$$\int_{\mathrm{Sec}'_\delta} \frac{dx}{|x+1/2|^2} = \frac{\pi}{2}\log(1/2\delta) + (\log 2)\theta_0$$

$$+ \int_0^{\theta_0} \log(1/\cos\theta)d\theta + \int_{\theta_0}^{\pi/2} \log(1/\sin\theta)d\theta.$$

Proof. As before, we have

$$\int_{\mathrm{Sec}'_\delta} = \int_{A_\delta} + \int_S .$$

For the integral over A_δ, we use polar coordinates, $x_1 + 1/2 = r\cos\theta$, $x_2 = r\sin\theta$. Then

$$\int_{A_\delta} \frac{dx}{|x+1/2|^2} = \int_0^{\pi/2}\int_\delta^{1/2} \frac{r\,dr\,d\theta}{r^2} = \frac{\pi}{2}\log(1/2\delta).$$

For the S-integral, with $0 \leqq \theta \leqq \theta_0$, the r-variable extends from 1/2 until $x_1 = 1/2$. Since $x_1 + 1/2 = r\cos\,\theta$, this means until $r = 1/\cos\,\theta$. So

$$\int_0^{\theta_0}\int_{1/2}^{1/\cos\theta} \frac{r dr d\theta}{r^2} = \int_0^{\theta_0} (\log 1/\cos\theta - \log\,1/2) d\theta$$

$$= \int_0^{\theta_0} \log(1/\cos\theta) d\theta + (\log 2)\theta_0.$$

For $\theta_0 \leqq \theta \leqq \pi/2$, the r-range extends to $x_2 = 1/2$. Since $x_2 = r\sin\,\theta$, r extends to $r = 1\sqrt{2}\sin\,\theta$. Then

$$\int_{\theta_0}^{\pi/2}\int_{1/2}^{1/2\sin\theta} \frac{r dr d\theta}{r^2} = \int_{\theta_0}^{\pi/2} \log(1/\sin\theta) d\theta.$$

Combining the two integrals proves Lemma 14.6.2.

Finally, the cases of the four corners will be proved in a corollary of the general Lemma 14.6.2. Let

$$C'' = (\log 2)\theta_0 + \int_0^{\theta_0} \log(1/\cos\theta) d\theta + \int_{\theta_0}^{\pi/2} \log(1/\sin\theta).$$

Corollary 14.6.3.

$$\int_{\mathrm{Sec}'_{-1,Y_0}} \frac{dx}{|x+1/2|^2} = \int_{\mathrm{Sec}'_{1,Y_0}} \frac{dx}{|x-1/2|^2} = \frac{\pi}{2}\log(4Y_0) + C'' \tag{14.36}$$

$$\int_{\mathrm{Sec}'_{1+\mathbf{i},Y_0}} \frac{dx}{|x+(1-\mathbf{i})/2|^2} = \int_{\mathrm{Sec}'_{-1+\mathbf{i},Y_0}} \frac{dx}{|x-1/2|^2} = \frac{\pi}{2}\log(2Y_0) + C''. \tag{14.37}$$

Proof. The first integral in (14.40) comes from Lemma 14.6.2 with $\delta = \delta_{-1,Y_0}$. The second integral in (14.40) follows from Lemma 6.3 under the change of variables $x_1 \mapsto -x_1$. The first integral in (14.41) follows from Lemma 14.6.2 with the change of variables $x_2 \mapsto 1/2 - x_2$ and $x_1 \mapsto -x_1$; and $\delta = \delta_{1+\mathbf{i},Y_0}$. The second integral in (14.41) follows from Lemma 14.6.2 with $\delta = \delta_{-1+\mathbf{i},Y_0}$, with the change of variables $x_2 \mapsto 1/2 - x_2$.

Putting the lemmas together, we get the value for C'_{Y_0} and C_1.

Proposition 14.6.4.

$$C'_{Y_0} = \pi \log Y_0 + \frac{5\pi}{4}\log 2 + (\log 2)\theta_0$$
$$+ \frac{1}{2}\int_0^{\pi/4} \log(1/\cos\theta)d\theta + \int_0^{\theta_0} \log(1/\cos\theta)d\theta + \int_{\theta_0}^{\pi/2} \log(1/\sin\theta)d\theta.$$

We write

$$C'_{Y_0} = \pi \log Y_0 + C',$$

where C' consists of all the above terms except the first term $\pi \log Y_0$. In light of the definition of the constant C_1 in Chapter 13, Corollary 13.6.8, we find the value for C_1, which is independent of Y_0.

Corollary 14.6.5. *Let* $C' = C'_{Y_0} - \pi \log Y_0$. *Then*

$$C_1 = 4\gamma_{\mathbf{Q(i)}} + \gamma^{\mathrm{Av},2}_{\mathrm{Hur},\mathbf{Z[i]}} - 2\pi \log(2) + C'.$$

Proof. The term $\pi \log Y_0$ cancels the $-\pi \log Y_0$ coming from the above-mentioned Corollary 13.6.8 of Chapter 13.

The constant $\gamma^{\mathrm{Av},2}_{\mathrm{Hur},\mathbf{Z[i]}}$ (average Hurwitz constant) is discussed in Section 14.4.

14.7 The Theta Inversion Formula

The theta inversion formula was stated in Chapter 12, Theorem 12.1.3:

$$e^{-2t}(4t)^{-1/2}\Theta^{\mathrm{NC}}(1/t) + \Theta^{\mathrm{Cus}}(1/t) = \theta_{\mathrm{Cus}}(t) + 1 + \theta_{\mathrm{Eis}}(t). \tag{14.38}$$

Now each term can be given explicitly. We know the series for $\theta_{\mathrm{Cus}}(t)$ from Section 12.1, (14.39). It is the theta series formed with the eigenvalues of Casimir on an orthonormal basis of $L^2_{\mathrm{cus}}(\Gamma\backslash G)$ consisting of eigenfunctions. The series for $\theta_{\mathrm{Eis}}(t)$ was given in Section 12.3, (7) and (8). The terms were computed explicitly in Sections 12.4 and 12.5, and we get the right side:

$$\theta_{\mathrm{Cus}}(t) + 1 + \theta_{\mathrm{Eis}}(t) = \sum_{n=1}^{\infty} e^{-\lambda_n t} + 1 - \frac{1}{2}e^{-2t} - \frac{1}{2\pi}\int_{-\infty}^{\infty} e^{-2(r^2+1)t\phi'/\phi(1+\mathbf{i}r)dr}, \tag{14.39}$$

where

$$\phi(s) = \Lambda\zeta_{\mathbf{Q(i)}}(2-s)/\Lambda\zeta_{\mathbf{Q(i)}}(s) \text{ and } \Lambda\zeta_{\mathbf{Q(i)}}(s) = \pi^{-s}\Gamma_{(s)}\zeta_{\mathbf{Q}(i)}(s).$$

As for the left side, the *noncuspidal term* is given in Chapter 5, Theorem 5.3.3, namely,

$$\Theta^{\mathrm{NC}}(1/t) = \sum_{c \in \mathrm{CC}_{\Gamma}(\Gamma')} a_c e^{-|\log b_c|^2/4t}, \tag{14.40}$$

where

$$a_c = \frac{1}{|\mathrm{Tor}(c)|} \log|\eta_{c,0}|^2 \frac{(2\pi)^{-3/2}}{|\eta_c - \eta_c^1|^2}$$

and

$|\mathrm{Tor}(c)|$ = the order of the torsion group of $\Gamma_{\gamma c}$;
η_c = eigenvalue of elements in c;
$\eta_{c,0}$ = eigenvalue of absolute value > 1 for a primitive element of $\Gamma_{\gamma c}$;
b_c^2 = polar component of $\gamma_c \gamma_c^*$, where γ_c is an element of c.

The *cuspidal term* $\Theta^{\mathrm{Cus}}(1/t)$ is partly given in Chapter 13, Corollary 13.6.8, some subterms being made explicit in the current Chapter 14. The bottom line is that

$$\Theta\mathrm{Cus}_{(1/t)} = \frac{4}{\pi^2} \zeta_{\mathbf{Q}(\mathbf{i})}(2) \frac{e^{-2t}}{(8\pi t)^{3/2}} + \int_0^{\infty} F_{\mathrm{cus}}(r) \mathbf{f}_t(r) dr, \tag{14.41}$$

where

$$F_{\mathrm{cus}}(r) = C_1 r + 3\pi r \log r + \pi \log\left(\frac{r^2+4}{4r}\right).$$

We recall:

$\mathbf{f}_t(r) = \mathbf{g}_t(u(r))$ with the heat Gaussian $\mathbf{g}_t$ (cf. Chapter 13, Lemma 13.2.1).
C_1 is the constant evaluated in Corollary 14.6.5 of the present Chapter 14.

The three integrals above are evaluated as far as possible in Corollaries 14.5.2, 14.5.3, and 14.5.4.

References

[Art 74] J. ARTHUR, The Selberg trace formula for groups of Gamma-rank one, *Ann. of Math.* **100** No. 2 (1974) pp. 326–385.

[Art 78] J. ARTHUR, A Trace Formula for Reductive Groups I: Terms Associated to Classes in G(Q), *Duke Math. J.* **45** No. 4 (1978) pp. 911–952.

[Art 80] J. ARTHUR, A Trace Formula for Reductive Groups II: Applications to a Truncation Operator, *Compositio Mathematica* **40** No. 1 (1980) pp. 87–121.

[Art 89] J. ARTHUR, The Trace Formula and Hecke Operators, *Number Theory, Trace Formulas, and Discrete Groups*, Academic Press (1989) pp. 11–27.

[Asa 70] T. ASAI, On a certain function analogous to log eta(z), *Nagoya Math. J.* **40** (1970) pp. 193–211.

[AtBP 73] M. ATIYAH, R. BOTT, V. PATODI, On the heat equation and the index theorem, *Inventiones Math.* **19** (1973) pp. 279–330.

[AtDS 83] M. ATIYAH, H. DONNELLY, I. SINGER, Eta invariants, signature defects of cusps, and values of L-functions, *Ann. of Math.* **118** (1983) pp. 131–177.

[BaM 83] D. BARBASCH, H. MOSCOVICI, L^2 index and the Selberg trace formula, *J. Functional Analysis* **53** (1983) pp. 151–201.

[Bea 83] A. F. BEARDON, *The Geometry of Discrete Groups*, Springer Verlag, GTM, 1983.

[BeG 04] J. BERNSTEIN and S. GELBART eds., *Introduction to the Langlands Program*, AMS 2004.

[Bor 62] A. BOREL, Arithmetic Properties of Linear Algebraic Groups, *Proceedings International Congress of Mathematicians*, Stockholm (1962) pp. 10–22.

[Bor 69] A. BOREL, *Introduction aux Groupes Arithmétiques*, Hermann, Paris, 1969.

[BoG 83] A. BOREL and H. GARLAND, Laplacian and the Discrete Spectrum of an Arithmetic Group, *Am. J. Math.* **105** No. 2 (1983) pp. 309–335.

[Bre 05] E. BRENNER, Thesis, Yale, 2005.

[BuO 94] U. BUNKE and M. OLBRICH, The wave kenel for the Laplacian on the classical locally symmetric spaces of rank one, theta functions, trace formulas and the Selbeg zeta function. With an appendix by Andreas Juhl. *Ann. Global Anal. Geom.* **12** No. 4 (1994) pp. 357–401. Appendix: 402–405 (1994).

[Cha 84] I. CHAVEL, *Eigenvalues in Riemannian Geometry*, Academic Press 1984.

[DeI 82] J.-M. DESHOUILLERS and H. IWANIEC, Kloosterman sums and Fourier coefficients of cusp forms, *Invent. Math.* **70** (1982) pp. 219–288.

[Dod 83] J. DODZIUK, Maximum Principle for Parabolic Inequalities and the Heat Flow on Open Manifolds, *Indiana U. Math. J.* **32** (1983), pp. 703–716.

[DoJ 98] J. DODZIUK and J. JORGENSON, *Spectral Asymptotics on Degenerating Hyperbolic 3-Manifolds*, Memoirs of the AMS No. 643, Vol. **135**, 1998.

[Efr 87] I. EFRAT, *The Selberg trace formula for* $\mathrm{PSL}_2(\mathbf{R})^n$, *Mem. AMS* **359** (1987).

[EfS 85] I. EFRAT and P. SARNAK, The determinant of the Eisenstein matrix and Hilbert class fields, *Trans. AMS* **290** (1985) pp. 815–824.

[EGM 85] J. ELSTRODT, E. GRUNEWALD, J. MENNICKE, Eisenstein series on three dimensional hyperbolic spaces and imaginary quadratic fields, *J. reine angew. Math.* **360** (1985) pp. 160–213.

[EGM 87] J. ELSTRODT, E. GRUNEWALD, J. MENNICKE, Zeta Functions of binary Hermitian forms and special values of Eisenstein series on three-dimensional hyperbolic space, *Math. Ann.* **277** (1987) pp. 655–708.

[EGM 98] J. ELSTRODT, E. GRUNEWALD, J. MENNICKE, *Groups Acting on Hyperbolic Space*, Monograph in Mathematics, Springer Verlag, 1998.

[FlJ 78] M. FLENSTED-JENSEN, Spherical functions on semisimple Lie groups: A method of reduction to the complex case, *J. Funct. Anal.* **30** (1978) pp. 106–146.

[FlJ 86] M. FLENSTED-JENSEN, *Analysis on Non-Riemannian Symmetric Spaces*, CBMS **61**, 1986.

[Fre 04] E. FRENKEL, Recent advances in the Langlands program, *Bull. AMS* **41** No. 2 (2004) pp. 151–184.

[Gaf 59] M. GAFFNEY, The conservation property of the heat equation on Riemannian manifolds, *Comm. Pure and Applied Math.* **12** (1959) pp. 1–11.

[Gan 68] R. GANGOLLI, Asymptotic Behaviour of Spectra of Compact Quotients of Certain Symmetric Spaces, *Acta Math.* **121** (1968) pp. 151–192.

[Gan 77] R. GANGOLLI, Zeta functions of Selberg's type for compact space forms of symmetric spaces of rank 1, *Illinois J. Math.* **21** (1977) pp. 1–42.

[GaV 88] R. GANGOLLI and V.S. VARADARAJAN, *Harmonic Analysis of Spherical Functions on Real Reductive Groups*, Ergebnisse Math. **101**, Springer-Verlag, 1988.

[GaW 80] R. GANGOLLI and G. WARNER, Zeta functions of Selberg's type for some noncompact quotients of symmetric spaces of rank one, *Nagoya Math J.* **78** (1980) pp. 1–44.

[Gar 60] L. GÅRDING, Vecteurs analytiques dans les représentations des groupes de Lie, *Bull. Soc. Math. France* **88** (1960) pp. 73–93.

[GeM 03] S. GELBART and S. MILLER, Riemann's zeta function and beyond, *Bulletin AMS* **41** No. 1 (2003) pp. 50–112.

[GGP 66] I. GELFAND, M. GRAEV, I. PIATETSKII-SHAPIRO, *Representation Theory and Automorphic Functions* (Generalized Functions vol. 6), Moscow 1966, Translation, Saunders, 1969.

[GeN 50/57] I. M. GELFAND and M. A. NAIMARK, *Unitäre Darstellungen der klassischen Gruppen*, Akademie Verlag, Berlin, 1957; German translation of *Unitary representations of the classical groups* (in Russian), Trudy Mat. Inst. Steklova **36** (1950) pp. 1–288.

[GePS 63] I. GELFAND, I. PIATETSKII-SHAPIRO, Representation theory and theory of automorphic functions, *Am. Math. Soc. Transl.* ser 2 **26** (1963) pp. 173–200.

[Gey 69] W-D. GEYER, Unendliche algebraische Zahlkörper, über denen jede Gleichung auflösbar von beschränkter Stufe ist, *Journal of Number Theory* **1** (1969) pp. 346–374.

[God 66] R. GODEMENT, The spectral decomposition of cusp forms, *Proc. Symp. Pure Math.* AMS **9** (1966) pp. 225–234.

[Gri 71] P. GRIFFITHS, Complex analytic properties of certain Zariski open sets on algebraic varieties, *Ann. of Math.* **94** (1971) pp. 21–51.

[HaL 21] G. H. HARDY and J. E. LITTLEWOOD, The zeros of Riemann's zeta-function on the critical line, *Math. Zeit.* **10** (1921) pp. 283–317.

[Har 54] HARISH-CHANDRA, Representations of semisimple lie groups III, *Trans. Am. Math. Soc.* **76** (1954) pp. 234–253.

[Har 58a] HARISH-CHANDRA, Spherical functions on a semisimple Lie group I, *Amer. J. Math.* **79** (1958) pp. 241–310.

[Har 58b] HARISH-CHANDRA, Spherical functions on a semisimple Lie group I, *Amer. J. Math.* **80** (1958) pp. 533–613.

[Har 65] HARISH-CHANDRA, Invariant distributions on semisimple Lie groups, Pub. IHES **27** (1965) pp. 5–54.

[Har 68] HARISH-CHANDRA, *Automorphic Forms on Semisimple Lie Groups*, Springer Lecture Notes **62** (1968); notes by J.G.M. Mars.

[Hel 59] S. HELGASON, Differential operators on homogeneous spaces, *Acta Math.* **102** (1959) pp. 239–299.

[Hel 84] S. HELGASON, *Groups and Geometric Analysis*, Academic Press, 1984.

[Hum 1884] G. HUMBERT, Sur la mesure de classes d'Hermite de discriminant donné dans un corps quadratique imaginaire, *C.R. Acad. Sci. Paris* Vol. **169** (1919) pp. 448–454.

[Iwa 95] H. IWANIEC, *Introduction to the Spectral Theory of Automorphic Forms*, Biblioteca de la Revista Matematica Iberoamericana, Madrid, 1995.

[Jor 80] C. JORDAN, Memoire sur l'Equivalence des formes, *J. Ec. Polytech.* XLVIII (1880) pp. 112–150.

[JoL 93] J. JORGENSON and S. LANG, *Basic Analysis of Regularized Series and Products*, Springer Lecture Notes **1564**, 1993.

[JoL 94] J. JORGENSON and S. LANG, *Explicit Formulas for Regularized Products and Series*, Springer Lecture Notes **1593**, 1993.

[JoL 96] J. JORGENSON and S. LANG, Extension of analytic number theory and the theory of regularized harmonic series from Dirichlet series to Bessel series, *Math. Ann.* **306** (1996) pp. 75–124.

[JoL 99] J. JORGENSON and S. LANG, Hilbert-Asai Eisenstein series, regularized products, and heat kernels, *Nagoya Math. J.* Vol. **153** (1999) pp. 155–188.

[JoL 01a] J. JORGENSON and S. LANG, *Spherical Inversion on* $SL_n(\mathbf{R})$, Springer-Verlag MIM, 2001.

[JoL 01b] J. JORGENSON and S. LANG, The Ubiquitous Heat Kernel, *Mathematics Unlimited: 2001 and Beyond*, Vol. I, Engquist and Schmid eds., Springer Verlag 2001, pp. 665–683.

[JoL 02] J. JORGENSON and S. LANG, *Heat Eisenstein Series on* $SL_n(C)$, to appear in Memoirs of the AMS.

[JoL 03a] J. JORGENSON and S. LANG, Spherical Inversion for $SL_2(C)$, in: *Heat Kernels and Analysis on Manifolds, Graphs, and Metric Spaces*, ed. P. Auscher, T. Coulhon, and A. Grigor'yan, *AMS Contemp. Math.* **338** (2003) pp. 241–270.

[JoL 03b] J. JORGENSON and S. LANG, A Gaussian Space of Test Functions, *Math. Nachr.* **278** (2005) pp. 824–832.

[JoL 04] J. JORGENSON and S. LANG, Heat Kernel and Theta Inversion on $SL_2(C)$, to appear.

[JoLS 03] J. JORGENSON, S. LANG, and A. SINTON, *Spherical Inversion and Totally Geodesic Embeddings of Non-compact G/K's*, in preparation.

[JoLu 95] J. JORGENSON and R. LUNDELIUS, Convergence of the heat kernel and the resolvant kernel on degenerating hyperbolic Riemann surfaces of finite volume, *Quaestiones Mathematicae* **18** (1995) pp. 345–363.

[JoLu 97a] J. JORGENSON and R. LUNDELIUS, Convergence of the normalized spectral function on degenerating hyperbolic Riemann surfaces of finite volume *J. Func. Analysis* **149** (1997) pp. 25–57.

[JoLu 97b] J. JORGENSON and R. LUNDELIUS, A regularized heat trace for hyperbolic Riemann surfaces of finite volume, *Comment. Math. Helv.* **72** (1997) pp. 636–659.

[Kat 92] S. KATOK, *Fuchsian Groups*, University of Chicago press, 1992.

[Kub 68] T. KUBOTA, Über diskontinuierlicher Gruppen Picardschen Typus und zugehörige Eisensteinsche Reihen, *Nagoya Math. J.* **32** (1968) pp. 259–271.

[Kub 73] T. KUBOTA, *Elementary Theory of Eisenstein Series*, Kodansha and John Wiley, Tokyo-New York 1973.

[Lan 73/87] S. LANG, *Elliptic Functions*, Addison Wesley, 1973; second edition, Springer Verlag, 1987.

[Lan 75/85] S. LANG, $SL_2(\mathbf{R})$, Addison Wesley 1973, Springer Verlag 1985.

[Lan 93] S. LANG, *Real and Functional Analysis*, Springer Verlag, 1993.

[Lan 70/94] S. LANG, *Algebraic Number Theory*, Addison Wesley 1970; second edition, Springer Verlag, 1994.

[Lan 97] S. LANG, *Undergraduate Analysis*, second edition, Springer Verlag 1997.

[Lan 99] S. LANG, *Math Talks for Undergraduates*, Springer Verlag, 1999.

[Lan 02] S. LANG, *Introduction to Differentiable Manifolds*, second edition, Springer Verlag 2002.

[Lgld 66] R. P. LANGLANDS, Eisenstein Series, Proc. Symposium in Pure Mathematics, AMS, Boulder Colorado 1966, *Algebraic Groups and Discontinuous Subgroups*, Borel and Mostow editors, pp. 235–252.

[Lgld 76] R. P. LANGLANDS, *On the Functional Equations Satisfied by Eisenstein Series*, Springer Lecture Notes **544**, 1976.

[Maa 49] H. MAASS, Über eine neue Art von Nichtanalytischen automorphen Funktionen und die Bestimmung Dirichletscher Reihen durch Funtionalgleichungen, *Math. Ann.* **121** (1949) pp. 141–183.

[McK 72] H. P. McKEAN, Selberg's Trace Formula as Applied to a Compact Riemann Surface, *Comm. Pure and Applied Math.* **XXV** (1972) pp. 225–246.

[Mul 83] W. MÜLLER, Spectral theory for Riemannian manifolds with cusps and a related trace formula, *Math. Nachrichten* **111** (1983) pp. 197–288.

[Mul 84] W. MÜLLER, Signature defects of cusps of Hilbert modular varieties and values of L-series at $s = 1$, *J. Diff. Geom.* **20** (1984) pp. 55–119.

[Mul 87] W. MÜLLER, *Manifolds with Cusps of Rank One*, Springer Lecture Notes **1244**, Springer Verlag 1987.

[Nel 59] E. NELSON, Analytic vectors, *Annals of Math.* **70** (1959) pp. 572–615.

[Pic 1884] E. PICARD, Sur un groupe de transformations des points de l'espace situés du même côté d'un plan, *Bull. Soc. Math. France* **12** (1884) pp. 43–47.

[Roe 56] W. ROELCKE, *Über die Wellengleichung bei Grenzkreisgruppen erster Art*, Sitz. Ber. Heidelberger Ak. der Wiss., Math. nat. Kl. 1956, **4** Abh.

[Roe 66] W. ROELCKE, Das Eigenwertproblem der automorphen Formen in der hyperbolischer Ebene I, *Math. Ann.* **167** (1966) pp. 292–337.

[Roe 67] W. ROELCKE, Das Eigenwertproblem der automorphen Formen in der hyperbolisches Ebene II, *Math. Ann.* **168** (1967) pp. 261–324.

[Sar 83] P. SARNAK, The arithmetic and geometry of some hyperbolic three manifolds, *Acta Math.* **151** (1983) pp. 253–295.

[Sar 03] P. SARNAK, Spectra of Hyperbolic Surfaces, *Bulletin of the AMS* **40** (2003) pp. 441–478.

[Sel 56] A. SELBERG, Harmonic Analysis and Discontinuous Groups in Weakly Symmetric Riemannian Spaces with Applications to Dirichlet Series, International Colloquium on Zeta Functions, *J. Indian Math. Soc.* (1956) pp. 47–87.

[Sel 62] A. SELBERG, Discontinuous Groups and Harmonic Analysis, *Proc. International Congress of Mathematicians, Stockholm* (1962) p. 177–189.

[Sel 89] A. SELBERG, Harmonic analysis. Introduction to the Goettingen lecture notes, *Collected Papers*, vol. I, Springer, 1989.

[Szm 83] J. SZMIDT, The Selberg trace formula for the Picard group $SL_2(\mathbf{Z}[\mathbf{i}])$, *Acta Arith.* **42** (1983) pp. 291–424.

[Szm 87] J. SZMIDT, *The Selberg trace formula and imaginary quadratic fields*, Schriftenreihe des Sonderforschungsbereichs Geometrie und Analysis #52, Mathematics, University of Göettingen, 1987.

[Tam 60] T. TAMAGAWA, On Selberg's trace formula, *J. Faculty of Science*, University of Tokyo, Sec. I, **VIII**, Part 2, pp. 363–386.

[Tan 77] Y. TANIGAWA, Selberg trace formula for Picard groups, *Int. Symp. Algebraic Number Theory*, Tokyo 1977.

[Tit 51] E. C. TITCHMARSH, *The Theory of the Riemann Zeta Function*, Oxford, 1951.

[Ven 73] A. B. VENKOV, Expansions in automorphic eigenfunctions of the Laplace-Beltrami operator in classical symmetric spaces of rank one and the Selberg trace formula, *Proceedings of the Steklov Institute of Mathematics* No. 125 (1973); AMS translation 1975, pp. 1–48.

[Wal 84] D. WALLACE, Conjugacy classes of hyperbolic matrices in SL(n, **Z**) and ideal classes in an order, *Trans. AMS* **283** No. 1 (1984) pp. 177–184.

[War 79] G. WARNER, *Selberg's trace formula for non-uniform lattices: the* **R***-rank one case*, Advance in Math. Studies **6** (1979) pp. 1–142.

[Wei 1885] K. WEIERSTRASS, Über die analytische Darstellbarkeit sogenannter willkürlicher Funktionen einer reellen Veränderlichen, *Sitzungsbericht Königl. Akad. Wiss.*, 2 and 30 July 1885 pp. 633–639 and 789–805.

[Yos 88] E. YOSHIDA, On an Application of Zagier's Method in the Theory of Selberg's Trace Formula, *Advanced Studies in Pure Mathematics* **13** (1988), Investigations in Number Theory, pp. 193–214.

[Zag 79] D. ZAGIER, Eisenstein series and the Selberg trace formula, in *Automorphic Forms, Representation Theory and Arithmetic*, Tata Institute, Bombay (1979) pp. 303–355.

[Zag 82] D. ZAGIER, The Rankin-Selberg method for automorphic functions which are not of rapid decay, *J. Fac. Sci. Univ. Tokyo* **I A 28** (1981) pp. 415–437.

[Zog 82] P. ZOGRAF, Selberg trace formula for the Hilbert modular group of a real quadratic number field, *J. Soviet Math.* **19** (1982) pp. 1637–1652.

Index